STUDENT'S SOLUTIONS MANUAL

WILLIAM CRAINE III

KIMBERLY SMITH

STATS: DATA AND MODELS

THIRD EDITION

Richard De Veaux
Williams College

Paul Velleman
Cornell University

David Bock
Cornell University

Addison-Wesley
is an imprint of

Reproduced by Pearson Addison-Wesley from electronic files supplied by the author.

Publishing as Addison-Wesley, 75 Arlington Street, Boston, MA 02116.

ISBN-13: 978-0-321-69349-5
ISBN-10: 0-321-69349-3

4 5 6 EBM 15 14 13 12

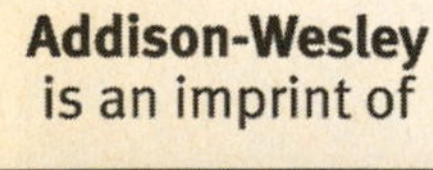

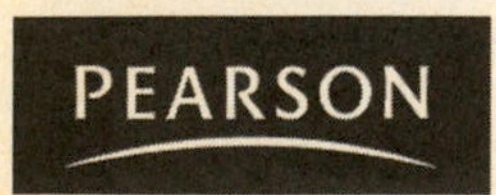

www.pearsonhighered.com

Contents

Chapter 2 - Data

1. **Voters.** The response is a categorical variable.

3. **Medicine.** The company is studying a quantitative variable.

5. **The News.** Answers will vary.

7. **Bicycle Safety.** *Who* – 2,500 cars. *What* – Distance from the bicycle to the passing car (in inches). *Population of interest* – All cars passing bicyclists.

9. **Honesty.** *Who* – Workers who buy coffee in an office. *What* – amount of money contributed to the collection tray. *Population of interest* – All people in honor system payment situations.

11. **Fitness.** *Who* – 25,892 men. *What* – Fitness level and cause of death. It is not clear what categories were used for fitness level. *Population of interest* – All men.

13. **Weighing bears.** *Who* – 54 bears. *What* – Weight, neck size, length (no specified units), and sex. *When* – Not specified. *Where* – Not specified. *Why* - Since bears are difficult to weigh, the researchers hope to use the relationships between weight, neck size, length, and sex of bears to estimate the weight of bears, given the other, more observable features of the bear. *How* – Researchers collected data on 54 bears they were able to catch. *Variables* – There are 4 variables; weight, neck size, and length are quantitative variables, and sex is a categorical variable. No units are specified for the quantitative variables. *Concerns* – The researchers are (obviously!) only able to collect data from bears they were able to catch. This method is a good one, as long as the researchers believe the bears caught are representative of all bears, in regard to the relationships between weight, neck size, length, and sex.

15. **Arby's menu.** *Who* – Arby's sandwiches. *What* – type of meat, number of calories (in calories), and serving size (in ounces). *When* – Not specified. *Where* – Arby's restaurants. *Why* – These data might be used to assess the nutritional value of the different sandwiches. *How* – Information was gathered from each of the sandwiches on the menu at Arby's, resulting in a census. *Variables* – There are three variables. Number of calories and serving size are quantitative variables, and type of meat is a categorical variable.

17. Babies. *Who* – 882 births. *What* – Mother's age (in years), length of pregnancy (in weeks), type of birth (caesarean, induced, or natural), level of prenatal care (none, minimal, or adequate), birth weight of baby (unit of measurement not specified, but probably pounds and ounces), gender of baby (male or female), and baby's health problems (none, minor, major).
When – 1998-2000. *Where* – Large city hospital. *Why* – Researchers were investigating the impact of prenatal care on newborn health. *How* – It appears that they kept track of all births in the form of hospital records, although it is not specifically stated. *Variables* – There are three quantitative variables: mother's age, length of pregnancy, and birth weight of baby. There are four categorical variables: type of birth, level of prenatal care, gender of baby, and baby's health problems.

19. Herbal medicine. *Who* – experiment volunteers. *What* – herbal cold remedy or sugar solution, and cold severity. *When* – Not specified. *Where* – Major pharmaceutical firm. *Why* – Scientists were testing the efficacy of an herbal compound on the severity of the common cold.
How – The scientists set up a controlled experiment. *Variables* – There are two variables. Type of treatment (herbal or sugar solution) is categorical, and severity rating is quantitative. *Concerns* – The severity of a cold seems subjective and difficult to quantify. Also, the scientists may feel pressure to report negative findings about the herbal product.

21. Streams. *Who* – Streams. *What* – Name of stream, substrate of the stream (limestone, shale, or mixed), acidity of the water (measured in pH), temperature (in degrees Celsius), and BCI (unknown units). *When* – Not specified. *Where* – Upstate New York. *Why* – Research is conducted for an Ecology class. *How* – Not specified. *Variables* – There are five variables. Name and substrate of the stream are categorical variables, and acidity, temperature, and BCI are quantitative variables.

23. Refrigerators. *Who* – 41 refrigerators. *What* – Brand, cost (probably in dollars), size (in cu. ft.), type, estimated annual energy cost (probably in dollars), overall rating, and repair history (in percent requiring repair over the past five years). *When* – 2006. *Where* – United States. *Why* – The information was compiled to provide information to the readers of *Consumer Reports*. *How* – Not specified. *Variables* – There are 7 variables. Brand, type, and overall rating are categorical variables. Cost, size, estimated energy cost, and repair history are quantitative variables.

25. Horse race 2008. *Who* – Kentucky Derby races. *What* – Date, winner, margin (in lengths), jockey, winner's payoff (in dollars), duration of the race (in minutes and seconds), and track condition. *When* – 1875 – 2008. *Where* – Churchill Downs, Louisville, Kentucky. *Why* – It is interesting to examine the trends in the Kentucky Derby. *How* – Official statistics are kept for the race each year. *Variables* – There are 7 variables. Winner, jockey, and track condition are categorical variables. Date, margin, winner's payoff, and duration are quantitative variables.

Chapter 3 – Displaying and Describing Categorical Data

1. **Graphs in the news.** Answers will vary.

3. **Tables in the news.** Answers will vary.

5. **Movie genres.**

   a) A pie chart seems appropriate from the movie genre data. Each movie has only one genre, and the 120 movies constitute a "whole". However, because of the percentages of each type of movie, it is difficult to compare the ratings. It's not really clear whether Action/Adventure or Comedy is the most common genre in this group of movies.

   b) Thriller/Horror is the least common genre. It has the smallest region in the chart.

7. **Genres, again.**

   a) Comedy has the highest bar, so it is the most common genre.

   b) This is easier to see on the bar chart. The percentages are so close that the difference is nearly indistinguishable in the pie chart.

9. **Magnet Schools.**

   There were 1,755 qualified applicants for the Houston Independent School District's magnet schools program. 53% were accepted, 17% were wait-listed, and the other 30% were turned away for lack of space.

11. **Causes of death 2006.**

   a) Yes, it is reasonable to assume that heart and respiratory disease caused approximately 32% of U.S. deaths in 2006, since there is no possibility for overlap. Each person could only have one cause of death.

   b) Since the percentages listed add up to 65.4%, other causes must account for 34.6% of US deaths.

   c) A pie chart is a good choice (with the inclusion of the "Other" category), since causes of US deaths represent parts of a whole. A bar chart would also be a good display.

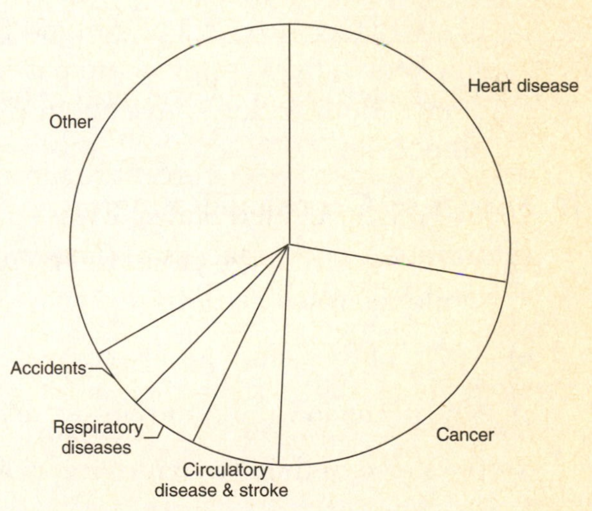

13. Oil spills 2008.

The bar chart shows that grounding is the most frequent cause of oil spillage for these 319 spills, and allows the reader to rank the other types as well. If being able to differentiate between these close counts is required, use the bar chart. The pie chart is also acceptable as a display, but it's difficult to tell whether, for example, there is a greater percentage of spills caused by fire and explosions or unknown causes. If you want to showcase the causes of oil spills as a fraction of all 319 spills, use the pie chart.

15. Global warming.

Perhaps the most obvious error is that the percentages in the pie chart only add up to 92%, when they should, of course, add up to 100%. Furthermore, the three-dimensional perspective view distorts the regions in the graph, violating the area principle. The regions corresponding to No Solid Evidence and Due to Natural Patterns should be roughly the same size, at 20% and 21% of respondents, respectively. However, the angle for the 21% region looks much bigger. Always use simple, two-dimensional graphs.

17. Teen smokers.

According to the Monitoring the Future study, teen smoking brand preferences differ somewhat by region. Although Marlboro is the most popular brand in each region, with about 58% of teen smokers preferring this brand in each region, teen smokers from the South prefer Newports at a higher percentage than teen smokers from the West, 22.5% to approximately 10%, respectively. Camels are more popular in the West, with 9.5% of teen smokers preferring this brand, compared to only 3.3% in the South. Teen smokers in the West are also more likely to have to particular brand than teen smokers in the South. 12.9% of teen smokers in the West have no particular brand, compared to only 6.7% in the South. Both regions have about 9% of teen smokers that prefer one of over 20 other brands.

19. Movies by Genre and Rating.

a) We can tell that the table uses column percents, since each column adds to 100%, while the rows do not.

b) 31.7% of these movies are comedies.

c) 60% of the PG-rated movies were comedies.

d) **i)** 35.7% of the PG-13 movies were comedies.

ii) You cannot determine this from the table.

iii) None (0%) of the dramas were G-rated.

iv) You cannot determine this from the table.

21. Seniors.

Plans	White	Minority	TOTAL
4-year college	198	44	242
2-year college	36	6	42
Military	4	1	5
Employment	14	3	17
Other	16	3	19
TOTAL	268	57	325

a) A table with marginal totals is above. There are 268 White graduates and 325 total graduates. 268/325 ≈ 82.5% of the graduates are White.

b) There are 42 graduates planning to attend 2-year colleges. 42/325 ≈ 12.9%

c) 36 white graduates are planning to attend 2-year colleges. 36/325 ≈ 11.1%

d) 36 white graduates are planning to attend 2-year colleges and there are 268 whites graduates. 36/268 ≈ 13.4%

e) There are 42 graduates planning to attend 2-year colleges. 36/42 ≈ 85.7%

23. More about seniors.

a) For white students, 73.9% plan to attend a 4-year college, 13.4% plan to attend a 2-year college, 1.5% plan on the military, 5.2% plan to be employed, and 6.0% have other plans.

b) For minority students, 77.2% plan to attend a 4-year college, 10.5% plan to attend a 2-year college, 1.8% plan on the military, 5.3% plan to be employed, and 5.3% have other plans.

c) A segmented bar chart is a good display of these data:

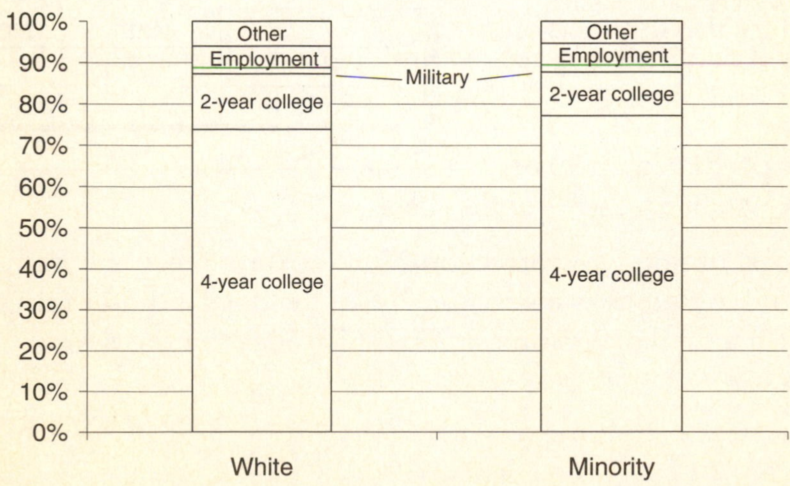

d) The conditional distributions of plans for Whites and Minorities are similar:
White – 74% 4-year college, 13% 2-year college, 2% military, 5% employment, 6% other.
Minority – 77% 4-year college, 11% 2-year college, 2% military, 5% employment, 5% other.
Caution should be used with the percentages for Minority graduates, because the total is so small. Each graduate is almost 2%. Still, the conditional distributions of plans are essentially the same for the two groups. There is little evidence of an association between race and plans for after graduation.

25. Magnet schools revisited.

a) There were 1755 qualified applicants to the Houston Independent School District's magnet schools program. Of those, 292, or about 16.6% were Asian.

b) There were 931 students accepted to the magnet schools program. Of those, 110, or about 11.8% were Asian.

c) There were 292 Asian applicants. Of those, 110, or about 37.7%, were accepted.

d) There were 1755 total applicants. Of those, 931, or about 53%, were accepted.

27. Back to school.

There were 1,755 qualified applicants for admission to the magnet schools program. 53% were accepted, 17% were wait-listed, and the other 30% were turned away. While the overall acceptance rate was 53%, 93.8% of Blacks and Hispanics were accepted, compared to only 37.7% of Asians, and 35.5% of whites. Overall, 29.5% of applicants were Black or Hispanics, but only 6% of those turned away were Black or Hispanic. Asians accounted for 16.6% of applicants, but 25.3% of those turned away. It appears that the admissions decisions were not independent of the applicant's ethnicity.

29. Weather forecasts.

a) The table shows the marginal totals. It rained on 34 of 365 days, or 9.3% of the days.

		Actual Weather		
		Rain	No Rain	**Total**
Forecast	Rain	27	63	90
	No Rain	7	268	275
	Total	34	331	365

b) Rain was predicted on 90 of 365 days. 90/365 ≈ 24.7% of the days.

c) The forecast of rain was correct on 27 of the days it actually rained and the forecast of No Rain was correct on 268 of the days it didn't rain. So, the forecast was correct a total of 295 times. 295/365 ≈ 80.8% of the days.

d) On rainy days, rain had been predicted 27 out of 34 times (79.4%). On days when it did not rain, forecasters were correct in their predictions 268 out of 331 times (81.0%). These two percentages are very close. There is no evidence of an association between the type of weather and the ability of the forecasters to make an accurate prediction.

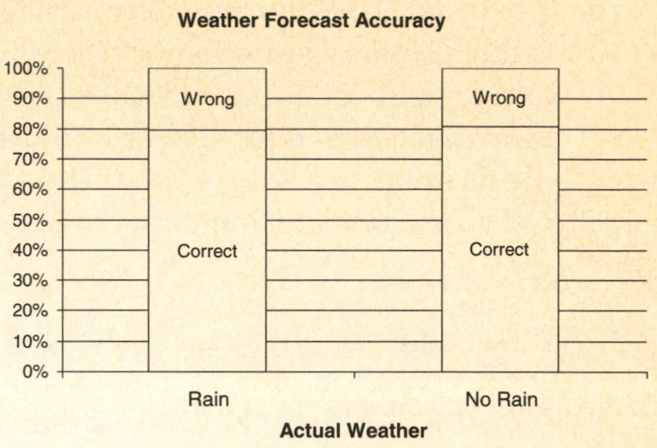

31. Blood pressure.

a) The marginal distribution of blood pressure for the employees of the company is the total column of the table, converted to percentages. 20% low, 49% normal and 31% high blood pressure.

Blood pressure	under 30	30 - 49	over 50	Total
low	27	37	31	95
normal	48	91	93	232
high	23	51	73	147
Total	98	179	197	474

b) The conditional distribution of blood pressure within each age category is:
Under 30 : 28% low, 49% normal, 23% high
30 – 49 : 21% low, 51% normal, 28% high
Over 50 : 16% low, 47% normal, 37% high

c) A segmented bar chart of the conditional distributions of blood pressure by age category is at the right.

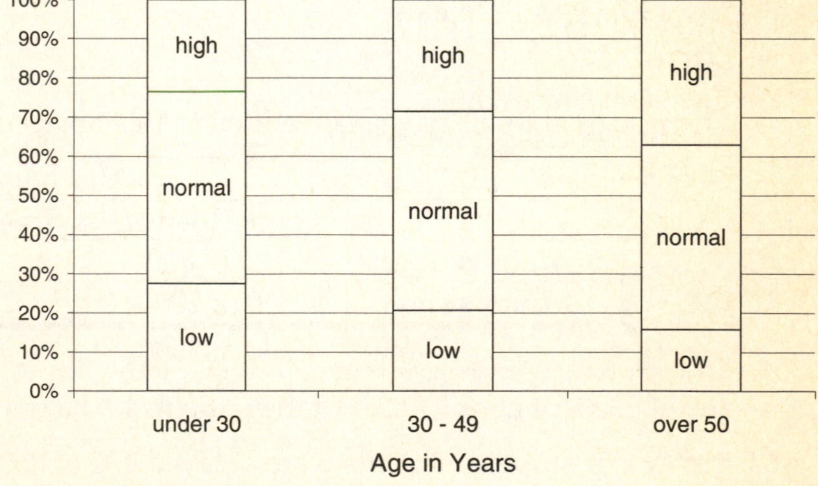

d) In this company, as age increases, the percentage of employees with low blood pressure decreases, and the percentage of employees with high blood pressure increases.

e) No, this does not prove that people's blood pressure increases as they age. Generally, an association between two variables does not imply a cause-and-effect relationship. Specifically, these data come from only one company and cannot be applied to all people. Furthermore, there may be some other variable that is linked to both age and blood pressure. Only a controlled experiment can isolate the relationship between age and blood pressure.

33. Anorexia.

These data provide no evidence that Prozac might be helpful in treating anorexia. About 71% of the patients who took Prozac were diagnosed as "Healthy", while about 73% of the patients who took a placebo were diagnosed as "Healthy". Even though the percentage was higher for the placebo patients, this does not mean that Prozac is hurting patients. The difference between 71% and 73% is not likely to be statistically significant.

35. Driver's licenses 2006.

a) There are 9,727,516 drivers under 20 and a total of 219,872,389 drivers in the U.S. That's about 4.4% of U.S. drivers under 20.

b) There are 109,720,479 males out of 219,872,389 total U.S. drivers, or about 49.9%.

c) Each age category appears to have about 50% male and 50% female drivers. There is a slight pattern in the deviations from 50%. At younger ages, males form the slight majority of drivers. This percentage shrinks until the percentages are 50% male and 50% for middle aged drivers. The percentage of male drivers continues to shrink until, at around age 45, female drivers hold a slight majority. This continues into the 85 and over category.

d) There appears to be a slight association between age and gender of U.S. drivers. Younger drivers are slightly more likely to be male, and older drivers are slightly more likely to be female.

37. Hospitals.

a) The marginal totals have been added to the table:

		Discharge delayed		
Procedure		**Large Hospital**	**Small Hospital**	**Total**
	Major surgery	120 of 800	10 of 50	130 of 850
	Minor surgery	10 of 200	20 of 250	30 of 450
	Total	130 of 1000	30 of 300	160 of 1300

160 of 1300, or about 12.3% of the patients had a delayed discharge.

b) Yes. Major surgery patients were delayed 130 of 850 times, or about 15.3% of the time.
Minor Surgery patients were delayed 30 of 450 times, or about 6.7% of the time.

c) Large Hospital had a delay rate of 130 of 1000, or 13%.
Small Hospital had a delay rate of 30 of 300, or 10%.
The small hospital has the lower overall rate of delayed discharge.

d) Large Hospital: Major Surgery 15% delayed and Minor Surgery 5% delayed.
Small Hospital: Major Surgery 20% delayed and Minor Surgery 8% delayed.
Even though small hospital had the lower overall rate of delayed discharge, the large hospital had a lower rate of delayed discharge for each type of surgery.

e) No. While the overall rate of delayed discharge is lower for the small hospital, the large hospital did better with *both* major surgery and minor surgery.

f) The small hospital performs a higher percentage of minor surgeries than major surgeries. 250 of 300 surgeries at the small hospital were minor (83%). Only 200 of the large hospital's 1000 surgeries were minor (20%). Minor surgery had a lower delay rate than major surgery (6.7% to 15.3%), so the small hospital's overall rate was artificially inflated. Simply put, it is a mistake to look at the overall percentages. The real truth is found by looking at the rates after the information is broken down by type of surgery, since the delay rates for each type of surgery are so different. The larger hospital is the better hospital when comparing discharge delay rates.

39. Graduate admissions.

Program	Males Accepted (of applicants)	Females Accepted (of applicants)	Total
1	511 of 825	89 of 108	600 of 933
2	352 of 560	17 of 25	369 of 585
3	137 of 407	132 of 375	269 of 782
4	22 of 373	24 of 341	46 of 714
Total	**1022 of 2165**	**262 of 849**	**1284 of 3014**

a) 1284 applicants were admitted out of 3014 applicants. 1284/3014 = 42.6%

b) 1022 of 2165 (47.2%) of males were admitted. 262 of 849 (30.9%) of females were admitted.

c) Since there are four comparisons to make, the table at the right organizes the percentages of males and females accepted in each program. Females are accepted at a higher rate in every program.

Program	Males	Females
1	61.9%	82.4%
2	62.9%	68.0%
3	33.7%	35.2%
4	5.9%	7%

d) The comparison of acceptance rate within each program is most valid. The overall percentage is an unfair average. It fails to take the different numbers of applicants and different acceptance rates of each program. Women tended to apply to the programs in which gaining acceptance was difficult for everyone. This is an example of Simpson's Paradox.

Chapter 4 – Displaying and Summarizing Quantitative Data

1. **Histogram.** Answers will vary.

3. **In the news.** Answers will vary.

5. **Thinking about shape.**

   a) The distribution of the number of speeding tickets each student in the senior class of a college has ever had is likely to be unimodal and skewed to the right. Most students will have very few speeding tickets (maybe 0 or 1), but a small percentage of students will likely have comparatively many (3 or more?) tickets.

   b) The distribution of player's scores at the U.S. Open Golf Tournament would most likely be unimodal and slightly skewed to the right. The best golf players in the game will likely have around the same average score, but some golfers might be off their game and score 15 strokes above the mean. (Remember that high scores are undesirable in the game of golf!)

   c) The weights of female babies in a particular hospital over the course of a year will likely have a distribution that is unimodal and symmetric. Most newborns have about the same weight, with some babies weighing more and less than this average. There may be slight skew to the left, since there seems to be a greater likelihood of premature birth (and low birth weight) than post-term birth (and high birth weight).

   d) The distribution of the length of the average hair on the heads of students in a large class would likely be bimodal and skewed to the right. The average hair length of the males would be at one mode, and the average hair length of the females would be at the other mode, since women typically have longer hair than men. The distribution would be skewed to the right, since it is not possible to have hair length less than zero, but it is possible to have a variety of lengths of longer hair.

7. **Sugar in cereals.**

   a) The distribution of the sugar content of breakfast cereals is bimodal, with a cluster of cereals with sugar content around 10% sugar and another cluster of cereals around 48% sugar. The lower cluster shows a bit of skew to the right. Most cereals in the lower cluster have between 0% and 10% sugar. The upper cluster is symmetric, with center around 45% sugar.

   b) There are two different types of breakfast cereals, those for children and those for adults. The children's cereals are likely to have higher sugar contents, to make them taste better (to kids, anyway!). Adult cereals often advertise low sugar content.

9. Vineyards.

a) There is information displayed about 36 vineyards and it appears that 28 of the vineyards are smaller than 60 acres. That's around 78% of the vineyards. (75% would be a good estimate!)

b) The distribution of the size of 36 Finger Lakes vineyards is skewed to the right. Most vineyards are smaller than 75 acres, with a few larger ones, from 90 to 160 acres. One vineyard was larger than all the rest, over 240 acres. The mode of the distribution is between 0 and 30 acres.

11. Heart attack stays.

a) The distribution of length of stays is skewed to the right, so the mean is larger than the median.

b) The distribution of the length of hospital stays of female heart attack patients is skewed to the right, with stays ranging from 1 day to 36 days. The distribution is centered around 8 days, with the majority of the hospital stays lasting between 1 and 15 days. There are a relatively few hospital stays longer than 27 days. Many patients have a stay of only one day, possibly because the patient died.

c) The median and IQR would be used to summarize the distribution of hospital stays, since the distribution is strongly skewed.

13. Super Bowl points.

a) The median number of points scored in the first 43 Super Bowl games is 45 points.

b) The first quartile of the number of points scored in the first 43 Super Bowl games is 37 points. The third quartile is 55 points.

c) In the first 43 Super Bowl games, the lowest number of points scored was 21, and the highest number of points scored was 75. The median number of points scored was 45, and the middle 50% of Super Bowls has between 37 and 55 points scored.

15. Summaries.

a) The mean price of the electric smoothtop ranges is $1001.50.

b) In order to find the median and the quartiles, the list must be ordered.
565 750 850 900 1000 1050 1050 1200 1250 1400
The median price of the electric ranges is $1025.
Quartile 1 = $850 and Quartile 3 = $1200.

c) The range of the distribution of prices is Max – Min = $1400 – $565 = $835.
The IQR = Q3 – Q1 = $1200 - $850 = $350.

17. Mistake.

a) As long as the boss's true salary of $200,000 is still above the median, the median will be correct. The mean will be too large, since the total of all the salaries will decrease by $2,000,000 - $200,000 = $1,800,000, once the mistake is corrected.

b) The range will likely be too large. The boss's salary is probably the maximum, and a lower maximum would lead to a smaller range. The IQR will likely be unaffected, since the new maximum has no effect on the quartiles. The standard deviation will be too large, because the $2,000,000 salary will have a large squared deviation from the mean.

19. Standard deviation I.

a) Set 2 has the greater standard deviation. Both sets have the same mean, 6, but set two has values that are generally farther away from the mean.
SD(Set 1) = 2.24   SD(Set 2) = 3.16

b) Set 2 has the greater standard deviation. Both sets have the same mean (15), maximum (20), and minimum (10), but 11 and 19 are farther from the mean than 14 and 16.
SD(Set 1) = 3.61   SD(Set 2) = 4.53

c) The standard deviations are the same. Set 2 is simply Set 1 + 80. Although the measures of center and position change, the spread is exactly the same.
SD(Set 1) = 4.24   SD(Set 2) = 4.24

21. Pizza prices.

The mean and standard deviation would be used to summarize the distribution of pizza prices, since the distribution is unimodal and symmetric.

23. Pizza prices again.

a) The mean pizza price is closest to $2.60. That's the balancing point of the histogram.

b) The standard deviation in pizza prices is closest to $0.15, since that is the typical distance to the mean. There are no pizza prices as far as $0.50 of $1.00.

25. Movie lengths.

a) A typical movie would be around 100 minutes long. This is near the center of the unimodal and slightly skewed histogram, with the outlier set aside.

b) You would be surprised to find that your movie ran for 150 minutes. Only 3 movies ran that long.

c) The mean run time would be higher, since the distribution of run times is skewed to the right, and also has a high outlier. The mean is be pulled towards this tail, while the median resistant.

27. Movie lengths II.

a) i) The distribution of movie running times is fairly consistent, with the middle 50% of running times between 97 and 119 minutes. The interquartile range is 22 minutes.

ii) The standard deviation of the distribution of movie running times is 19.6 minutes, which indicates that movies typically have running times fairly close to the mean running time.

b) Since the distribution of movie running times is skewed to the right and contains an outlier, the standard deviation is a poor choice of numerical summary for the spread. The interquartile range is better, since it is resistant to outliers.

29. Movie budgets.

The industry publication is using the median, while the watchdog group is using the mean. It is likely that the mean is pulled higher by a few very expensive movies.

31. Payroll.

a) The mean salary is $\frac{(1200+700+6(400)+4(500))}{12}=\$525$.

The median salary is the middle of the ordered list:
400 400 400 400 400 400 500 500 500 500 700 1200
The median is $450.

b) Only two employees, the supervisor and the inventory manager, earn more than the mean wage.

c) The median better describes the wage of the typical worker. The mean is affected by the two higher salaries.

d) The IQR is the better measure of spread for the payroll distribution. The standard deviation and the range are both affected by the two higher salaries.

33. Gasoline.

a)

Gasoline Prices

Stem	Leaf
2.4	56
2.4	
2.3	68
2.3	23
2.2	677789
2.2	1234

Key:
2.2 | 1 = $2.21/gal

b) The distribution of gas prices is unimodal and skewed to the right, centered around \$2.27 per gallon, with most stations charging between \$2.26 and \$2.33 per gallon. The lowest and highest prices were \$2.27 and \$2.46 per gallon.

c) There is a gap in the distribution of gasoline prices. There were no stations that charged between \$2.40 and \$2.44.

35. States.

a) The distribution of state populations is skewed heavily to the right. Therefore, the median and IQR are the appropriate measures of center and spread.

b) The mean population must be larger than the median population. The extreme values on the right affect the mean greatly and have no effect on the median.

c) There are 51 entries in the stemplot, so the 26th entry must be the median. Counting in the ordered stemplot gives median = 4 million people. The middle of the lower 50% of the list (26 state populations) is between the 13th and 14th population, or 1.5 million people. The middle of the upper half of the list (26 state populations) is between the 13th and 14th population from the top, or 6 million people. The IQR = Q3 – Q1 = 6 – 1.5 = 4.5 million people.

d) The distribution of population for the 50 U.S. States and Washington, D.C. is skewed heavily to the right. The median population is 4 million people, with 50% of states having populations between 1 and 6 million people. There is one outlier, a state with 34 million people. The next highest population is only 21 million.

37. A-Rod 2009.

The distribution of the number of homeruns hit by Alex Rodriguez during the 1994 – 2009 seasons is reasonably symmetric, with a typical number of homeruns per season in around 40. With the exception of 2 seasons in which Rodriguez hit 0 and 5 homeruns, his total number of homeruns per season was between 20 and the maximum of 57.

39. Hurricanes 2006.

a) A dotplot of the number of hurricanes each year from 1944 through 2006 is displayed. Each dot represents a year in which there were that many hurricanes.

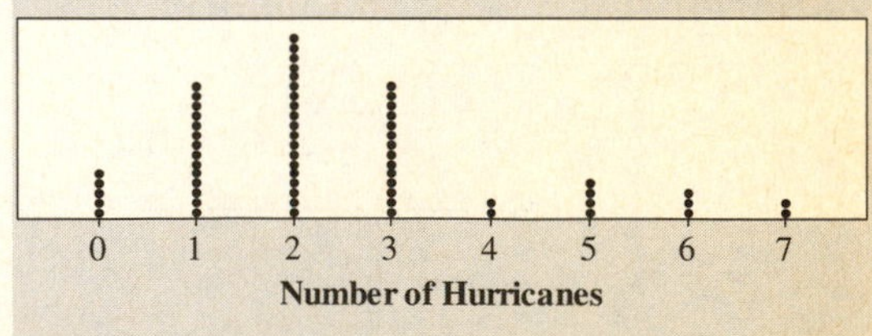

b) The distribution of the number of hurricanes per year is unimodal and skewed to the right, with center around 2 hurricanes per year. The number of hurricanes per year ranges from 0 to 7. There are no outliers. There may be a second mode at 5 hurricanes per year, but since there were only 4 years in which 5 hurricanes occurred, it is unlikely that this is anything other than natural variability.

41. A-Rod again.

a) This is not a histogram. The horizontal axis should the number of home runs per year, split into bins of a convenient width. The vertical axis should show the frequency; that is, the number years in which Rodriguez hit a number of home runs within the interval of each bin. The display shown is a bar chart/time plot hybrid that simply displays the data table visually. It is of no use in describing the shape, center, spread, or unusual features of the distribution of home runs hit per year by Rodriguez.

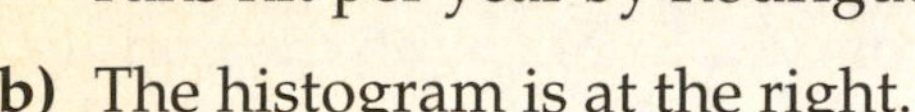

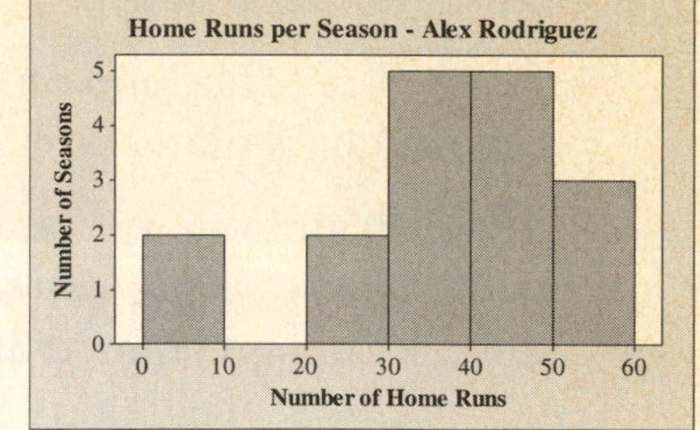

b) The histogram is at the right.

43. Acid rain.

The distribution of the pH readings of water samples in Allegheny County, Penn. is bimodal. A roughly uniform cluster is centered around a pH of 4.4. This cluster ranges from pH of 4.1 to 4.9. Another smaller, tightly packed cluster is centered around a pH of 5.6. Two readings in the middle seem to belong to neither cluster.

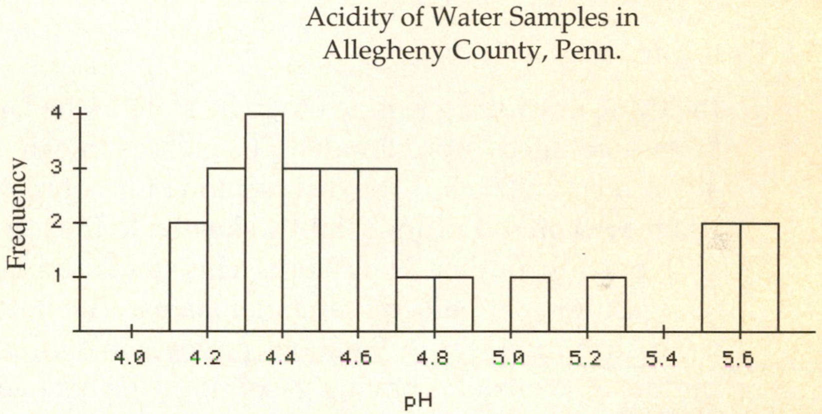

45. Final grades.

The width of the bars is much too wide to be of much use. The distribution of grades is skewed to the left, but not much more information can be gathered.

47. Zip codes.

Even though zip codes are numbers, they are not quantitative in nature. Zip codes are categories. A histogram is not an appropriate display for categorical data. The histogram the Holes R Us staff member displayed doesn't take into account that some 5-digit numbers do not correspond to zip codes or that zip codes falling into the same classes may not even represent similar cities or towns.

The employee could design a better display by constructing a bar chart that groups together zip codes representing areas with similar demographics and geographic locations.

49. Math scores 2005.

a) Median: 239
IQR: 9
Mean: 237.6
Standard deviation: 5.7

b) Since the distribution of Math scores is skewed to the left, it is probably better to report the median and IQR.

c) The distribution of average math achievement scores for eighth graders in the United States is skewed slightly to the left, and roughly unimodal. The distribution is centered at 239. Scores range from 224 to 247, with the middle 50% of the scores falling between 233 and 242. Several low scores, namely New Mexico, Alabama, and Mississippi, pull down the mean, making the median a better measure of center.

51. Gasoline usage 2004.

In 2004, per capita gasoline usage by state in the United States averaged approximately 500 gallons (mean 488.7, median 500.5). The distribution of gasoline usage was bimodal, and slightly skewed to the left, with one low outlier, New York. This state used much less gasoline per person than other states. The IQR of the distribution was 96.9 gallons per person, with the middle 50% of states having gasoline usage between 447.5 and 544.4 gallons per person.

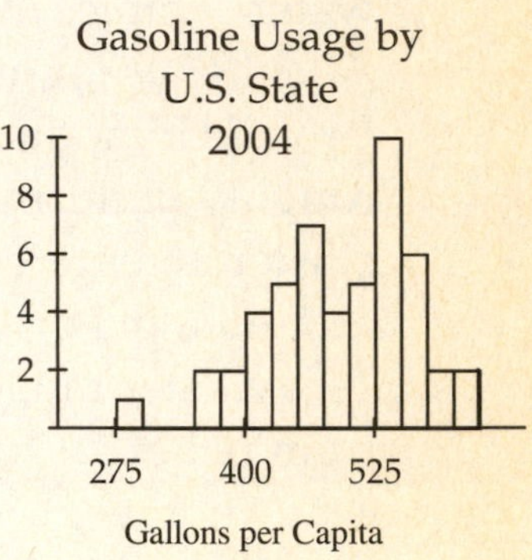

Chapter 5 – Understanding and Comparing Distributions

1. **In the news.** Answers will vary.

3. **Time on the Internet.** Answers will vary.

5. **Pizza prices.**

   a) Pizza prices appear to be both higher on average, and more variable, in Baltimore than in the other three cities. Prices in Chicago may be slightly higher on average than in Dallas and Denver, but the difference is small.

   b) There are low outliers in the distribution of pizza prices in Baltimore and Chicago. There is one high outlier in the distribution of pizza prices in Dallas. These outliers do not affect the overall conclusions reached in the previous part.

7. **Rock concert accidents.**

   a) The histogram and boxplot of the distribution of "crowd crush" victims' ages both show that a typical crowd crush victim was approximately 18 - 20 years of age, that the range of ages is 36 years, that there are two outliers, one victim at age 36 - 38 and another victim at age 46 – 48.

   b) This histogram shows that there may have been two modes in the distribution of ages of "crowd crush" victims, one at 18 - 20 years of age and another at 22 – 24 years of age. Boxplots, in general, can show symmetry and skewness, but not features of shape like bimodality or uniformity.

   c) Median is the better measure of center, since the distribution of ages has outliers. Median is more resistant to outliers than the mean.

   d) IQR is a better measure of spread, since the distribution of ages has outliers. IQR is more resistant to outliers than the standard deviation.

9. **Cereals.**

   a) The maximum sugar content is approximately 60% and the minimum sugar content is approximately 1%, so the range of sugar contents is about 60 – 1 = 59%.

   b) The distribution of sugar content of cereals is bimodal, with modes centered around 5% and 45% sugar by weight.

   c) Some cereals are healthy, low-sugar brands, and others are very sugary.

   d) Yes. The minimum sugar content in the children's cereals is about 35% and the maximum sugar content of adult cereals is only 34%.

   e) The range of sugar contents is about the same for the two types of cereals, approximately 28%, but the IQR is larger for the adult cereals. This is an indication of more variability in the sugar content of the middle 50% of adult cereals.

11. Population growth.

a) Comparative boxplots are at the right.

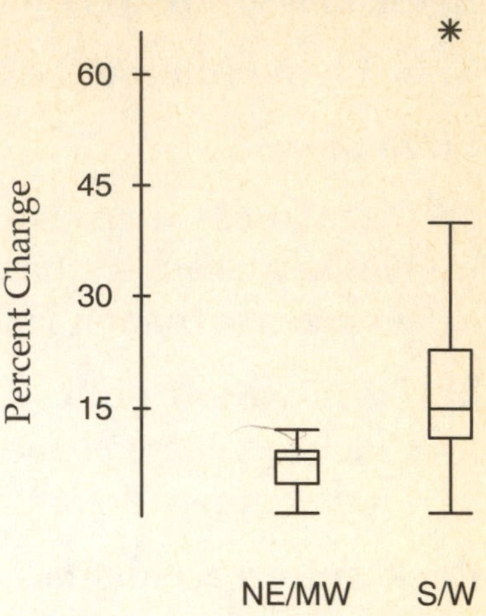

b) The distribution of population growth in NE/MW states is unimodal, symmetric and tightly clustered around 5% growth. The distribution of population growth in S/W states is much more spread out, with most states having population growth between 5% and 30%. A typical state had about 15% growth. There were two outliers, Arizona and Nebraska, with 40% and 66% growth, respectively. Generally, the growth rates in the S/W states were higher and more variable than the rates in the NE/MW states.

13. Hospital stays.

a) The histograms of male and female hospital stay durations would be easier to compare if the were constructed with the same scale, perhaps from 0 to 20 days.

b) The distribution of hospital stays for men is skewed to the right, with many men having very short stays of about 1 or 2 days. The distribution tapers off to a maximum stay of approximately 25 days. The distribution of hospital stays for women is skewed to the right, with a mode at approximately 5 days, and tapering off to a maximum stay of approximately 22 days. Typically, hospital stays for women are longer than those for men.

c) The peak in the distribution of women's hospital stays can be explained by childbirth. This time in the hospital increases the length of a typical stay for women, and not for men.

15. Women's basketball.

a) Both girls have a median score of about 17 points per game, but Scyrine is much more consistent. Her IQR is about 2 points, while Alexandra's is over 10.

b) If the coach wants a consistent performer, she should take Scyrine. She'll almost certainly deliver somewhere between 15 and 20 points. But, if she wants to take a chance and needs a "big game", she should take Alexandra. Alex scores over 24 points about a quarter of the time. On the other hand, she scores under 11 points about as often.

17. Marriage age.

The distribution of marriage age of U.S. men is skewed right, with a typical man (as measured by the median) first marrying at around 24 years old. The middle 50% of male marriage ages is between about 23 and 26 years. For U.S. women, the distribution of marriage age is also skewed right, with median of around 21 years. The middle 50% of female marriage age is between about 20 and 23 years. When comparing the two distributions, the most striking feature is that the distributions are nearly identical in spread, but have different centers. Females typically seem to marry earlier than males. In fact, between 50% and 75% of the women marry at a younger age than *any* man.

19. Fuel economy II.

(Note: numerical details may vary.) In general, fuel economy is higher in cars than in either SUVs or vans. There are numerous outliers on both ends for cars and a few high outliers for SUVs. The top 50% of cars get higher fuel economy than 75% of SUVs and nearly all vans. On average, SUVs and vans get about the same fuel economy, although the distribution for vans shows less spread. The range from vans is about 10 mpg, while for SUVS it is nearly 30 mpg.

21. Test scores.

Class A is Class 1. The median is 60, but has less spread than Class B, which is Class 2. Class C is Class 3, since it's median is higher, which corresponds to the skew to the left.

23. Graduation?

a) The distribution of the percent of incoming college freshman who graduate on time is roughly symmetric. The mean and the median are reasonably close to one another and the quartiles are approximately the same distance from the median.

b) Upper Fence: Q3+1.5(IQR)=74.75+1.5(74.75-59.15)
=74.75+23.4
=98.15

Lower Fence: Q1-1.5(IQR)=59.15-1.5(74.75-59.15)
=59.15-23.4
=35.75

Since the maximum value of the distribution of the percent of incoming freshmen who graduate on time is 87.4% and the upper fence is 98.15%, there are no high outliers. Likewise, since the minimum is 43.2% and the lower fence is 35.75%, there are no low outliers. Since the minimum and maximum percentages are within the fences, all percentages must be within the fences.

c) A boxplot of the distribution of the percent of freshmen who graduate on time is at the right.

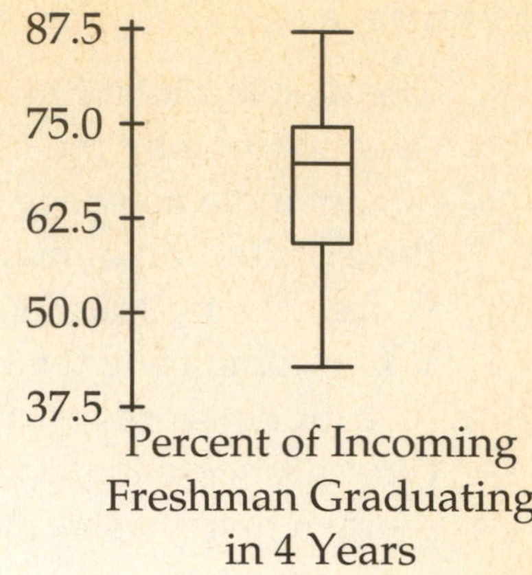

d) The distribution of the percent of incoming freshmen who graduate on time is roughly symmetric, with mean of approximately 68% of freshmen graduating on time. Universities surveyed had between 43.2% and 87.4% of students graduating on time, with the middle 50% of universities reporting between 59.15% and 74.75% graduating on time.

25. Caffeine.

a) *Who* – 45 volunteeers. *What* – Level of caffeine consumption and memory test score.
When – Not specified. *Where* – Not specified. *Why* – The student researchers want to see the possible effects of caffeine on memory. *How* – It appears that the researchers imposed the treatment of level of caffeine consumption in an experiment. However, this point is not clear. Perhaps they allowed the subjects to choose their own level of caffeine.

b) *Variables* – Caffeine level is a categorical variable with three levels: no caffeine, low caffeine, and high caffeine. Test score is a quantitative variable, measured in number of items recalled correctly.

c)

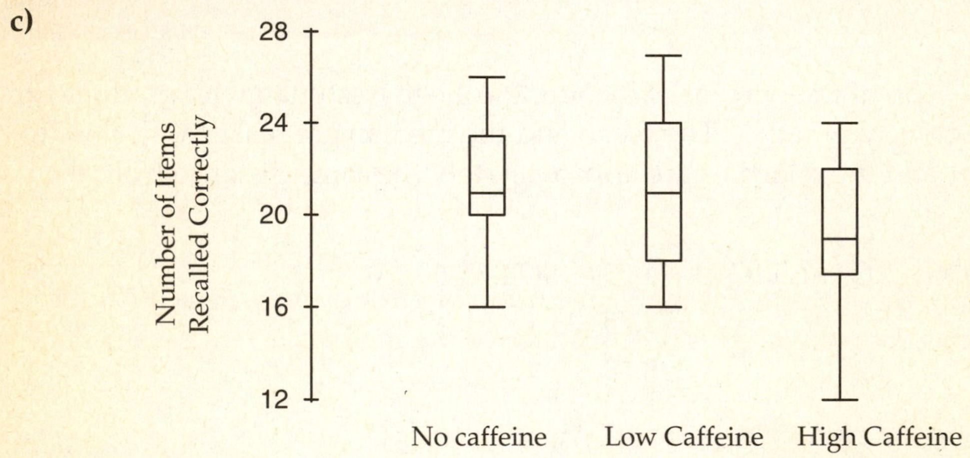

d) The groups consuming no caffeine and low caffeine had comparable memory test scores. A typical score from these groups was around 21. However, the scores of the group consuming no caffeine were more consistent, with a smaller range and smaller interquartile range than the scores of the group consuming low caffeine. The group consuming high caffeine had lower memory scores in general, with a median score of about 19. No one in the high caffeine group scored above 24, but 25% of each of the other groups scored above 24.

27. Derby speeds 2007.

a) The median speed is the speed at which 50% of the winning horses ran slower. Find 50% on the left, move straight over to the graph and down to a speed of about 36 mph.

b) Quartile 1 is at 25% on the left, and Quartile 3 is at 75% on the left. Matching these to the ogive, Q1 = 34.5 mph and Q3 = 36.5 mph.

c) Range = Max – Min = 38 – 31 = 7 mph
IQR = Q3 – Q1 = 36.5 – 34.5 = 2 mph

d) An approximate boxplot of winning Kentucky Derby Speeds is at the right.

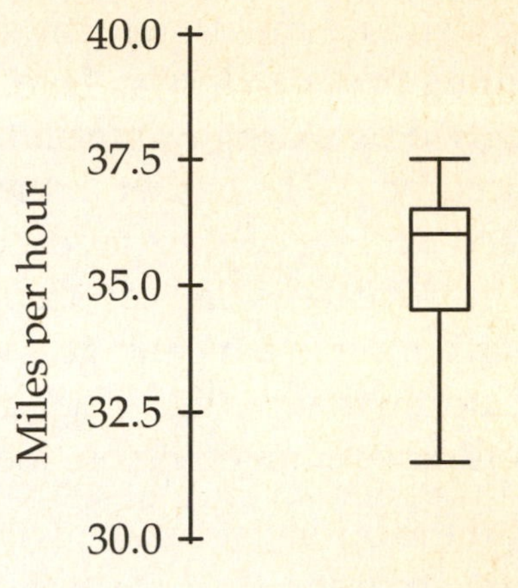

e) The distribution of winning speeds in the Kentucky Derby is skewed to the left. The lowest winning speed is just under 31 mph and the fastest speed is about 37.5 mph. The median speed is approximately 36 mph, and 75% of winning speeds are above 34.5 mph. Only a few percent of winners have had speeds below 33 mph.

29. Reading scores.

a) The highest score for boys was 6, which is higher than the highest score for girls, 5.9.

b) The range of scores for boys is greater than the range of scores for girls.
Range = Max – Min   Range(Boys) = 4   Range(Girls) = 3.1

c) The girls had the greater IQR.
IQR = Q3 – Q1   IQR(Boys) = 4.9 – 3.9 = 1   IQR(Girls) = 5.2 – 3.8 = 1.4

d) The distribution of boys' scores is more skewed. The quartiles are not the same distance from the median. In the distribution of girls' scores, Q1 is 0.7 units below the median, while Q3 is 0.7 units above the median.

e) Overall, the girls did better on the reading test. The median, 4.5, was higher than the median for the boys, 4.3. Additionally, the upper quartile score was higher for girls than boys, 5.2 compared to 4.9. The girls' lower quartile score was slightly lower than the boys' lower quartile score, 3.8 compared to 3.9.

f) The overall mean is calculated by weighting each mean by the number of students. $\frac{14(4.2)+11(4.6)}{25} = 4.38$

31. Industrial experiment.

First of all, there is an extreme outlier in the distribution of distances for the slow speed drilling. One hole was drilled almost an inch away from the center of the target! If that distance is correct, the engineers at the computer production plant should investigate the slow speed drilling process closely. It may be plagued by extreme, intermittent inaccuracy. The outlier in the slow speed drilling process is so extreme that no graphical display can display the distribution in a meaningful way while including that outlier. That distance should be removed before looking at a plot of the drilling distances.

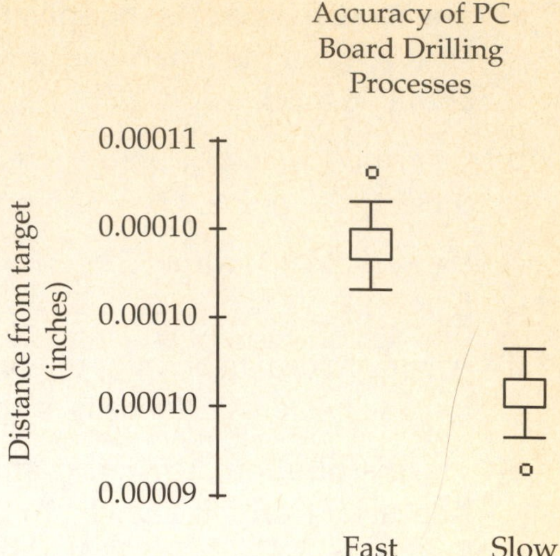

With the outlier set aside, we can determine that the slow drilling process is more accurate. The greatest distance from the target for the slow drilling process, 0.000098 inches, is still more accurate than the smallest distance for the fast drilling process, 0.000100 inches.

33. MPG.

a) A back-to-back stemplot of these data is shown at the right. A plot with with tens and units digits for stems and tenths for leaves would have been quite long, but still useable. A plot with tens as stems and rounded units as leaves would have been too compact. This plot has tens as the stems, but the stems are split 5 ways. The uppermost 2 stem displays 29 and 28, the next 2 stem displays 27 and 26, and so on. The key indicates the rounding used, as well as the accuracy of the original data. In this case, the mileages were given to the nearest tenth.

US Cars		Other Cars
	3	7
44	3	45
	3	222
01	3	01
89	2	8
777	2	7
	2	
2	2	222
11	2	0
888999	1	
6777	1	67

KEY:
2 | 5 = 24.5 – 25.4 mpg

b) In general, the Other cars got better gas mileage than the US Cars, although both distributions were highly variable. The distribution of US cars was bimodal and skewed to the right, with many cars getting mileages in the high teens and low twenties, and another group of cars whose mileages were in the high twenties and low thirties. Two high outliers had mileages of 34 miles per gallon. The distribution of Other cars, in contrast, was bimodal and skewed to the left. Most cars had mileages in the high twenties and thirties, with a small group of cars whose mileages were in the low twenties. Two low outliers had mileages of 16 and 17 miles per gallon.

35. Fruit Flies.

a) The most fruit flies died around day 16. Over 60,000 fruit flies died that day.

b) The largest proportion of fruit flies died around day 65. About 20% of that previously surviving population died that day.

c) At around day 50, the number of fruit flies wasn't changing by very much day to day.

37. Assets.

a) The distribution of assets of 79 companies chosen from the *Forbes* list of the nation's top corporations is skewed so heavily to the right that the vast majority of companies have assets represented in the first bar of the histogram, 0 to 10 billion dollars. This makes meaningful discussion of center and spread impossible.

b) Re-expressing these data by, for example, logs or square roots might help make the distribution more nearly symmetric. Then a meaningful discussion of center might be possible.

39. Assets again.

a) The distribution of logarithm of assets is preferable, because it is roughly unimodal and symmetric. The distribution of the square root of assets is still skewed right, with outliers.

b) If $\sqrt{Assets} = 50$, then the companies assets are approximately $50^2 = 2500$ million dollars.

c) If $\log(Assets) = 3$, then the companies assets are approximately $10^3 = 1000$ million dollars.

41. Stereograms.

a) The two variables discussed in the description are fusion time and treatment group.

b) Fusion time is a quantitative variable, measured in seconds. Treatment group is a categorical variable, with subjects either receiving verbal clues only, or visual and verbal clues.

c) Generally, the Visual/Verbal group had shorter fusion times than the No/Verbal group. The median for the Visual/Verbal group was approximately the same as the lower quartile for the No/Verbal group. The No/Verbal Group also had an extreme outlier, with at least one subject whose fusion time was approximately 50 seconds. There is evidence that visual information may reduce fusion time.

Chapter 6 – The Standard Deviation as a Ruler and the Normal Model

1. **Shipments.**

   a) Adding 4 ounces will affect only the median. The new median will be 68 +4 = 72 ounces, and the IQR will remain at 40 ounces.

   b) Changing the units will affect both the median and IQR. The median will be 72/16 = 4.5 pounds and the IQR will be 40/16 = 2.5 pounds.

3. **Payroll.**

   a) The distribution of salaries in the company's weekly payroll is skewed to the right. The mean salary, \$700, is higher than the median, \$500.

   b) The IQR, \$600, measures the spread of the middle 50% of the distribution of salaries.

$$\begin{aligned} Q3 - Q1 &= IQR \\ Q3 &= Q1 + IQR \\ Q3 &= \$350 + \$600 \\ Q3 &= \$950 \end{aligned}$$

   50% of the salaries are found between \$350 and \$950.

   c) If a \$50 raise were given to each employee, all measures of center or position would increase by \$50. The minimum would change to \$350, the mean would change to \$750, the median would change to \$550, and the first quartile would change to \$400. Measures of spread would not change. The entire distribution is simply shifted up \$50. The range would remain at \$1200, the IQR would remain at \$600, and the standard deviation would remain at \$400.

   d) If a 10% raise were given to each employee, all measures of center, position, and spread would increase by 10%.

Minimum = \$330	Mean = \$770	Median = \$550	Range = \$1320
IQR = \$660	First Quartile = \$385	St. Dev. = \$440	

5. **SAT or ACT?**

   Measures of center and position (lowest score, top 25% above, mean, and median) will be multiplied by 40 and increased by 150 in the conversion from ACT to SAT by the rule of thumb. Measures of spread (standard deviation and IQR) will only be affected by the multiplication.

Lowest score = 910	Mean = 1230	Standard deviation = 120
Top 25% above = 1350	Median = 1270	IQR = 240

7. Stats test.

A z-score of 2.20 means that your score was 2.20 standard deviations above the mean.

9. Stats test, part II.

Gregor scored 65 points.

$$z = \frac{y - \mu}{\sigma}$$

$$-2 = \frac{y - 75}{5}$$

$$y = 65$$

11. Temperatures.

In January, with mean temperature 36° and standard deviation in temperature 10°, a high temperature of 55° is almost 2 standard deviations above the mean. In July, with mean temperature 74° and standard deviation 8°, a high temperature of 55° is more than two standard deviations below the mean. A high temperature of 55° is less likely to happen in July, when 55° is farther away from the mean.

13. Combining test scores.

The z-scores, which account for the difference in the distributions of the two tests, are 1.5 and 0 for Derrick and 0.5 and 2 for Julie. Derrick's total is 1.5 which is less than Julie's 2.5.

15. Final Exams.

a) Anna's average is $\frac{83+83}{2} = 83$. Megan's average is $\frac{77+95}{2} = 86$.

Only Megan qualifies for language honors, with an average higher than 85.

b) On the French exam, the mean was 81 and the standard deviation was 5. Anna's score of 83 was 2 points, or 0.4 standard deviations, above the mean. Megan's score of 77 was 4 points, or 0.8 standard deviations below the mean.

On the Spanish exam, the mean was 74 and the standard deviation was 15. Anna's score of 83 was 9 points, or 0.6 standard deviations, above the mean. Megan's score of 95 was 21 points, or 1.4 standard deviations, above the mean.

Measuring their performance in standard deviations is the only fair way in which to compare the performance of the two women on the test.

Anna scored 0.4 standard deviations above the mean in French and 0.6 standard deviations above the mean in Spanish, for a total of 1.0 standard deviation above the mean.

Megan scored 0.8 standard deviations below the mean in French and 1.4 standard deviations above the mean in Spanish, for a total of only 0.6 standard deviations above the mean.

Anna did better overall, but Megan had the higher average. This is because Megan did very well on the test with the higher standard deviation, where it was comparatively easy to do well.

17. Cattle.

a) A steer weighing 1000 pounds would be about 1.81 standard deviations below the mean weight.

$$z = \frac{y-\mu}{\sigma} = \frac{1000-1152}{84} \approx -1.81$$

b) A steer weighing 1000 pounds is more unusual. Its z-score of –1.81 is further from 0 than the 1250 pound steer's z-score of 1.17.

19. More cattle.

a) The new mean would be 1152 – 1000 = 152 pounds. The standard deviation would not be affected by subtracting 1000 pounds from each weight. It would still be 84 pounds.

b) The mean selling price of the cattle would be 0.40(1152) = \$460.80. The standard deviation of the selling prices would be 0.40(84) = \$33.60.

21. Cattle, part III.

Generally, the minimum and the median would be affected by the multiplication and subtraction. The standard deviation and the IQR would only be affected by the multiplication.

Minimum = 0.40(980) – 20 = \$372.00
Median = 0.40(1140) – 20 = \$436
Standard deviation = 0.40(84) = \$33.60
IQR = 0.40(102) = \$40.80

23. Professors.

The standard deviation of the distribution of years of teaching experience for college professors must be 6 years. College professors can have between 0 and 40 (or possibly 50) years of experience. A workable standard deviation would cover most of that range of values with ±3 standard deviations around the mean. If the standard deviation were 6 months ($\frac{1}{2}$ year), some professors would have years of experience 10 or 20 standard deviations away from the mean, whatever it is. That isn't possible. If the standard deviation were 16 years, ±2 standard deviations would be a range of 64 years. That's way too high. The only reasonable choice is a standard deviation of 6 years in the distribution of years of experience.

25. Guzzlers?

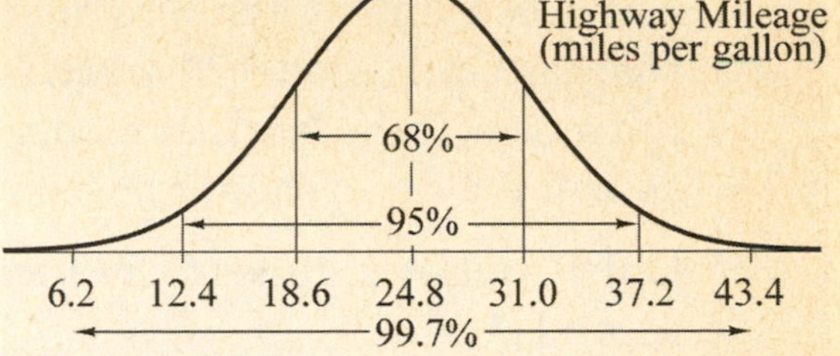

a) The Normal model for auto fuel economy is at the right.

b) Approximately 68% of the cars are expected to have highway fuel economy between 18.6 mpg and 31.0 mpg.

c) Approximately 16% of the cars are expected to have highway fuel economy above 31 mpg.

d) Approximately 13.5% of the cars are expected to have highway fuel economy between 31 mpg and 37 mpg.

e) The worst 2.5% of cars are expected to have fuel economy below approximately 12.4 mpg.

27. Small steer.

Any weight more than 2 standard deviations below the mean, or less than 1152 – 2(84) = 984 pounds might be considered unusually low.. We would expect to see a steer below
1152 – 3(84) = 900 very rarely.

29. Trees.

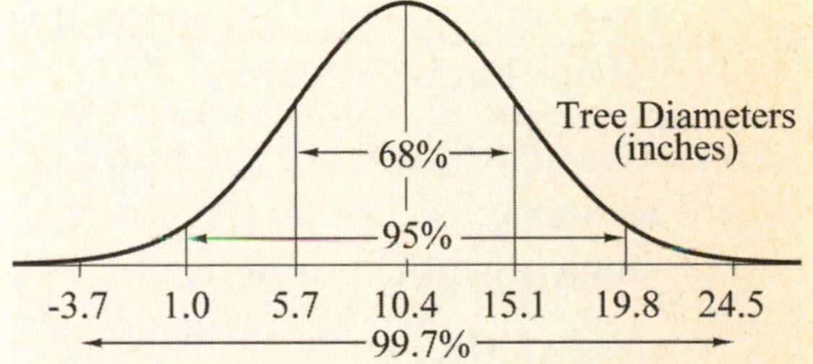

a) The Normal model for the distribution of tree diameters is at the right.

b) Approximately 95% of the trees are expected to have diameters between 1.0 inch and 19.8 inches.

c) Approximately 2.5% of the trees are expected to have diameters less than an inch.

d) Approximately 34% of the trees are expected to have diameters between 5.7 inches and 10.4 inches.

e) Approximately 16% of the trees are expected to have diameters over 15 inches.

31. Trees, part II.

The use of the Normal model requires a distribution that is unimodal and symmetric. The distribution of tree diameters is neither unimodal nor symmetric, so use of the Normal model is not appropriate.

33. Winter Olympics 2006 downhill.

a) The 2006 Winter Olympics downhill times have mean of 113.02 seconds and standard deviation 3.24 seconds. 109.78 seconds is 1 standard deviation below the mean. If the Normal model is appropriate, 16% of the times should be below 99.7 seconds.

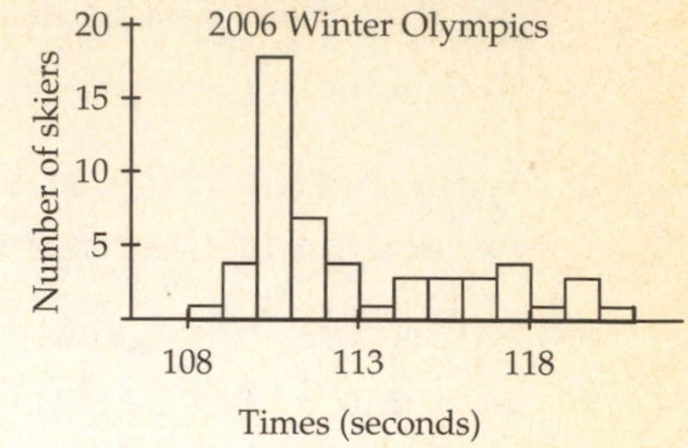

b) Only 2 out of 53 times (3.8%) are below 109.78 seconds.

c) The percentages in parts a and b do not agree because the Normal model is not appropriate in this situation.

d) The histogram of 2006 Winter Olympic Downhill times is skewed to the right, and is skewed to the right. The Normal model is not appropriate for the distribution of times, because the distribution is symmetric.

35. Receivers.

a) Approximately 2.5% of the receivers are expected to gain fewer yards than 2 standard deviations below the mean number of yards gained.

b) The distribution of the number of yards gained has mean 435 yards and standard deviation 384 yards. According to the Normal model, receivers who run below 2 standard deviations below the mean number of yards are expected to run fewer than –333 yards. Of course, it is impossible to run fewer than 0 yards, let alone fewer than –333 yards.

c) The distribution of the number of yards run by wide receivers is skewed heavily to the right. Use of the Normal model is not appropriate for this distribution, since it is not symmetric.

37. Normal cattle.

a)

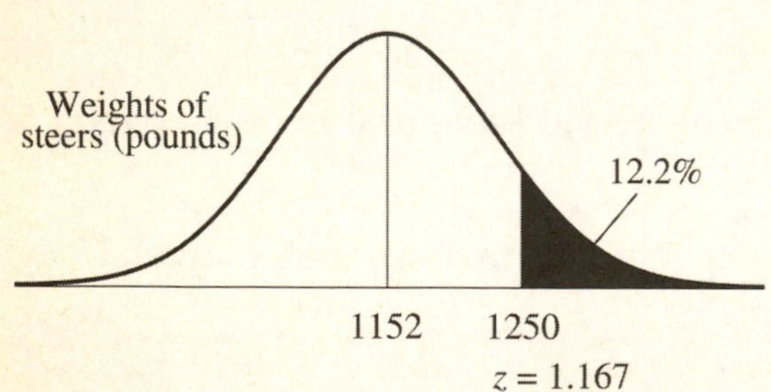

$$z = \frac{y - \mu}{\sigma}$$

$$z = \frac{1250 - 1152}{84}$$

$$z \approx 1.167$$

According to the Normal model, 12.2% of steers are expected to weigh over 1250 pounds.

b)

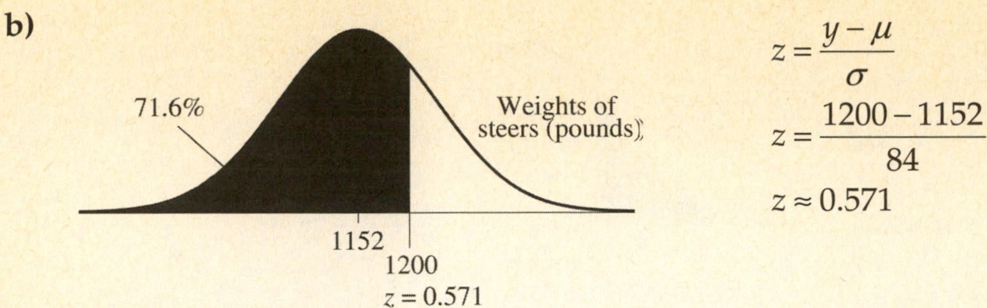

$$z = \frac{y - \mu}{\sigma}$$

$$z = \frac{1200 - 1152}{84}$$

$$z \approx 0.571$$

According to the Normal model, 71.6% of steers are expected to weigh under 1200 pounds.

c)

$$z = \frac{y - \mu}{\sigma} \qquad z = \frac{y - \mu}{\sigma}$$

$$z = \frac{1000 - 1152}{84} \qquad z = \frac{1100 - 1152}{84}$$

$$z \approx -1.810 \qquad z \approx -0.619$$

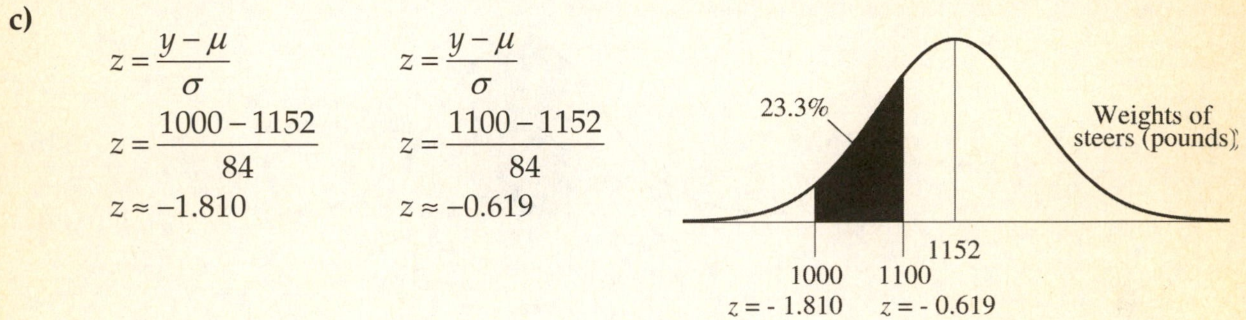

According to the Normal model, 23.3% of steers are expected to weigh between 1000 and 1100 pounds.

39. More cattle.

a)

Weights of steers (pounds)
10%
1152
$z = 1.282$
$y = 1259.7$ pounds

$$z = \frac{y - \mu}{\sigma}$$

$$1.282 = \frac{y - 1152}{84}$$

$$y \approx 1259.7$$

According to the Normal model, the highest 10% of steer weights are expected to be above approximately 1259.7 pounds.

b)

$$z = \frac{y-\mu}{\sigma}$$
$$-0.842 = \frac{y-1152}{84}$$
$$y \approx 1081.3$$

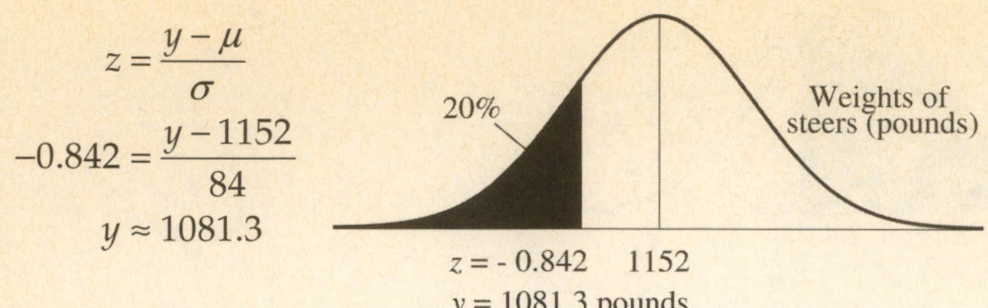

According to the Normal model, the lowest 20% of weights of steers are expected to be below approximately 1081.3 pounds.

c)

$$z = \frac{y-\mu}{\sigma}$$
$$-0.524 = \frac{y-1152}{84}$$
$$y \approx 1108.0$$

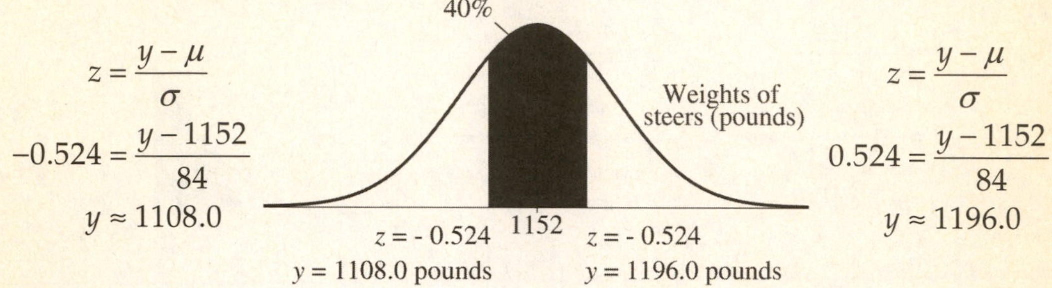

$$z = \frac{y-\mu}{\sigma}$$
$$0.524 = \frac{y-1152}{84}$$
$$y \approx 1196.0$$

According to the Normal model, the middle 40% of steer weights is expected to be between about 1108.0 pounds and 1196.0 pounds.

41. Cattle, finis.

a)

$$z = \frac{y-\mu}{\sigma}$$
$$-0.253 = \frac{y-1152}{84}$$
$$y \approx 1130.7$$

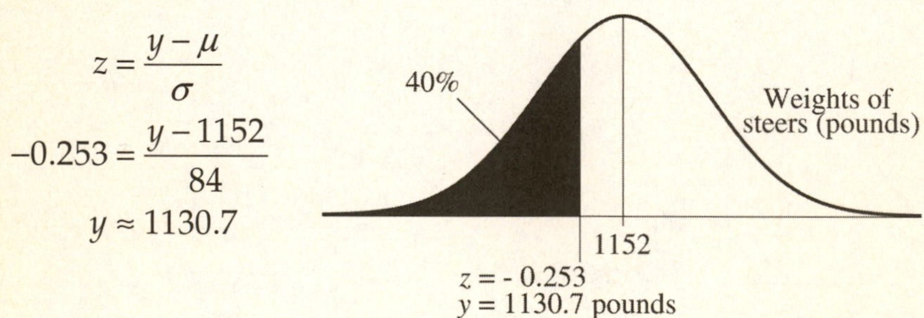

According to the Normal model, the weight at the 40th percentile is 1130.7 pounds. This means that 40% of steers are expected to weigh less than 1130.7 pounds.

b)

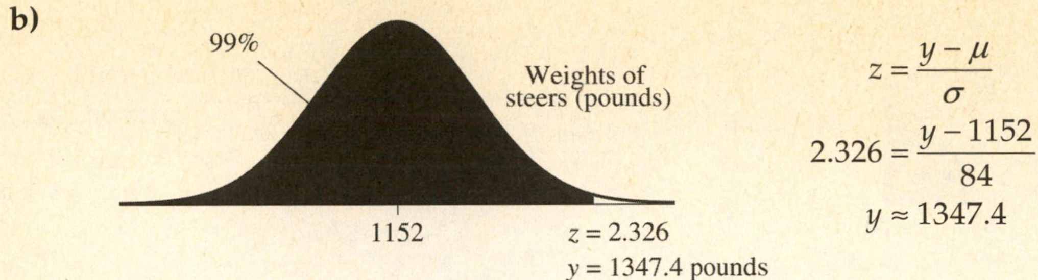

According to the Normal model, the weight at the 99th percentile is 1347.4 pounds. This means that 99% of steers are expected to weigh less than 1347.4 pounds.

c)

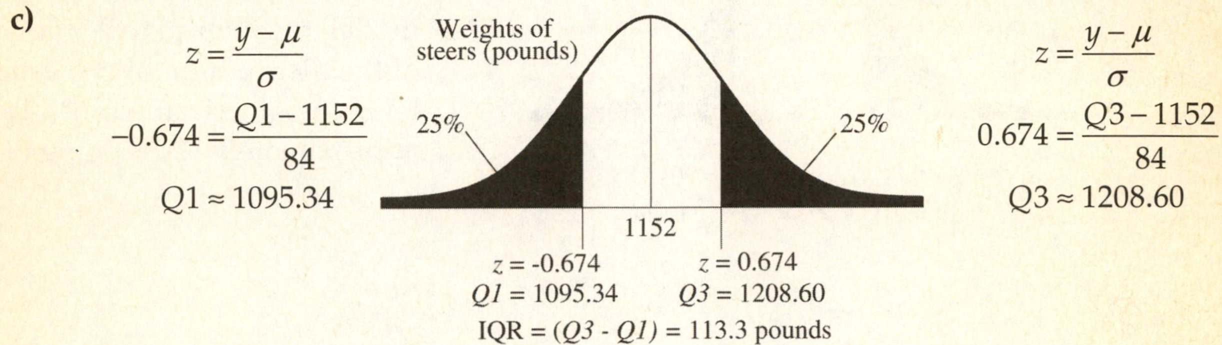

According to the Normal model, the IQR of the distribution of weights of Angus steers is about 113.3 pounds.

43. Cholesterol.

a) The Normal model for cholesterol levels of adult American women is at the right.

b)

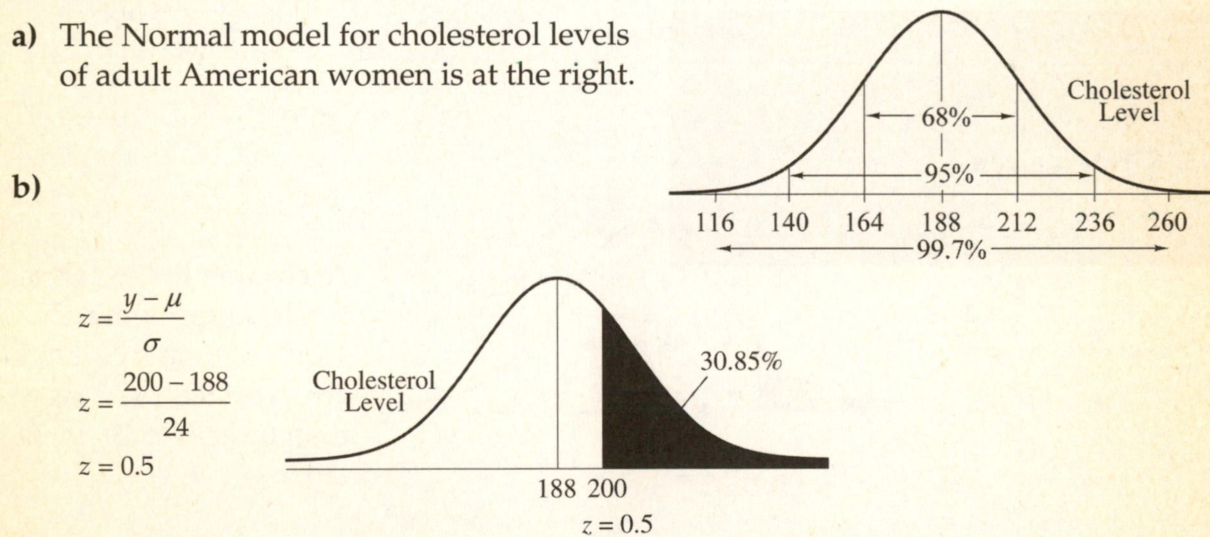

According to the Normal model, 30.85% of American women are expected to have cholesterol levels over 200.

c)

17%

Cholesterol Level

150 170 188

$z = -1.583$ $z = -0.75$

According to the Normal model, 17.00% of American women are expected to have cholesterol levels between 150 and 170.

d)

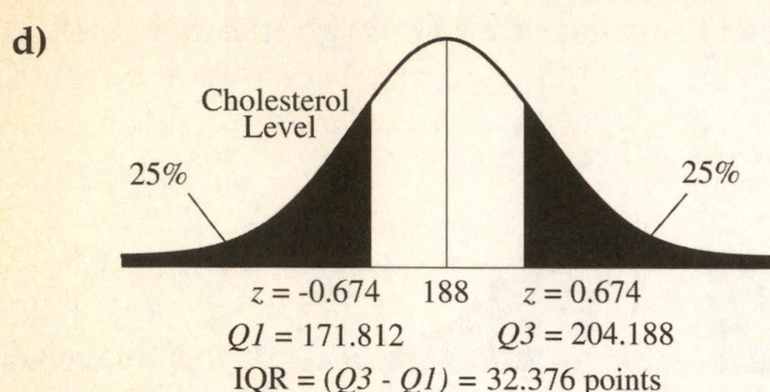

According to the Normal model, the interquartile range of the distribution of cholesterol levels of American women is approximately 32.38 points.

e)

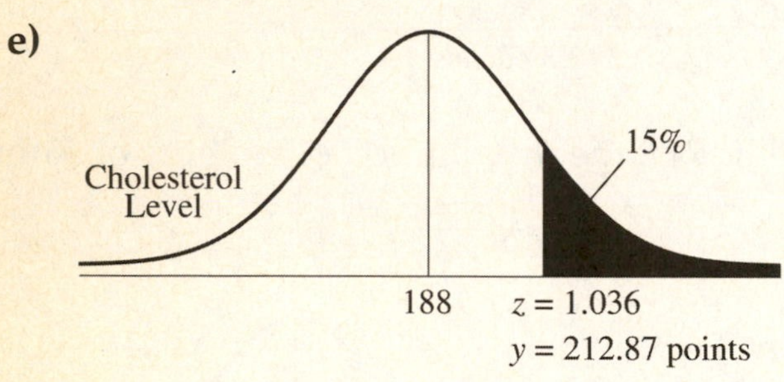

$$z = \frac{y - \mu}{\sigma}$$

$$1.036 = \frac{y - 188}{24}$$

$$y = 212.87$$

According to the Normal model, the highest 15% of women's cholesterol levels are above approximately 212.87.

45. Kindergarten.

a)

$$z = \frac{y - \mu}{\sigma}$$

$$z = \frac{36 - 38.2}{1.8}$$

$$z = -1.222$$

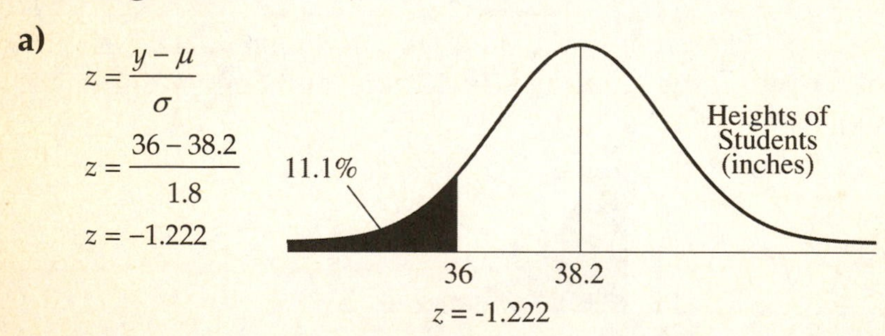

According to the Normal model, approximately 11.1% of kindergarten kids are expected to be less than three feet (36 inches) tall.

b)

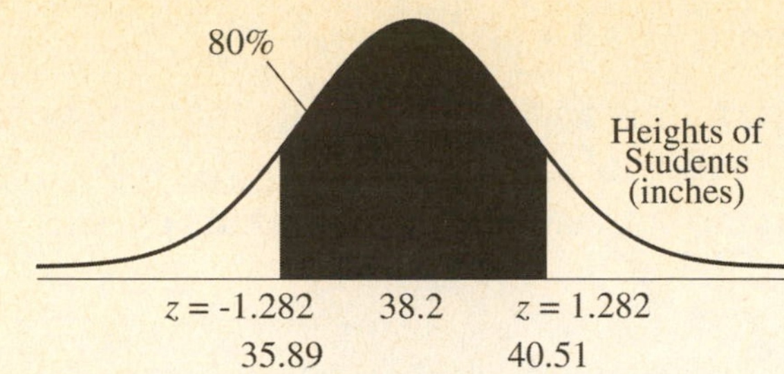

$$z = \frac{y - \mu}{\sigma}$$
$$-1.282 = \frac{y_1 - 38.2}{1.8}$$
$$y_1 = 35.89$$

$$z = \frac{y - \mu}{\sigma}$$
$$1.282 = \frac{y_2 - 38.2}{1.8}$$
$$y_2 = 40.51$$

According to the Normal model, the middle 80% of kindergarten kids are expected to be between 35.89 and 40.51 inches tall. (The appropriate values of $z = \pm 1.282$ are found by using right and left tail percentages of 10% of the Normal model.)

c)

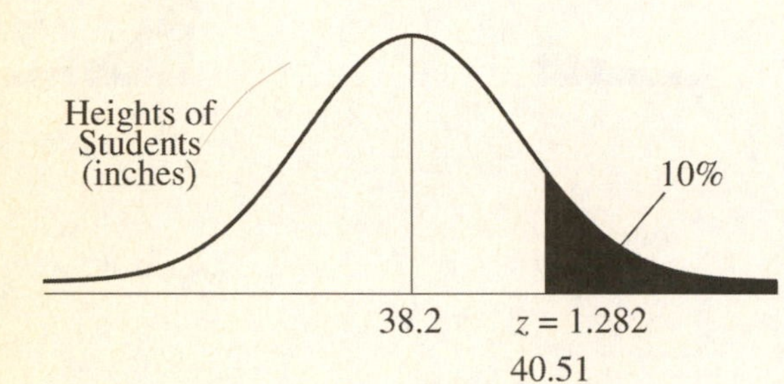

$$z = \frac{y - \mu}{\sigma}$$
$$1.282 = \frac{y - 38.2}{1.8}$$
$$y = 40.51$$

According to the Normal model, the tallest 10% of kindergarteners are expected to be at least 40.51 inches tall.

47. Eggs.

a)

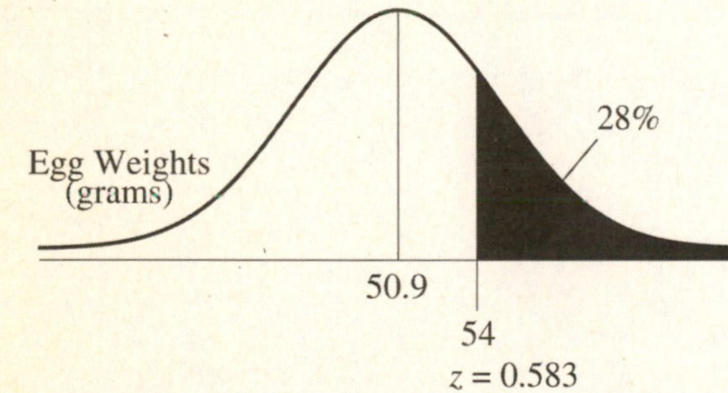

$$z = \frac{y - \mu}{\sigma}$$
$$0.583 = \frac{54 - 50.9}{\sigma}$$
$$0.583\sigma = 3.1$$
$$\sigma = 5.317$$

According to the Normal model, the standard deviation of the egg weights for young hens is expected to be 5.3 grams.

b)

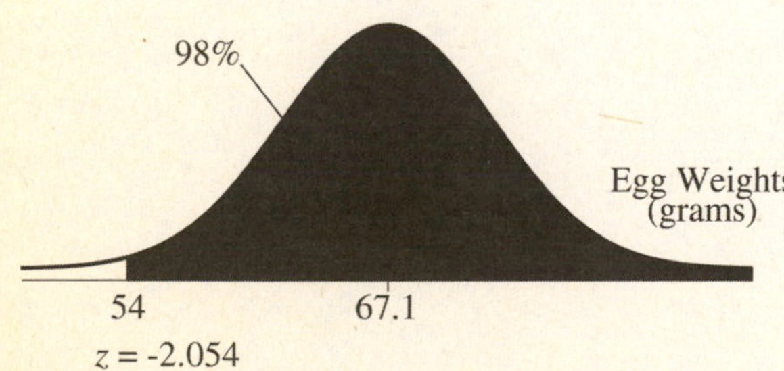

$$z = \frac{y - \mu}{\sigma}$$
$$-2.054 = \frac{54 - 67.1}{\sigma}$$
$$-2.054\sigma = -13.1$$
$$\sigma = 6.377$$

According to the Normal model, the standard deviation of the egg weights for older hens is expected to be 6.4 grams.

c) The younger hens lay eggs that have more consistent weights than the eggs laid by the older hens. The standard deviation of the weights of eggs laid by the younger hens is lower than the standard deviation of the weights of eggs laid by the older hens.

d) A good way to visualize the solution to this problem is to look at the distance between 54 and 70 grams in two different scales. First, 54 and 70 grams are 16 grams apart. When measured in standard deviations, the respective *z*-scores, -1.405 and 1.175, are 2.580 standard deviations apart. So, 16 grams must be the same as 2.580 standard deviations.

According to the Normal model, the mean weight of the eggs is 62.7 grams, with a standard deviation of 6.2 grams.

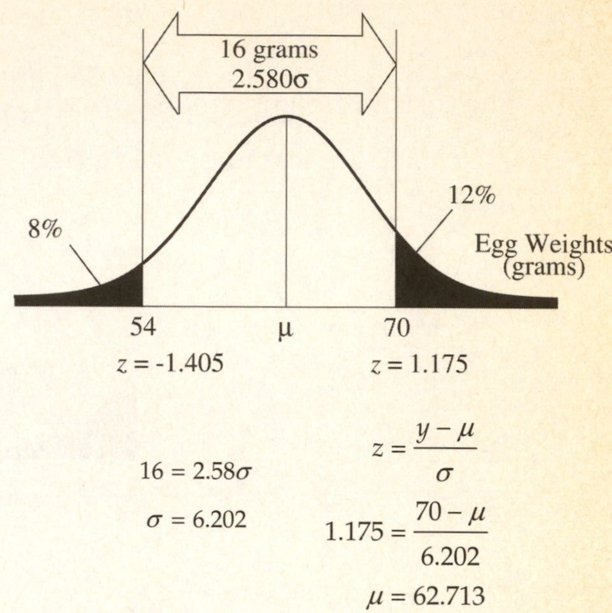

Review of Part I – Exploring and Understanding Data

1. **Bananas.**

   a) A histogram of the prices of bananas from 15 markets, as reported by the USDA, appears at the right.

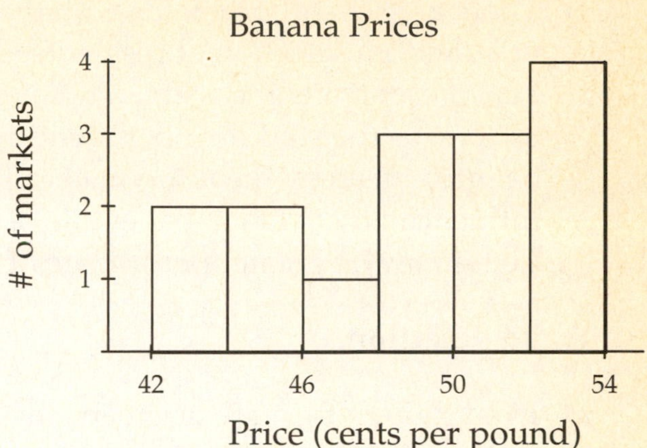

   b) The distribution of banana prices is skewed to the left, so median and IQR are appropriate measures of center and spread. Median = 49 cents per pound
   IQR = 7 cents per pound

   c) The distribution of the prices of bananas from 15 markets, as reported by the USDA, is unimodal and skewed to the left. The center of the distribution is approximately 50 cents, with the lowest price 42 cents per pound and the highest price 53 cents per pound.

3. **Singers.**

   a) The two statistics could be the same if there were many sopranos of that height.

   b) The distribution of heights of each voice part is roughly symmetric. The basses and tenors are generally taller than the altos and sopranos, with the basses being slightly taller than the tenors. The sopranos and altos have about the same median height. Heights of basses and sopranos are more consistent than altos and tenors.

5. **Beanstalks.**

   a) The greater standard deviation for the distribution of women's heights means that their heights are more variable than the heights of men.

   b) The z-score for women to qualify is 2.4 compared with 1.75 for men, so it is harder for women to qualify.

7. **State University.**

   a) *Who* – Local residents near State University. *What* – Age, whether or not the respondent attended college, and whether or not the respondent had a favorable opinion of State University. *When* – Not specified. *Where* – Region around State University. *Why* – The information will be included in a report to the University's directors. *How* – 850 local residents were surveyed by phone.

b) There is one quantitative variable, age, probably measured in years. There are two categorical variables, college attendance (yes or no), and opinion of State University (favorable or unfavorable).

c) There are several problems with the design of the survey. No mention is made of a random selection of residents. Furthermore, there may be a non-response bias present. People with an unfavorable opinion of the university may hang up as soon as the staff member identifies himself or herself. Also, response bias may be introduced by the interviewer. The responses of the residents may be influenced by the fact that employees of the university are asking the questions. There may be greater percentage of favorable responses than truly exist.

9. Fraud detection.

a) Even though they are numbers, the SIC code is a categorical variable. A histogram is a quantitative display, so it is not appropriate.

b) The Normal model will not work at all. The Normal model is for modeling distributions of unimodal and symmetric quantitative variables. SIC code is a categorical variable.

11. Cramming.

a) Comparitive boxplots of the distributions of Friday and Monday scores are at the right.

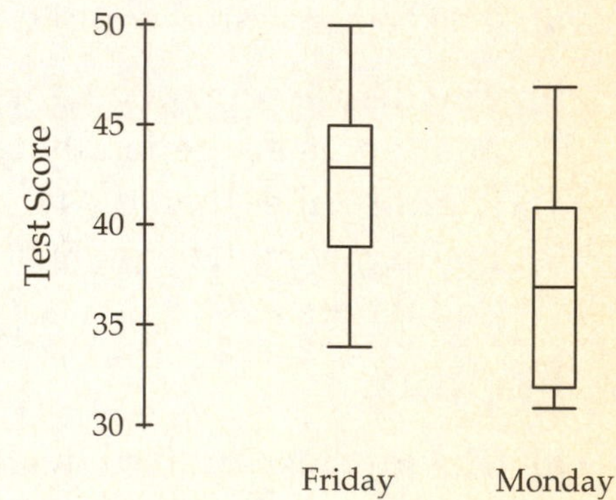

b) The distribution of scores on Friday was generally higher by about 5 points. Students fared worse on Monday after preparing for the test on Friday. The spreads are about the same, but the scores on Monday are slightly skewed to the right.

c) A histogram of the distribution of change in test score is at the right.

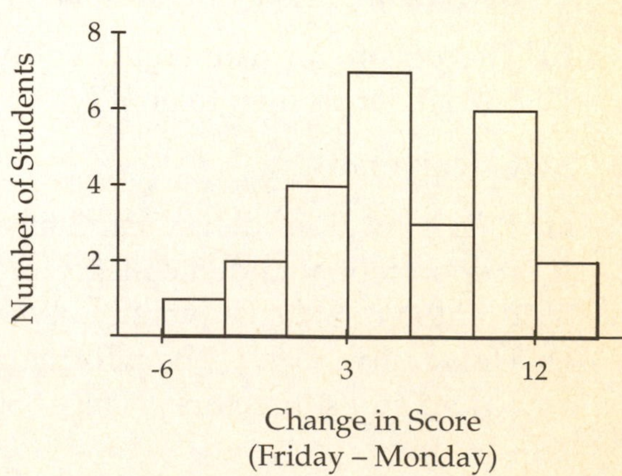

d) The distribution of changes in score is roughly unimodal and symmetric, and is centered near 4 points. Changes ranged from a student who scored 5 points higher on Monday, to two students who each scored 14 points higher on Friday. Only three students did better on Monday.

13. Let's play cards.

a) Suit is a categorical variable.

b) In the game of Go Fish, the denomination is not ordered. Numbers are merely matched with one another. You may have seen children's Go Fish decks that have symbols or pictures on the cards instead of numbers. These work just fine.

c) In the game of Gin Rummy, the order of the cards is important. During the game, ordered "runs" of cards are assembled (with Jack = 11, Queen = 12, King = 13), and at the end of the hand, points are totaled from the denomination of the card (face cards = 10 points). However, even in Gin Rummy, the denomination of the card sometimes behaves like a categorical variable. When you are collecting 3s, for example, order doesn't matter.

15. Hard water.

a) The variables in this study are both quantitative. Annual mortality rate for males is measured in deaths per 100,000. Calcium concentration is measured in parts per million.

b) The distribution of calcium concentration is skewed right, with many towns having concentrations below 25 ppm. The rest of the towns have calcium concentrations which are distributed in a fairly uniform pattern from 25 ppm to 100 ppm, tapering off to a maximum concentration around 150 ppm.
The distribution of mortality rates is unimodal and symmetric, with center approximately 1500 deaths per 100,000. The distribution has a range of 1000 deaths per 100,000, from 1000 to 2000 deaths per 100,000.

17. Seasons.

a) The two histograms have different horizontal and vertical scales. This makes a quick comparison impossible.

b) The center of the distribution of average temperatures in January in is the low 30s, compared to a center of the distribution of July temperatures in the low 70s. The January distribution is also much more spread out than the July distribution. The range is over 50 degrees in January, compared to a range of over 20 in July. The distribution of average temperature in January is skewed slightly to the right, while the distribution of average temperature in July is roughly symmetric.

c) The distribution of difference in average temperature (July – January) for 60 large U.S. cities is slightly skewed to the left, with median at approximately 44 degrees. There are several low outliers, cities with very little difference between their average July and January temperatures. The single high outlier is a city with a large difference in average temperature between July and January. The middle 50% of differences are between approximately 38 and 46 degrees.

19. Old Faithful?

a) The distribution of duration of the 222 eruptions is bimodal, with modes at approximately 2 minutes and 4.5 minutes. The distribution is fairly symmetric around each mode.

b) The bimodal shape of the distribution of duration of the 222 eruptions suggests that there may be two distinct groups of eruption durations. Summary statistics would try to summarize these two groups as a single group, which wouldn't make sense.

c) The intervals between eruptions are generally longer for long eruptions than the intervals for short eruptions. Over 75% of the short eruptions had intervals of approximately 60 minutes or less, while almost all of the long eruptions had intervals of more than 60 minutes.

21. Liberty's nose.

a) The distribution of the ratio of arm length to nose length of 18 girls in a statistics class is unimodal and roughly symmetric, with center around 15. There is one low outlier, a ratio of 11.8. A boxplot is provided at the right. A histogram or stemplot is also an appropriate display.

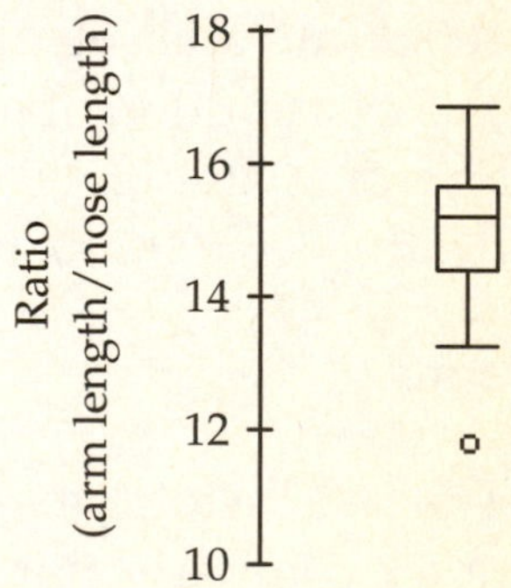

b) In the presence of an outlier, the 5-number summary is the appropriate choice for summary statistics. The 5-number summary is 11.8, 14.4, 15.25, 15.7, 16.9. The IQR is 1.3.

c) The ratio of 9.3 for the Statue of Liberty is very low, well below the lowest ratio in the statistics class, 11.8, which is already a low outlier. Compared to the girls in the statistics class, the Statue of Liberty's nose is very long in relation to her arm.

23. Sample.

Overall, the follow-up group was insured only 11.1% of the time as compared to 16.6% for the not traced group. At first, it appears that group is associated with presence of health insurance. But for blacks, the follow-up group was quite close (actually slightly higher) in terms of being insured: 8.9% to 8.7%. The same is true for whites. The follow-up group was insured 83.3% of the time, compared to 82.5% of the not traced group. When broken down by race, we see that group is not associated with presence of health insurance for either race. This demonstrates Simpson's paradox, because the overall percentages lead us to believe that there is an association between health insurance and group, but we see the truth when we examine the situation more carefully.

25. Be quick!

a) The Normal model for the distribution of reaction times is at the right.

b) The distribution of reaction times is unimodal and symmetric, with mean 1.5 seconds, and standard deviation 0.18 seconds. According to the Normal model, 95% of drivers are expected to have reaction times between 1.14 seconds and 1.86 seconds.

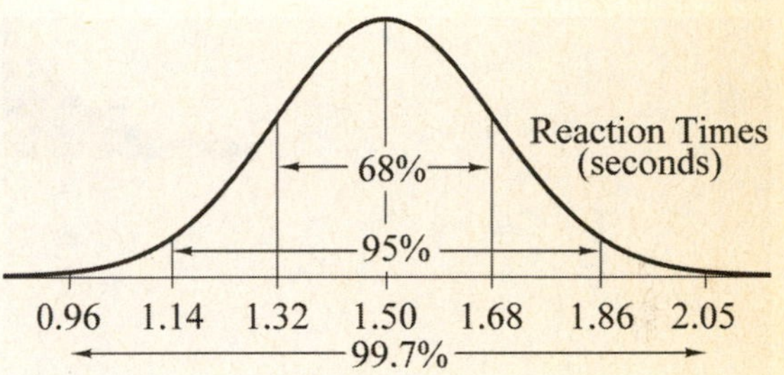

c)

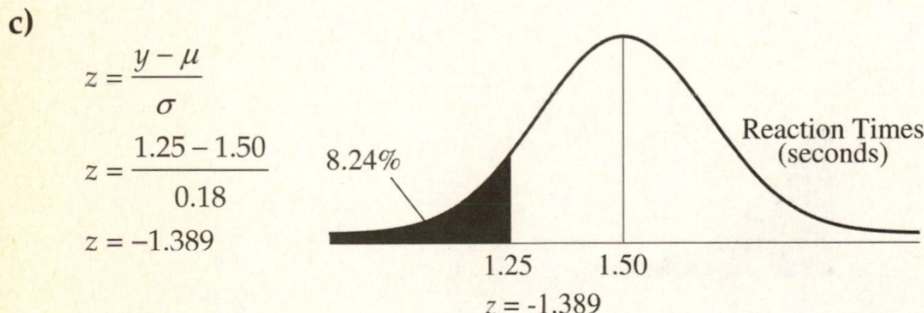

According to the Normal model, 8.24% of drivers are expected to have reaction times below 1.25 seconds.

d)

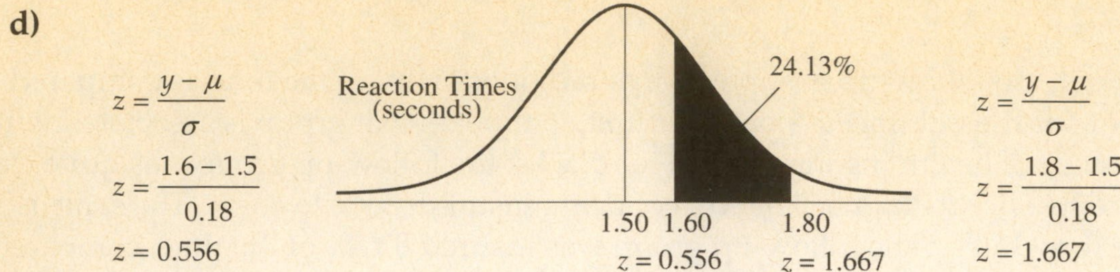

According to the Normal model, 24.13% of drivers are expected to have reaction times between 1.6 seconds and 1.8 seconds.

e)

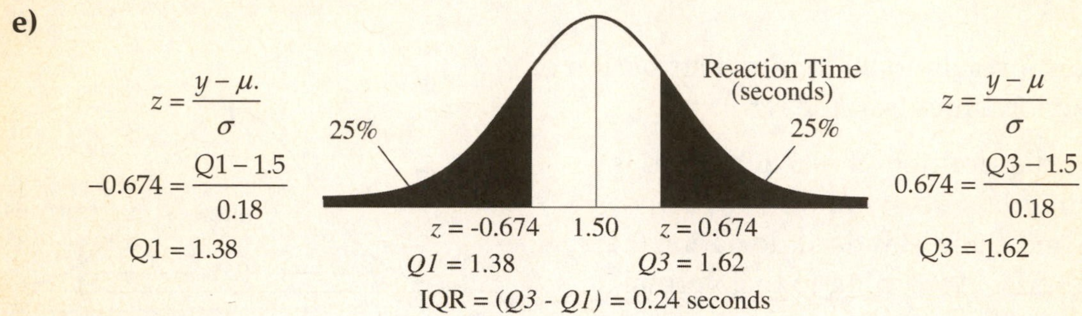

According to the Normal model, the interquartile range of the distribution of reaction times is expected to be 0.24 seconds.

f)

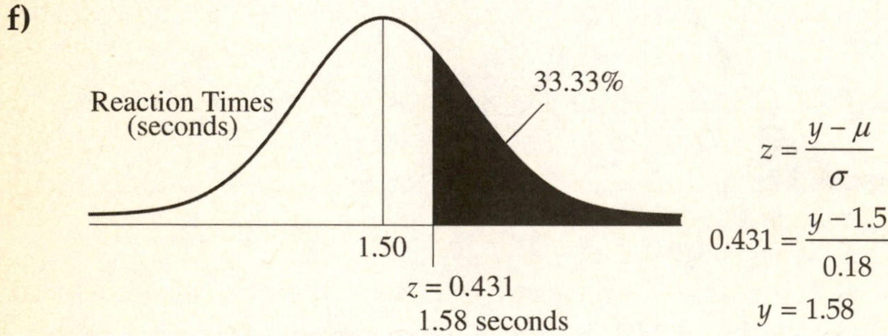

$$z = \frac{y-\mu}{\sigma}$$
$$0.431 = \frac{y-1.5}{0.18}$$
$$y = 1.58$$

According to the Normal model, the slowest 1/3 of all drivers are expected to have reaction times of 1.58 seconds or more. (Remember that a high reaction time is a SLOW reaction time!)

27. Mail.

a) A histogram of the number of pieces of mail received at a school office is at the right.

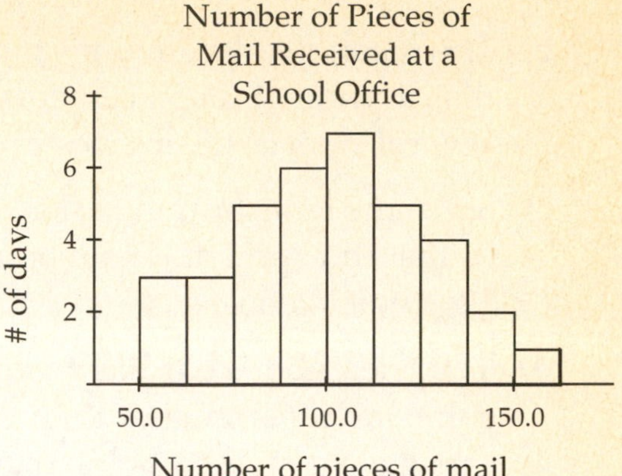

b) Since the distribution of number of pieces of mail is unimodal and symmetric, the mean and standard deviation are appropriate measures of center and spread. The mean number of pieces of mail is 100.25, and the standard deviation is 25.54 pieces.

c) The distribution of the number of pieces of mail received at the school office is unimodal and symmetric, with mean 100.25 and standard deviation 25.54. The lowest number of pieces of mail received in a day was 52 and the highest was 151.

d) 23 of the 36 days (64%) had a number of pieces of mail received within one standard deviation of the mean, or within the interval 74.71 - 125.79. This is fairly close to the 68% predicted by the Normal model. The Normal model may be useful for modeling the number of pieces of mail received by this school office.

29. Herbal medicine.

a) *Who* – 100 customers. *What* – Researchers asked whether or not the customer had taken the cold remedy and had customers rate the effectiveness of the remedy on a scale from 1 to 10. *When* – Not specified. *Where* – Store where natural health products are sold. *Why* – The researchers were from the Herbal Medicine Council, which sounds suspiciously like a group that might be promoting the use of herbal remedies. *How* – Researchers conducted personal interviews with 100 customers. No mention was made of any type of random selection.

b) "Have you taken the cold remedy?" is a categorical variable. Effectiveness on a scale of 1 to 10 is a categorical variable, as well, with respondents rating the remedy by placing it into one of 10 categories.

c) Very little confidence can be placed in the Council's conclusions. Respondents were people who already shopped in a store that sold natural remedies. They may be pre-disposed to thinking that the remedy was effective. Furthermore, no attempt was made to randomly select respondents in a representative manner. Finally, the Herbal Medicine Council has an interest in the success of the remedy.

31. Engines.

a) The count of cars is 38.

b) The mean displacement is higher than the median displacement, indicating a distribution of displacements that is skewed to the right. There are likely to be several very large engines in a group that consists of mainly smaller engines.

c) Since the distribution is skewed, the median and IQR are useful measures of center and spread. The median displacement is 148.5 cubic inches and the IQR is 126 cubic inches.

d) Your neighbor's car has an engine that is bigger than the median engine, but 227 cubic inches is smaller than the third quartile of 231, meaning that at least 25% of cars have a bigger engine than your neighbor's car. Don't be impressed!

e) Using the Outlier Rule (more than 1.5 IQRs beyond the quartiles) to find the fences:

Upper Fence: Q3 + 1.5(IQR) = 231 + 1.5(126) = 420 cubic inches.

Lower Fence: Q1 – 1.5(IQR) = 105 – 1.5(126) = - 84 cubic inches.

Since there are certainly no engines with negative displacements, there are no low outliers. Q1 + Range = 105 + 275 = 380 cubic inches. This means that the maximum must be less than 380 cubic inches. Therefore, there are no high outliers (engines over 420 cubic inches).

f) It is not reasonable to expect 68% of the car engines to measure within one standard deviation of the mean. The distribution engine displacements is skewed to the right, so the Normal model is not appropriate.

g) Multiplying each of the engine displacements by 16.4 to convert cubic inches to cubic centimeters would affect measures of position and spread. All of the summary statistics (except the count!) could be converted to cubic centimeters by multiplying each by 16.4.

33. Age and party 2007.

a) 1101 of 4002, or approximately 27.5%, of all voters surveyed were Republicans.

b) This was a representative telephone survey conducted by Gallup, a reputable polling firm. It is likely to be a reasonable estimate of the percentage of all voters who are Republicans.

c) 1001 + 1004 = 2005 of 4002, or approximately 50.1%, of all voters surveyed were under 30 or over 65 years old.

d) 409 of 4002, or approximately 10.2%, of all voters surveyed were Independents under the age of 30.

e) 409 of the 1497 Independents surveyed, or approximately 27.3%, were under the age of 30.

f) 409 of the 1001 respondents under 30, or approximately 40.9%, were Independents.

35. Age and party II.

a) The marginal distribution of party affiliation is:
Republican – 27.5% Democrat – 35.1% Independent – 37.4%
(As counts: Republican – 1101 Democrat – 1404 Independent – 1497)

b)

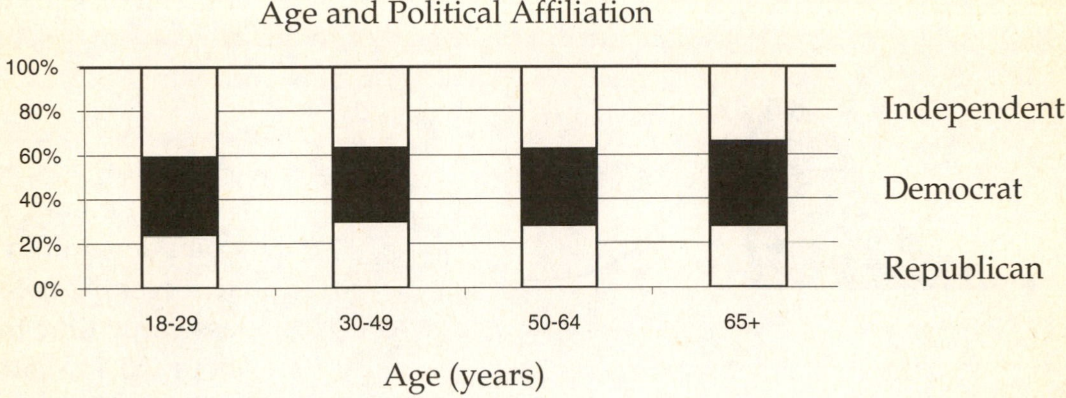

c) Political affiliation appears to be largely unrelated to age. According to The Gallup Poll, the percentages of Independents, Democrats and Republicans within four age categories are roughly the same, with approximately 35-40% Independent, 33-38% Democrat, and 25-30% Republican. However, there is some evidence that younger voters are more likely to be Independent than older voters.

d) The percentages of Independents, Democrats, and Republicans are roughly the same within each age category. Age and political affiliation appear to be independent. At the very least, there is no evidence of a strong association between the two.

37. Some assembly required.

a)

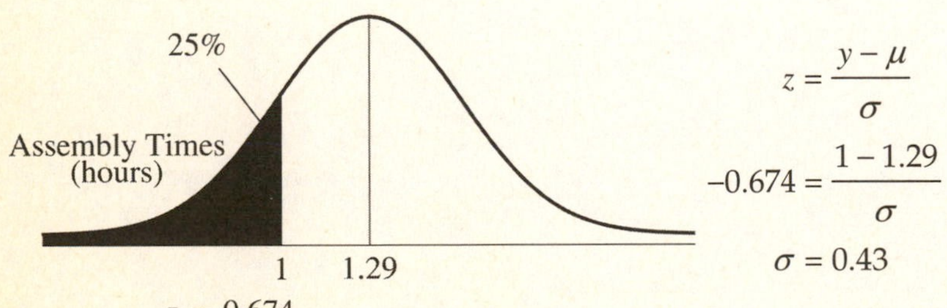

$$z = \frac{y - \mu}{\sigma}$$

$$-0.674 = \frac{1 - 1.29}{\sigma}$$

$$\sigma = 0.43$$

According to the Normal model, the standard deviation is 0.43 hours.

b)

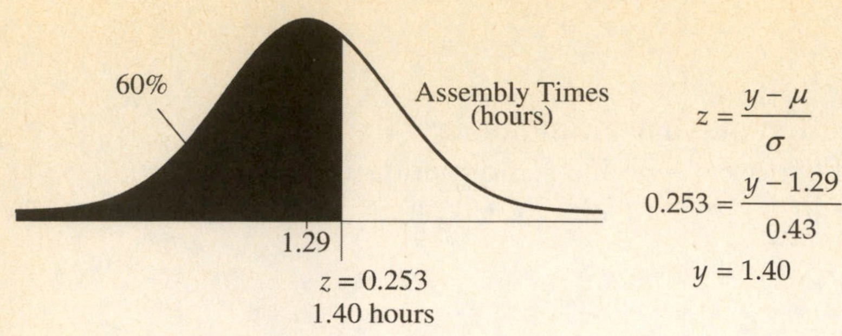

$$z = \frac{y - \mu}{\sigma}$$

$$0.253 = \frac{y - 1.29}{0.43}$$

$$y = 1.40$$

According to the Normal model, the company would need to claim that the desk takes "less than 1.40 hours to assemble", not the catchiest of slogans!

c)

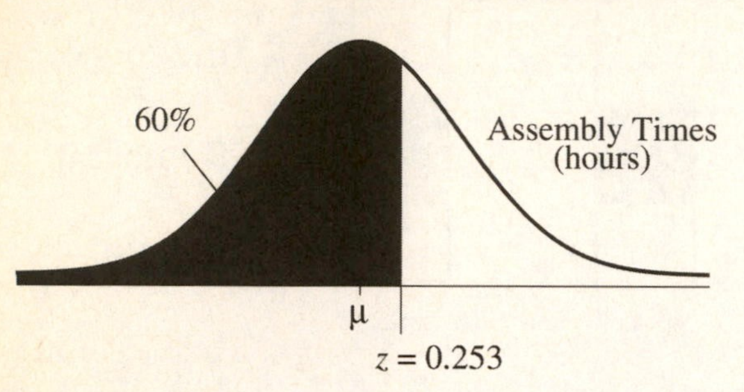

$$z = \frac{y - \mu}{\sigma}$$

$$0.253 = \frac{1 - \mu}{0.43}$$

$$\mu = 0.89$$

According to the Normal model, the company would have to lower the mean assembly time to 0.89 hour (53.4 minutes).

d) The new instructions and part-labeling may have helped lower the mean, but it also may have changed the standard deviation, making the assembly times more consistent as well as lower.

Chapter 7 – Scatterplots, Association, and Correlation

1. **Association.**

   a) Either weight in grams or weight in ounces could be the explanatory or response variable. Greater weights in grams correspond with greater weights in ounces. The association between weight of apples in grams and weight of apples in ounces would be positive, straight, and perfect. Each apple's weight would simply be measured in two different scales. The points would line up perfectly.

   b) Circumference is the explanatory variable, and weight is the response variable, since one-dimensional circumference explains three-dimensional volume (and therefore weight). For apples of roughly the same size, the association would be positive, straight, and moderately strong. If the sample of apples contained very small and very large apples, the association's true curved form would become apparent.

   c) There would be no association between shoe size and GPA of college freshmen.

   d) Number of miles driven is the explanatory variable, and gallons remaining in the tank is the response variable. The greater the number of miles driven, the less gasoline there is in the tank. If a sample of different cars is used, the association is negative, straight, and moderate. If the data is gathered on different trips with the same car, the association would be strong.

3. **Association.**

   a) Altitude is the explanatory variable, and temperature is the response variable. As you climb higher, the temperature drops. The association is negative, straight, and strong.

   b) At first, it appears that there should be no association between ice cream sales and air conditioner sales. When the lurking variable of temperature is considered, the association becomes more apparent. When the temperature is high, ice cream sales tend to increase. Also, when the temperature is high, air conditioner sales tend to increase. Therefore, there is likely to be an increase in the sales of air conditioners whenever there is an increase in the sales of ice cream. The association is positive, straight, and moderate. Either one of the variables could be used as the explanatory variable.

   c) Age is the explanatory variable, and grip strength is the response variable. The association is neither negative nor positive, but is curved, and moderate in strength, due to the variability in grip strength among people in general. The very young would have low grip strength, and grip strength would increase as age increased. After reaching a maximum (at whatever age physical prowess peaks), grip strength would decline again, with the elderly having low grip strengths.

d) Blood alcohol content is the explanatory variable, and reaction time is the response variable. As blood alcohol level increase, so does the time it takes to react to a stimulus. The association is positive, probably curved, and strong. The scatterplot would probably be almost linear for low concentrations of alcohol in the blood, and then begin to rise dramatically, with longer and longer reaction times for each incremental increase in blood alcohol content.

5. Scatterplots.

a) None of the scatterplots show little or no association, although # 4 is very weak.

b) #3 and #4 show negative association. Increases in one variable are generally related to decreases in the other variable.

c) #2, #3, and #4 each show a straight association.

d) #2 shows a moderately strong association.

e) #1 and #3 each show a very strong association. #1 shows a curved association and #3 shows a straight association.

7. Performance IQ scores *vs.* brain size.

The scatterplot of IQ scores *vs.* Brain Sizes is scattered, with no apparent pattern. There appears to be little or no association between the IQ scores and brain sizes displayed in this scatterplot.

9. Firing pottery.

a) A histogram of the number of broken pieces is at the right.

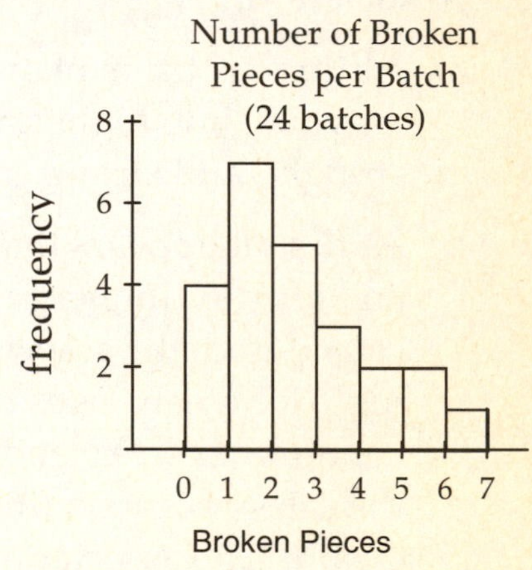

b) The distribution of the number broken pieces per batch of pottery is skewed right, centered around 1 broken piece per batch. Batches had from 0 and 6 broken pieces. The scatterplot does not show the center or skewness of the distribution.

c) The scatterplot shows that the number of broken pieces increases as the batch number increases. If the 8 daily batches are numbered sequentially, this indicates that batches fired later in the day generally have more broken pieces. This information is not visible in the histogram.

11. Matching.

a) 0.006 **b)** 0.777 **c)** - 0.923 **d)** - 0.487

13. Politics.

The candidate might mean that there is an **association** between television watching and crime. The term correlation is reserved for describing linear associations between quantitative variables. We don't know what type of variables "television watching" and "crime" are, but they seem categorical. Even if the variables are quantitative (hours of TV watched per week, and number of crimes committed, for example), we aren't sure that the relationship is linear. The politician also seems to be implying a cause-and-effect relationship between television watching and crime. Association of any kind does not imply causation.

15. Roller Coasters.

a) It is appropriate to calculate correlation. Both height of the drop and speed are quantitative variables, the scatterplot shows an association that is straight enough, and there are not outliers.

b) There is a strong, positive, straight association between drop and speed; the greater the height of the initial drop, the higher the top speed.

17. Hard water.

a) It is not appropriate to summarize the strength of the association between water hardness and pH with a correlation, since the association is curved, not Straight Enough.

b) Kendall's tau would be a more appropriate measure. It will not be affected by curvature in the association.

19. Cold nights.

The correlation is between the number of days since January 1 and temperature is likely to be near zero. We expect the temperature to be low in January, increase through the spring and summer, then decrease again. The relationship is not Straight Enough, so correlation is not an appropriate measure of strength.

21. Prediction units.

The correlation between prediction error and year would not change, since the correlation is based on z-scores. The z-scores are the same whether the prediction errors are measured in nautical miles or miles.

23. Correlation errors.

a) If the association between GDP and infant mortality is linear, a correlation of –0.772 shows a moderate, negative association. Generally, as GDP increases, infant mortality rate decreases.

b) Continent is a categorical variable. Correlation measures the strength of linear associations between quantitative variables.

25. Height and reading.

a) Actually, this *does* mean that taller children in elementary school are better readers. However, this does *not* mean that height causes good reading ability.

b) Older children are generally both taller and are better readers. Age is the lurking variable.

27. Correlations conclusions I.

a) No. We don't know this from correlation alone. The relationship between age and income may be non-linear, or the relationship may contain outliers.

b) No. We can't tell the form of the relationship between age and income. We need to look at the scatterplot.

c) No. The correlation between age and income doesn't tell us anything about outliers.

d) Yes. Correlation is based on z-scores, and is unaffected by changes in units.

29. Baldness and heart disease.

Even though the variables baldness and heart disease were assigned numerical values, they are categorical. Correlation is only an appropriate measure of the strength of linear association between quantitative variables. Their conclusion is meaningless.

31. Income and housing.

a) There is a positive, moderately strong, linear relationship between *Housing Cost Index* and *Median Family Income*, with several states whose *Housing Cost Index* seems high for their *Median Family Income*, and one state whose *Housing Cost Index* seems low for their *Median Family Income*.

b) Correlation is based on z-scores. The correlation would still be 0.65.

c) Correlation is based on z-scores, and is unaffected by changes in units. The correlation would still be 0.65.

d) Washington D.C. would be a moderately high outlier, with *Housing Cost Index* high for its *Median Family Income*. Since it doesn't fit the pattern, the correlation would decrease slightly if Washington D.C. were included.

e) No. We can only say that higher *Housing Cost Index* scores are associated with higher *Median Family Income*, but we don't know why. There may be other variables at work.

f) With Kendall's tau equal to 0.51, we can only say that higher *Housing Cost Index* scores are associated with higher *Median Family Income,* and that the association is of moderate strength. This, by no means, implies a cause-and-effect relationship.

33. Fuel economy 2007.

a) A scatterplot of average fuel economy vs. horsepower ratings is at the right.

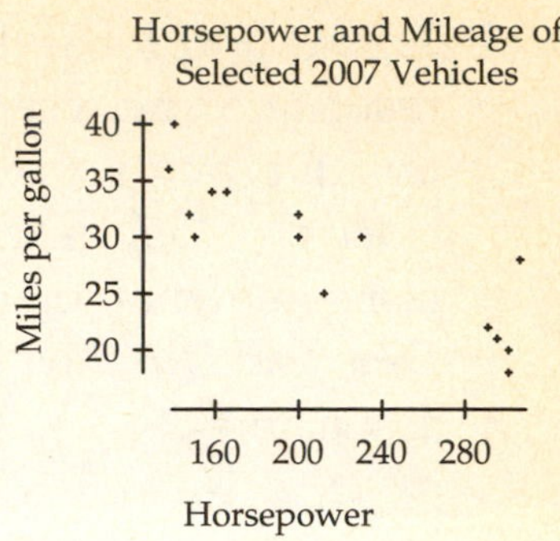

b) There is a strong, negative, straight association between horsepower and mileage of the selected vehicles. There don't appear to be any outliers. All of the cars seem to fit the same pattern. Cars with more horsepower tend to have lower mileage.

c) Since the relationship is linear, with no outliers, correlation is an appropriate measure of strength. The correlation between horsepower and mileage of the selected vehicles is $r = -0.869$.

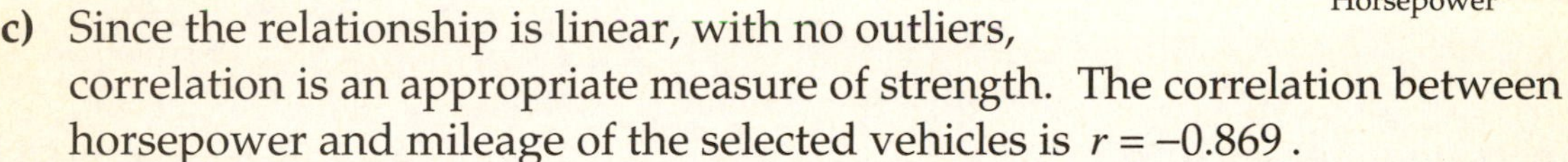

d) There is a strong linear relationship in the negative direction between horsepower and highway gas mileage. Lower fuel efficiency is associated with higher horsepower.

35. Burgers.

a) There is no apparent association between the number of grams of fat and the number of milligrams of sodium in several brands of fast food burgers. The correlation is only $r = 0.199$, which is close to zero, an indication of no association. One burger had a much lower fat content than the other burgers, at 19 grams of fat, with 920 milligrams of sodium. Without this (comparatively) low fat burger, the correlation would have been even lower.

b) Spearman's rho is –0.018. As with the Pearson correlation found in part a, it shows no evidence of an association between the number of fat grams and the number of milligrams of sodium in several brands of fast food burgers. In fact, Spearman's rho is closer to zero than the Pearson correlation, since the outlier has less influence in Spearman's rho.

37. Attendance 2006.

a) Number of runs scored and attendance are quantitative variables, the relationship between them appears to be straight, and there are no outliers, so calculating a correlation is appropriate.

b) The association between attendance and runs scored is positive, straight, and moderate in strength. Generally, as the number of runs scored increases, so does attendance.

c) There is evidence of an association between attendance and runs scored, but a cause-and-effect relationship between the two is not implied. There may be lurking variables that can account for the increases in each. For example, perhaps winning teams score more runs and also have higher attendance. We don't have any basis to make a claim of causation.

39. Thrills.

The scatterplot at the right shows that the association between *Drop* and *Duration* is straight, positive, and weak, with no outliers. Generally, rides on coasters with a greater initial drop tend to last somewhat longer. The correlation between *Drop* and *Duration* is 0.35, indicating a weak association.

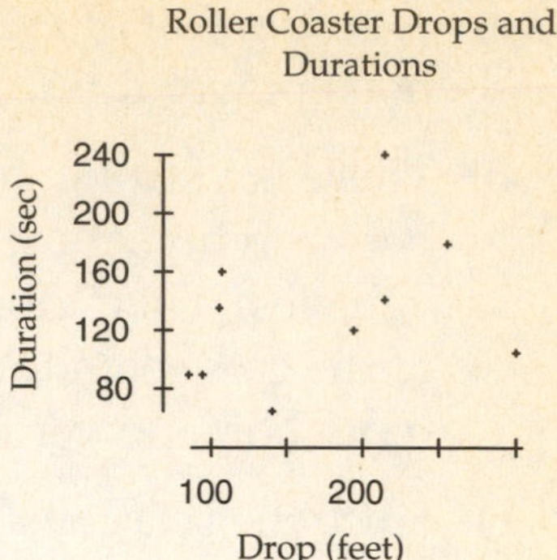

41. Planets (more or less).

a) The association between Position Number of each planet and its distance from the sun (in millions of miles) is very strong, positive and curved. The scatterplot is at the right.

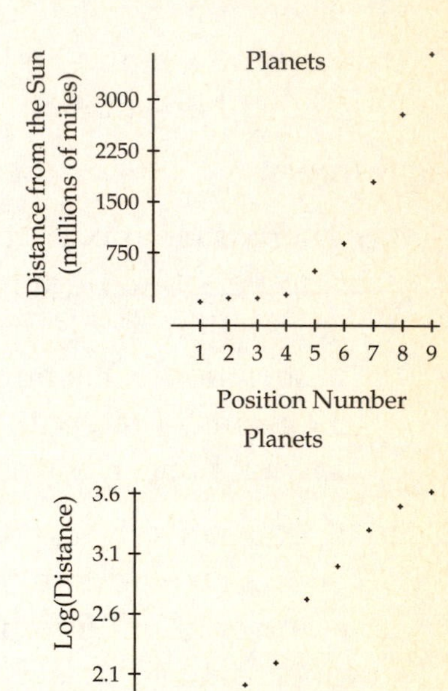

b) The relationship between Position Number and distance from the sun is not linear. Correlation is a measure of the degree of *linear* association between two variables.

c) The scatterplot of the logarithm of distance versus Position Number (shown at the right) still shows a strong, positive relationship, but it is straighter than the previous scatterplot. It still shows a curve in the scatterplot, but it is straight enough that correlation may now be used as an appropriate measure of the strength of the relationship between logarithm of distance and Position Number, which will in turn give an indication of the strength of the association.

d) Kendall's tau is 1, since the association is always increasing. The slope between any pair of points is always positive.

Chapter 8 – Linear Regression

1. Cereals.

$\widehat{Potassium} = 38 + 27Fiber = 38 + 27(9) = 281$ mg. According to the model, we expect cereal with 9 grams of fiber to have 281 milligrams of potassium.

3. More cereal.

A negative residual means that the potassium content is actually lower than the model predicts for a cereal with that much fiber.

5. Another bowl.

The model predicts that cereals will have approximately 27 more milligrams of potassium for each additional gram of fiber.

7. Cereal again.

$R^2 = (0.903)^2 \approx 0.815$. About 81.5% of the variability in potassium content is accounted for by the model.

9. Last bowl!

True potassium content of cereals vary from the predicted values with a standard deviation of 30.77 milligrams.

11. Regression equations.

$\bar{x}$	s_x	$\bar{y}$	s_y	r	$\hat{y} = b_0 + b_1 x$
a) 10	2	20	3	0.5	$\hat{y} = 12.5 + .75x$
b) 2	0.06	7.2	1.2	–0.4	$\hat{y} = 23.2 - 8x$
c) 12	6	152	30	–0.8	$\hat{y} = 200 - 4x$
d) 2.5	1.2	25	100	0.6	$\hat{y} = -100 + 50x$

a)
$$b_1 = \frac{rs_y}{s_x} \qquad b_1 = \frac{(0.5)(3)}{2} \qquad b_1 = 0.75$$
$$\hat{y} = b_0 + b_1 x \qquad \bar{y} = b_0 + b_1 \bar{x} \qquad 20 = b_0 + 0.75(10) \qquad b_0 = 12.5$$

b)
$$b_1 = \frac{rs_y}{s_x} \qquad b_1 = \frac{(-0.4)(1.2)}{0.06} \qquad b_1 = -8$$
$$\hat{y} = b_0 + b_1 x \qquad \bar{y} = b_0 + b_1 \bar{x} \qquad 7.2 = b_0 - 8(2) \qquad b_0 = 23.2$$

c)
$$\hat{y} = b_0 + b_1 x \qquad \bar{y} = b_0 + b_1 \bar{x} \qquad \bar{y} = 200 - 4(12) \qquad \bar{y} = 152$$
$$b_1 = \frac{rs_y}{s_x} \qquad -4 = \frac{-0.8 s_y}{6} \qquad s_y = 30$$

d)
$$\hat{y} = b_0 + b_1 x \qquad \bar{y} = b_0 + b_1 \bar{x} \qquad \bar{y} = -100 + 50(2.5) \qquad \bar{y} = 25$$
$$b_1 = \frac{rs_y}{s_x} \qquad 50 = \frac{r(100)}{1.2} \qquad r = 0.6$$

13. Residuals.

a) The scattered residuals plot indicates an appropriate linear model.

b) The curved pattern in the residuals plot indicates that the linear model is not appropriate. The relationship is not linear.

c) The fanned pattern indicates that the linear model is not appropriate. The model's predicting power decreases as the values of the explanatory variable increase.

15. Real estate.

a) The explanatory variable (x) is size, measured in square feet, and the response variable (y) is price measured in thousands of dollars.

b) The units of the slope are thousands of dollars per square foot.

c) The slope of the regression line predicting price from size should be positive. Bigger homes are expected to cost more.

17. What slope?

The only slope that makes sense is 300 pounds per foot. 30 pounds per foot is too small. For example, a Honda Civic is about 14 feet long, and a Cadillac DeVille is about 17 feet long. If the slope of the regression line were 30 pounds per foot, the Cadillac would be predicted to outweigh the Civic by only 90 pounds! (The real difference is about 1500 pounds.) Similarly, 3 pounds per foot is too small. A slope of 3000 pounds per foot would predict a weight difference of 9000 pounds (4.5 tons) between Civic and DeVille. The only answer that is even reasonable is 300 pounds per foot, which predicts a difference of 900 pounds. This isn't very close to the actual difference of 1500 pounds, but at least it is in the right ballpark.

19. Real estate again.

71.4% of the variability in price can be accounted for by variability in size. (In other words, 71.4% of the variability in price can be accounted for by the linear model.)

21. Misinterpretations.

a) R^2 is an indication of the strength of the model, not the appropriateness of the model. A scattered residuals plot is the indicator of an appropriate model.

b) Regression models give predictions, not actual values. The student should have said, "The model predicts that a bird 10 inches tall is expected to have a wingspan of 17 inches."

23. Real estate redux.

a) The correlation between size and price is $r = \sqrt{R^2} = \sqrt{0.714} = 0.845$. The positive value of the square root is used, since the relationship is believed to be positive.

b) The price of a home that is one standard deviation above the mean size would be predicted to be 0.845 standard deviations (in other words ***r*** standard deviations) above the mean price.

c) The price of a home that is two standard deviations below the mean size would be predicted to be 1.69 (or 2×0.845) standard deviations below the mean price.

25. ESP.

a) First, since no one has ESP, you must have scored 2 standard deviations above the mean by chance. On your next attempt, you are unlikely to duplicate the extraordinary event of scoring 2 standard deviations above the mean. You will likely "regress" towards the mean on your second try, getting a lower score. If you want to impress your friend, don't take the test again. Let your friend think you can read his mind!

b) Your friend doesn't have ESP, either. No one does. Your friend will likely "regress" towards the mean score on his second attempt, meaning his score will probably go up. If the goal is to get a higher score, your friend should try again.

27. More real estate.

a) According to the linear model, the price of a home is expected to increase \$61 (0.061 thousand dollars) for each additional square-foot in size.

b) $\widehat{Price} = 47.82 + 0.061(Sqrft)$

$\widehat{Price} = 47.82 + 0.061(3000)$

$\widehat{Price} = 230.82$

According to the linear model, a 3000 square-foot home is expected to have a price of \$230,820.

c) $\widehat{Price} = 47.82 + 0.061(Sqrft)$

$\widehat{Price} = 47.82 + 0.061(1200)$

$\widehat{Price} = 121.02$

According to the linear model, a 1200 square-foot home is expected to have a price of \$121,020. The asking price is \$121,020 - \$6000 = \$115,020. \$6000 is the (negative) residual.

29. Cigarettes.

a) A linear model is probably appropriate. The residuals plot shows some initially low points, but there is not clear curvature.

b) 92.4% of the variability in nicotine level is accounted for by variability in tar content. (In other words, 92.4% of the variability in nicotine level is accounted for by the linear model.)

31. Another cigarette.

a) The correlation between tar and nicotine is $r = \sqrt{R^2} = \sqrt{0.924} = 0.961$. The positive value of the square root is used, since the relationship is believed to be positive. Evidence of the positive relationship is the positive coefficient of tar in the regression output.

b) The average nicotine content of cigarettes that are two standard deviations below the mean in tar content would be expected to be about 1.922 (2×0.961) standard deviations below the mean nicotine content.

c) Cigarettes that are one standard deviation above average in nicotine content are expected to be about 0.961 standard deviations (in other words, ***r*** standard deviations) above the mean tar content.

33. Last cigarette.

a) $\widehat{Nicotine} = 0.15403 + 0.065052(Tar)$ is the equation of the regression line that predicts nicotine content from tar content of cigarettes.

b)

$$\widehat{Nicotine} = 0.15403 + 0.065052(Tar)$$
$$\widehat{Nicotine} = 0.15403 + 0.065052(4)$$
$$\widehat{Nicotine} = 0.414$$

The model predicts that a cigarette with 4 mg of tar will have about 0.414 mg of nicotine.

c) For each additional mg of tar, the model predicts an increase of 0.065 mg of nicotine.

d) The model predicts that a cigarette with no tar would have 0.154 mg of nicotine.

e)

$$\widehat{Nicotine} = 0.15403 + 0.065052(Tar)$$
$$\widehat{Nicotine} = 0.15403 + 0.065052(7)$$
$$\widehat{Nicotine} = 0.6094$$

The model predicts that a cigarette with 7 mg of tar will have 0.6094 mg of nicotine. If the residual is –0.5, the cigarette actually had 0.1094 mg of nicotine.

35. Income and housing revisited.

a) Yes. Both housing cost index and median family income are quantitative. The scatterplot is Straight Enough, although there may be a few outliers. The spread increases a bit for states with large median incomes, but we can still fit a regression line.

b) Using the summary statistics given in the problem, calculate the slope and intercept:

$b_1 = \frac{rs_{HCI}}{s_{MFI}}$

$b_1 = \frac{(0.65)(116.55)}{7072.47}$

$b_1 = 0.0107$

$\hat{y} = b_0 + b_1 x$

$\bar{y} = b_0 + b_1 \bar{x}$

$338.2 = b_0 + 0.0107(46234)$

$b_0 = -156.50$

The regression equation that predicts HCI from MFI is $\widehat{HCI} = -156.50 + 0.0107MFI$

(If you went back to Chapter 7, and found the regression equation from the original data, the equation is $\widehat{HCI} = -157.64 + 0.0107MFI$, not a huge difference!)

c)

$\widehat{HCI} = -156.50 + 0.0107MFI$

$\widehat{HCI} = -156.50 + 0.0107(44993)$

$\widehat{HCI} = 324.93$

The model predicts that a state with median family income of $44993 have an average housing cost index of 324.93

(Using the regression equation calculated from the actual data would give an average housing cost index of approximately 324.87.)

d) The prediction is 223.09 too low. Washington has a positive residual. (223.15 from the equation generated from the original data.)

e) The correlation is the slope of the regression line that relates z-scores, so the regression equation would be $\hat{z}_{HCI} = 0.65z_{MFI}$.

f) The correlation is the slope of the regression line that relates z-scores, so the regression equation would be $\hat{z}_{MFI} = 0.65z_{HCI}$.

37. Online clothes.

a) Using the summary statistics given in the problem, calculate the slope and intercept:

$b_1 = \frac{rs_{Total}}{s_{Age}}$

$b_1 = \frac{(0.037)(253.62)}{8.51}$

$b_1 = 1.1027$

$\hat{y} = b_0 + b_1 x$

$\bar{y} = b_0 + b_1 \bar{x}$

$572.52 = b_0 + 1.1027(29.67)$

$b_0 = 539.803$

The regression equation that predicts total online clothing purchase amount from age is $\widehat{Total} = 539.803 + 1.103Age$

b) Yes. Both total purchases and age are quantitative variables, and the scatterplot is Straight Enough, even though it is quite flat. There are no outliers and the plot does not spread throughout the plot.

c)

$\widehat{Total} = 539.803 + 1.103Age$
$\widehat{Total} = 539.803 + 1.103(18)$
$\widehat{Total} = 559.66$

The model predicts that an 18 year old will have $559.66 in total yearly online clothing purchases.

$\widehat{Total} = 539.803 + 1.103Age$
$\widehat{Total} = 539.803 + 1.103(50)$
$\widehat{Total} = 594.95$

The model predicts that a 50 year old will have $594.95 in total yearly online clothing purchases.

d) $R^2 = (0.037)^2 \approx 0.0014 = 0.14\%$..

e) This model would not be useful to the company. The scatterplot is nearly flat. The model accounts for almost none of the variability in total yearly purchases.

39. SAT scores.

a) The association between SAT Math scores and SAT Verbal Scores was linear, moderate in strength, and positive. Students with high SAT Math scores typically had high SAT Verbal scores.

b) One student got a 500 Verbal and 800 Math. That set of scores doesn't seem to fit the pattern.

c) $r = 0.685$ indicates a moderate, positive association between SAT Math and SAT Verbal, but only because the scatterplot shows a linear relationship. Students who scored one standard deviation above the mean in SAT Math were expected to score 0.685 standard deviations above the mean in SAT Verbal. Additionally, $R^2 = (0.685)^2 = 0.469225$, so 46.9% of the variability in math score was accounted for by variability in verbal score.

d) The scatterplot of verbal and math scores shows a relationship that is straight enough, so a linear model is appropriate.

$$b_1 = \frac{rs_{Math}}{s_{Verbal}}$$
$$b_1 = \frac{(0.685)(96.1)}{99.5}$$
$$b_1 = 0.661593$$

$$\hat{y} = b_0 + b_1 x$$
$$\bar{y} = b_0 + b_1 \bar{x}$$
$$612.2 = b_0 + 0.661593(596.3)$$
$$b_0 = 217.692$$

The equation of the least squares regression line for predicting SAT Math score from SAT Verbal score is $\widehat{Math} = 217.692 + 0.662(Verbal)$.

e) For each additional point in verbal score, the model predicts an increase of 0.662 points in math score. A more meaningful interpretation might be scaled up. For each additional 10 points in verbal score, the model predicts an increase of 6.62 points in math score.

f)

$$\widehat{Math} = 217.692 + 0.662(Verbal)$$
$$\widehat{Math} = 217.692 + 0.662(500)$$
$$\widehat{Math} = 548.692$$

According to the model, a student with a verbal score of 500 was expected to have a math score of 548.692.

g)

$$\widehat{Math} = 217.692 + 0.662(Verbal)$$
$$\widehat{Math} = 217.692 + 0.662(800)$$
$$\widehat{Math} = 747.292$$

According to the model, a student with a verbal score of 800 was expected to have a math score of 747.292. She actually scored 800 on math, so her residual was 800 – 747.292 = 52.708 points

41. SAT, take 2.

a) $r = 0.685$. The correlation between SAT Math and SAT Verbal is a unitless measure of the degree of linear association between the two variables. It doesn't depend on the order in which you are making predictions.

b) The scatterplot of verbal and math scores shows a relationship that is straight enough, so a linear model is appropriate.

$$b_1 = \frac{rs_{Verbal}}{s_{Math}}$$
$$b_1 = \frac{(0.685)(99.5)}{96.1}$$
$$b_1 = 0.709235$$

$$\hat{y} = b_0 + b_1 x$$
$$\bar{y} = b_0 + b_1 \bar{x}$$
$$596.3 = b_0 + 0.709235(612.2)$$
$$b_0 = 162.106$$

The equation of the least squares regression line for predicting SAT Verbal score from SAT Math score is: $\widehat{Verbal} = 162.106 + 0.709(Math)$

c) A positive residual means that the student's actual verbal score was higher than the score the model predicted for someone with the same math score.

d)

$$\widehat{Verbal} = 162.106 + 0.709(Math)$$
$$\widehat{Verbal} = 162.106 + 0.709(500)$$
$$\widehat{Verbal} = 516.606$$

According to the model, a person with a math score of 500 was expected to have a verbal score of 516.606 points.

e)

$$\widehat{Math} = 217.692 + 0.662(Verbal)$$
$$\widehat{Math} = 217.692 + 0.662(516.606)$$
$$\widehat{Math} = 559.685$$

According to the model, a person with a verbal score of 516.606 was expected to have a math score of 559.685 points.

f) The prediction in part e) does not cycle back to 500 points because the regression equation used to predict math from verbal is a different equation than the regression equation used to predict verbal from math. One was generated by minimizing squared residuals in the verbal direction, the other was generated by minimizing squared residuals in the math direction. If a math score is one standard deviation above the mean, its predicted verbal score regresses toward the mean. The same is true for a verbal score used to predict a math score.

43. Wildfires 2008.

a) The scatterplot shows a roughly linear relationship between the year and the number of wildfires, so the linear model is appropriate.

b) The model predicts an increase of an average of 296.665 wildfires per year.

c) It seems reasonable to interpret the intercept. The model predicts 73,790.7 wildfires in 1985, which is within the scope of the data, although it isn't very useful since we know the actual number of wildfires in 1985. There isn't much need for a prediction.

d) The standard deviation of the residuals is 12,323 fires. That's a large residual, considering that these years show between 50,000 and 100,000 fires per year. The association just isn't very strong.

e) The model only accounts for about 2.9% of the variability in the number of fires each year. The rest of the variability is due to other factors.

45. Used cars 2007.

a) We are attempting to predict the price in dollars of used Toyota Corollas from their age in years. A scatterplot of the relationship is at the right.

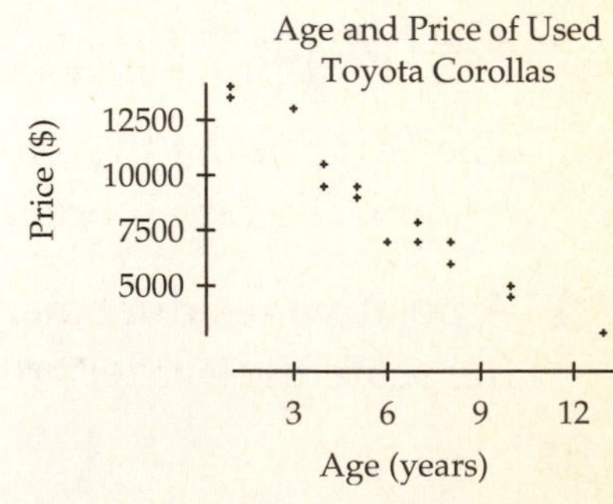

b) There is a strong, negative, linear association between price and age of used Toyota Corollas.

c) The scatterplot provides evidence that the relationship is Straight Enough. A linear model will likely be an appropriate model.

d) Since $R^2 = 0.944$, simply take the square root to find r. $\sqrt{0.944} = 0.972$. Since association between age and price is negative, $r = -0.972$.

e) 94.4% of the variability in price of a used Toyota Corolla can be accounted for by variability in the age of the car.

f) The relationship is not perfect. Other factors, such as options, condition, and mileage explain the rest of the variability in price.

47. More used cars 2007.

a) The scatterplot from the previous exercise shows that the relationship is straight, so the linear model is appropriate. The regression equation to predict the price of a used Toyota Corolla from its age is $\widehat{Price} = 14286 - 959(Years)$. The computer regression output used is at the right.

Dependent variable is: **Price ($)**
No Selector
R squared = 94.4% R squared (adjusted) = 94.0%
s = 816.2 with 15 - 2 = 13 degrees of freedom

Source	Sum of Squares	df	Mean Square	F-ratio
Regression	146917777	1	146917777	221
Residual	8660659	13	666205	

Variable	Coefficient	s.e. of Coeff	t-ratio	prob
Constant	14285.9	448.7	31.8	< 0.0001
Age (yr)	-959.046	64.58	-14.9	< 0.0001

b) According to the model, for each additional year in age, the car is expected to drop $959 in price.

c) The model predicts that a new Toyota Corolla (0 years old) will cost $14,286.

d)

$\widehat{Price} = 14286 - 959(Years)$

$\widehat{Price} = 14286 - 959(7)$

$\widehat{Price} = 7573$

According to the model, an appropriate price for a 7-year old Toyota Corolla is $7573.

e) Buy the car with the negative residual. Its actual price is lower than predicted.

f)

$\widehat{Price} = 14286 - 959(Years)$

$\widehat{Price} = 14286 - 959(10)$

$\widehat{Price} = 4696$

According to the model, a 10-year-old Corolla is expected to cost $4696. The car has an actual price of $3500, so its residual is $3500 — $4696 = — $1196

g) The model would not be useful for predicting the price of a 20-year-old Corolla. The oldest car in the list is 13 years old. Predicting a price after 20 years would be an extrapolation.

49. Burgers.

a) The scatterplot of calories vs. fat content in fast food hamburgers is at the right. The relationship appears linear, so a linear model is appropriate.

Fat and Calories of Fast Food Burgers

Dependent variable is: **Calories**
No Selector
R squared = 92.3%   R squared (adjusted) = 90.7%
s = 27.33 with 7 - 2 = 5 degrees of freedom

Source	Sum of Squares	df	Mean Square	F-ratio
Regression	44664.3	1	44664.3	59.8
Residual	3735.73	5	747.146	

Variable	Coefficient	s.e. of Coeff	t-ratio	prob
Constant	210.954	50.10	4.21	0.0084
Fat	11.0555	1.430	7.73	0.0006

b) From the computer regression output, $R^2 = 92.3\%$. 92.3% of the variability in the number of calories can be explained by the variability in the number of grams of fat in a fast food burger.

c) From the computer regression output, the regression equation that predicts the number of calories in a fast food burger from its fat content is:
$\widehat{Calories} = 210.954 + 11.0555(Fat)$

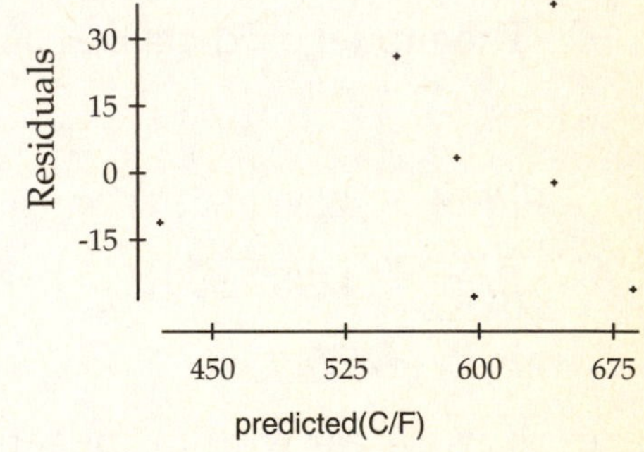

d) The residuals plot at the right shows no pattern. The linear model appears to be appropriate.

e) The model predicts that a fat free burger would have 210.954 calories. Since there are no data values close to 0, this extrapolation isn't of much use.

f) For each additional gram of fat in a burger, the model predicts an increase of 11.056 calories.

g) $\widehat{Calories} = 210.954 + 11.056(Fat) = 210.954 + 11.0555(28) = 520.508$

The model predicts a burger with 28 grams of fat will have 520.508 calories. If the residual is +33, the actual number of calories is 520.508 + 33 ≈ 553.5 calories.

51. A second helping of burgers.

a) The model from Exercise 49 was for predicting number of calories from number of grams of fat. In order to predict grams of fat from the number of calories, a new linear model needs to be generated.

b) The scatterplot at the right shows the relationship between number fat grams and number of calories in a set of fast food burgers. The association is strong, positive, and linear. Burgers with higher numbers of calories typically have higher fat contents. The relationship is straight enough to apply a linear model.

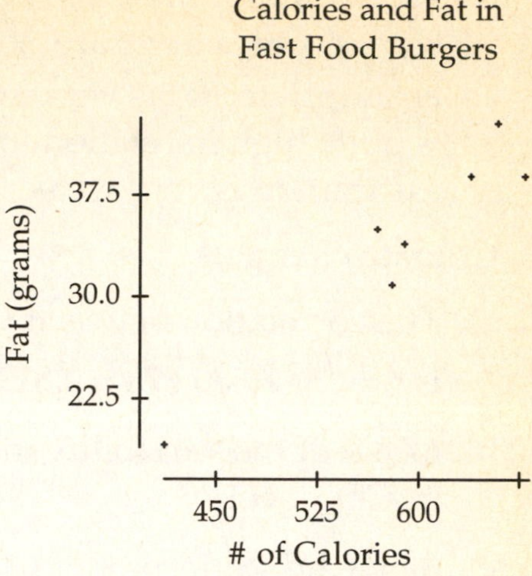

Dependent variable is: Fat
No Selector
R squared = 92.3% R squared (adjusted) = 90.7%
s = 2.375 with 7 - 2 = 5 degrees of freedom

Source	Sum of Squares	df	Mean Square	F-ratio
Regression	337.223	1	337.223	59.8
Residual	28.2054	5	5.64109	

Variable	Coefficient	s.e. of Coeff	t-ratio	prob
Constant	-14.9622	6.433	-2.33	0.0675
Calories	0.083471	0.0108	7.73	0.0006

The linear model for predicting fat from calories is:
$\widehat{Fat} = -14.9622 + 0.083471(Calories)$

The model predicts that for every additional 100 calories, the fat content is expected to increase by about 8.3 grams.

The residuals plot shows no pattern, so the model is appropriate. $R^2 = 92.3\%$, so 92.3% of the variability in fat content can be accounted for by the model.

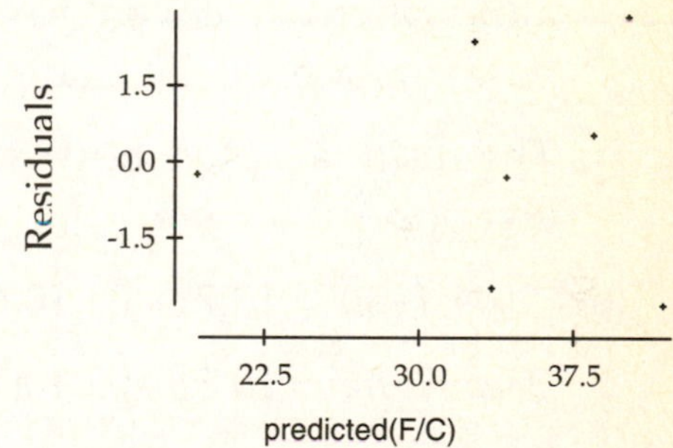

$\widehat{Fat} = -14.9622 + 0.083471(Calories)$
$\widehat{Fat} = -14.9622 + 0.083471(600)$
$\widehat{Fat} \approx 35.1$

According to the model, a burger with 600 calories is expected to have 35.1 grams of fat.

53. New York bridges

a) Overall, the model predicts the condition score of new bridges to be 4.95, a little below the cutoff of 5. Of course, looking at the scatterplot, we can see that the cluster of new bridges all have scores above 5. Still, the model is not a very encouraging one in regards to the conditions of New York City bridges.

b) According to the model, the condition of the bridges in New York City is decreasing by an average of 0.0048 per year.

c) We shouldn't place too much faith in the model. First of all, the scatterplot shows some curvature, so the linear model may not be appropriate. Even if it is appropriate, R^2 is very low, and the standard deviation of the residuals, 0.6708, is quite high in relation to the scope of the data values themselves. This association is very weak.

55. Climate change.

a) The correlation between CO_2level and mean temperature is $r = \sqrt{R^2} = \sqrt{0.678} = 0.8234$.

b) 67.8% of the variability in mean temperature can be accounted for by variability in CO_2 level.

c) Since the scatterplot of CO_2 level and mean temperature shows a relationship that is straight enough, use of the linear model is appropriate. The linear regression model that predicts mean temperature from CO_2 level is:
$\widehat{MeanTemp} = 10.71 + 0.010(CO_2)$

d) The model predicts that an increase in CO_2 level of 1 ppm is associated with an increase of 0.010 °C in mean temperature.

e) According to the model, the mean temperature is predicted to be 10.71 °C when there is no CO_2 in the atmosphere. This is an extrapolation outside of the range of data, and isn't very meaningful in context, since there is always CO_2 in the atmosphere. We want to use this model to study the change in CO_2 level and how it relates to the change in temperature.

f) The residuals plot shows no apparent patterns. The linear model appears to be an appropriate one.

g) $\widehat{MeanTemp} = 10.71 + 0.010(CO_2)$

$\widehat{MeanTemp} = 10.71 + 0.010(390)$

$\widehat{MeanTemp} = 14.61$

According to the model, the temperature is predicted to be 14.61 °C when the CO_2 level is 390 ppm.

57. Body fat.

a) The scatterplot of % body fat and weight of 20 male subjects, at the right, shows a strong, positive, linear association. Generally, as a subject's weight increases, so does % body fat. The association is straight enough to justify the use of the linear model.

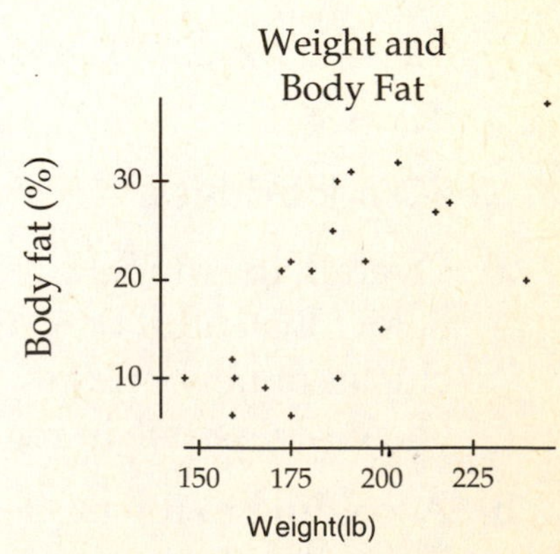

The linear model that predicts % body fat from weight is: $\%\widehat{Fat} = -27.3763 + 0.249874(Weight)$

b) The residuals plot, at the right, shows no apparent pattern. The linear model is appropriate.

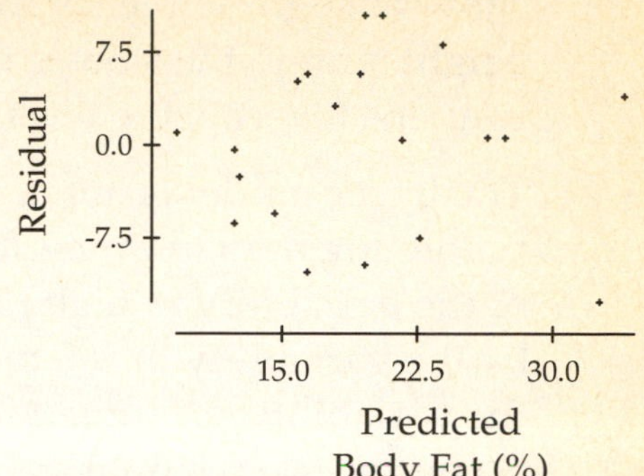

c) According to the model, for each additional pound of weight, body fat is expected to increase by about 0.25%.

d) Only 48.5% of the variability in % body fat can be accounted for by the model. The model is not expected to make predictions that are accurate.

e)

$$\%\hat{Fat} = -27.3763 + 0.249874(Weight)$$
$$\%\hat{Fat} = -27.3763 + 0.249874(190)$$
$$\%\hat{Fat} = 20.09976$$

According to the model, the predicted body fat for a 190-pound man is 20.09976%.
The residual is 21—20.09976 ≈ 0.9%.

59. Heptathlon 2004.

a) Both high jump height and 800 meter time are quantitative variables, the association is straight enough to use linear regression.

Dependent variable is: **High Jump**
No Selector
R squared = 16.4% R squared (adjusted) = 12.9%
s = 0.0617 with 26 - 2 = 24 degrees of freedom

Source	Sum of Squares	df	Mean Square	F-ratio
Regression	0.017918	1	0.017918	4.71
Residual	0.091328	24	0.003805	

Variable	Coefficient	s.e. of Coeff	t-ratio	prob
Constant	2.68094	0.4225	6.35	≤ 0.0001
800m	-6.71360e-3	0.0031	-2.17	0.0401

Heptathlon Results

The regression equation to predict high jump from 800m results is: $\widehat{Highjump} = 2.681 - 0.00671(800m)$.

According to the model, the predicted high jump decreases by an average of 0.00671 meters for each additional second in 800 meter time.

b) $R^2 = 16.4\%$. This means that 16.4% of the variability in high jump height is accounted for by the variability in 800 meter time.

c) Yes, good high jumpers tend to be fast runners. The slope of the association is negative. Faster runners tend to jump higher, as well.

d) The residuals plot is fairly patternless. The scatterplot shows a slight tendency for less variation in high jump height among the slower runners than the faster runners. Overall, the linear model is appropriate.

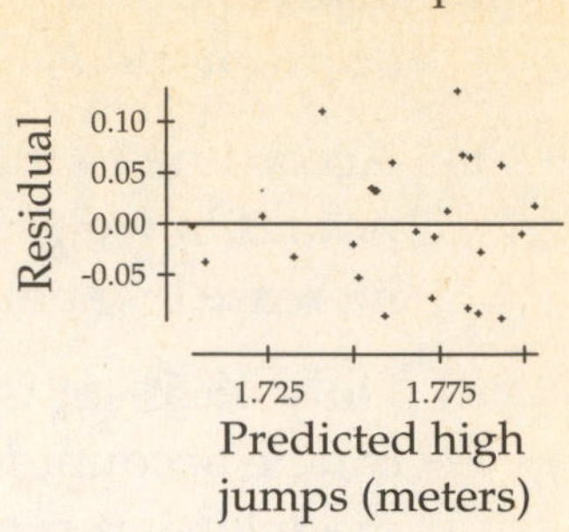

e) The linear model is not particularly useful for predicting high jump performance. First of all, 16.4% of the variability in high jump height is accounted for by the variability in 800 meter time, leaving 83.6% of the variability accounted for by other variables. Secondly, the residual standard deviation is 0.062 meters, which is not much smaller than the standard deviation of all high jumps, 0.066 meters. Predictions are not likely to be accurate.

61. Hard water.

a) There is a fairly strong, negative, linear relationship between calcium concentration (in ppm) in the water and mortality rate (in deaths per 100,000). Towns with higher calcium concentrations tended to have lower mortality rates.

b) The linear regression model that predicts mortality rate from calcium concentration is $\widehat{Mortality} = 1676 - 3.23(Calcium)$.

c) The model predicts a decrease of 3.23 deaths per 100,000 for each additional ppm of calcium in the water. For towns with no calcium in the water, the model predicts a mortality rate of 1676 deaths per 100,000 people.

d) Exeter had 348.6 fewer deaths per 100,000 people than the model predicts.

e) $\widehat{Mortality} = 1676 - 3.23(Calcium)$

$\widehat{Mortality} = 1676 - 3.23(100)$

$\widehat{Mortality} = 1353$

The town of Derby is predicted to have a mortality rate of 1353 deaths per 100,000 people.

f) 43% of the variability in mortality rate can be explained by the model.

63. Least squares.

If the 4 x-values are plugged into $\hat{y} = 7 + 1.1x$, the 4 predicted values are $\hat{y} = 18$, 29, 51 and 62, respectively. The 4 residuals are –8, 21, –31, and 18. The squared residuals are 64, 441, 961, and 324, respectively. The sum of the squared residuals is 1790. Least squares means that no other line has a sum lower than 1790. In other words, it's the best fit.

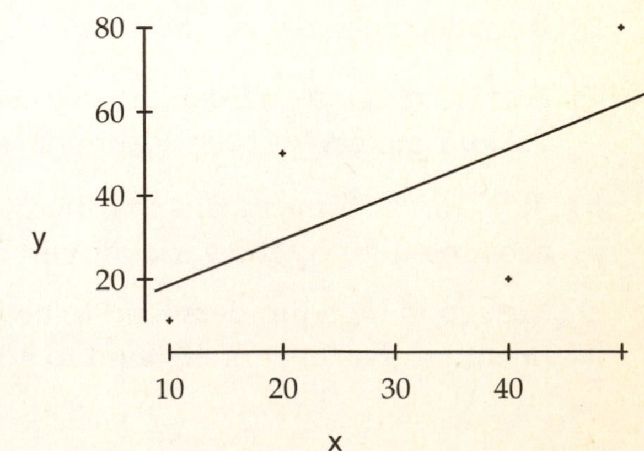

Chapter 9 – Regression Wisdom

1. Marriage age 2007.

a) The trend in age at first marriage for American women is very strong over the entire time period recorded on the graph, but the direction and form are different for different time periods. The trend appears to be somewhat linear, and consistent at around 22 years, up until about 1940, when the age seemed to drop dramatically, to under 21. From 1940 to about 1970, the trend appears non-linear and slightly positive. From 1975 to the present, the trend again appears linear and positive. The marriage age rose rapidly during this time period.

b) The association between age at first marriage for American women and year is strong over the entire time period recorded on the graph, but some time periods have stronger trends than others.

c) The correlation, or the measure of the degree of linear association is not high for this trend. The graph, as a whole, is non-linear. However, certain time periods, like 1975 to present, have a high correlation.

d) Overall, the linear model is not appropriate. The scatterplot is not "straight enough" to satisfy the condition. You could fit a linear model to the time period from 1975 to 1995, but this seems unnecessary. The ages for each year are reported, and, given the fluctuations in the past, extrapolation seems risky.

3. Human Development Index.

a) Fitting a linear model to the association between HDI and GDPPC would be misleading, since the relationship is not straight.

b) If you fit a linear model to these data, the residuals plot will be curved downward.

c) Setting aside the single data point corresponding to Luxembourg will not improve the model. The relationship will still be curved.

5. Good model?

a) The student's reasoning is not correct. A scattered residuals plot, not high R^2, is the indicator of an appropriate model. Once the model is deemed appropriate, R^2 is used as a measure of the strength of the model.

b) The model may not allow the student to make accurate predictions. The data may be curved, in which case the linear model would not fit well.

7. Movie Dramas.

a) The units for the slopes of these lines are millions of dollars per minutes of running time.

b) The slopes of the regression lines are the same. Dramas and movies from other genres have costs for longer movies that increase at the same rate.

c) The regression line for dramas has a lower *y*-intercept. Regardless of running time, dramas cost about 20 million dollars less than other genres of movies of the same running time.

9. Oakland passengers.

a) According to the linear model, the use of the Oakland airport has been increasing by about 59,700 passengers per year, starting at about 282,000 passengers in 1990.

b) About 71% of the variability in the number of passengers can be accounted for by the model.

c) Errors in prediction based on the model have a standard deviation of 104,330 passengers.

d) No, the model would not be useful in predicting the number of passengers in 2010. This year would be an extrapolation too far from the years we have observed.

e) The negative residual is September of 2001. Air traffic was artificially low following the attacks on 9/11/2001.

11. Unusual points.

a) **1)** The point has high leverage and a small residual.
2) The point is not influential. It has the *potential* to be influential, because its position far from the mean of the explanatory variable gives it high leverage. However, the point is not *exerting* much influence, because it reinforces the association.
3) If the point were removed, the correlation would become weaker. The point heavily reinforces the positive association. Removing it would weaken the association.
4) The slope would remain roughly the same, since the point is not influential.

b) **1)** The point has high leverage and probably has a small residual.
2) The point is influential. The point alone gives the scatterplot the appearance of an overall negative direction, when the points are actually fairly scattered.
3) If the point were removed, the correlation would become weaker. Without the point, there would be very little evidence of linear association.
4) The slope would increase, from a negative slope to a slope near 0. Without the point, the slope of the regression line would be nearly flat.

c) 1) The point has moderate leverage and a large residual.
2) The point is somewhat influential. It is well away from the mean of the explanatory variable, and has enough leverage to change the slope of the regression line, but only slightly.
3) If the point were removed, the correlation would become stronger. Without the point, the positive association would be reinforced.
4) The slope would increase slightly, becoming steeper after the removal of the point. The regression line would follow the general cloud of points more closely.

d) 1) The point has little leverage and a large residual.
2) The point is not influential. It is very close to the mean of the explanatory variable, and the regression line is anchored at the point $(\bar{x}, \bar{y})$, and would only pivot if it were possible to minimize the sum of the squared residuals. No amount of pivoting will reduce the residual for the stray point, so the slope would not change.
3) If the point were removed, the correlation would become slightly stronger, decreasing to become more negative. The point detracts from the overall pattern, and its removal would reinforce the association.
4) The slope would remain roughly the same. Since the point is not influential, its removal would not affect the slope.

13. The extra point.

1) Point e is very influential. Its addition will give the appearance of a strong, negative correlation like $r = -0.90$.

2) Point d is influential (but not as influential as point e). Its addition will give the appearance of a weaker, negative correlation like $r = -0.40$.

3) Point c is directly below the middle of the group of points. Its position is directly below the mean of the explanatory variable. It has no influence. Its addition will leave the correlation the same, $r = 0.00$.

4) Point b is almost in the center of the group of points, but not quite. Its addition will give the appearance of a very slight positive correlation like $r = 0.05$.

5) Point a is very influential. Its addition will give the appearance of a strong, positive correlation like $r = 0.75$.

15. What's the cause?

1) High blood pressure may cause high body fat.

2) High body fat may cause high blood pressure.

3) Both high blood pressure and high body fat may be caused by a lurking variable, such as a genetic or lifestyle trait.

17. Reading.

a) The principal's description of a strong, positive trend is misleading. First of all, "trend" implies a change over time. These data were gathered during one year, at different grade levels. To observe a trend, one class's reading scores would have to be followed through several years. Second, the strong, positive relationship only indicates the yearly improvement that would be expected, as children get older. For example, the 4th graders are reading at approximately a 4th grade level, on average. This means that the school's students are progressing adequately in their reading, not extraordinarily. Finally, the use of average reading scores instead of individual scores increases the strength of the association.

b) The plot appears very straight. The correlation between grade and reading level is very high, probably between 0.9 and 1.0.

c) If the principal had made a scatterplot of all students' scores, the correlation would have likely been lower. Averaging reduced the scatter, since each grade level has only one point instead of many, which inflates the correlation.

d) If a student is reading at grade level, then that student's reading score should equal his or her grade level. The slope of that relationship is 1. That would be "acceptable", according to the measurement scale of reading level. Any slope greater than 1 would indicate above grade level reading scores, which would certainly be acceptable as well. A slope less than 1 would indicate below grade level average scores, which would be unacceptable.

19. Heating.

a) The model predicts a decrease in \$2.13 in heating cost for an increase in temperature of 1° Fahrenheit. Generally, warmer months are associated with lower heating costs.

b) When the temperature is 0° Fahrenheit, the model predicts a monthly heating cost of \$133.

c) When the temperature is around 32° Fahrenheit, the predictions are generally too high. The residuals are negative, indicating that the actual values are lower than the predicted values.

d)

$$\hat{C} = 133 - 2.13(Temp)$$
$$\hat{C} = 133 - 2.13(10)$$
$$\hat{C} = \$111.70$$

According to the model, the heating cost in a month with average daily temperature 10° Fahrenheit is expected to be \$111.70.

e) The residual for a 10° day is approximately –\$6, meaning that the actual cost was \$6 less than predicted, or \$111.70 – \$6 = \$105.70.

f) The model is not appropriate. The residuals plot shows a definite curved pattern. The association between monthly heating cost and average daily temperature is not linear.

g) A change of scale from Fahrenheit to Celsius would not affect the relationship. Associations between quantitative variables are the same, no matter what the units.

21. Interest rates.

a) $r = \sqrt{R^2} = \sqrt{0.774} = 0.88$. The correlation between rate and year is +0.88, since the scatterplot shows a positive association.

b) According to the model, interest rates during this period increased at about 0.25% per year, starting from an interest rate of about 0.64%in 1950.

c) The linear regression equation predicting interest rate from year is

$\widehat{Rate} = 0.640282 + 0.247637(Year - 1950)$

$\widehat{Rate} = 0.640282 + 0.247637(50)$

$\widehat{Rate} = 13.022$

According to the model, the interest rate is predicted to be about 13% in the year 2000.

d) This prediction is not likely to have been a good one. Extrapolating 20 years beyond the final year in the data would be risky, and unlikely to be accurate.

23. Interest rates revisited.

a) The values of R^2 are approximately the same, so the models fit comparably well, but they have very different slopes.

b) The model that predicts the interest rate on 3-month Treasury bills from the number of years since 1950 is $\widehat{Rate} = 21.0688 - 0.356578(Year - 1950)$. This model predicts the interest rate to be 3.24%, a rate much lower than the prediction from the other model.

c) We can trust the newer prediction, since it is in the middle of the data used to generate the model. Additionally, the model accounts for 74.5% of the variability in interest rate.

d) Since 2020 is at least 15 years after the last year included in the newer model. It would be extremely risky to use this, or any, model to make a prediction that far into the future.

25. Gestation.

a) The association would be stronger if humans were removed. The point on the scatterplot representing human gestation and life expectancy is an outlier from the overall pattern and detracts from the association. Humans also represent an influential point. Removing the humans would cause the slope of the linear regression model to increase, following the pattern of the non-human animals much more closely.

b) The study could be restricted to non-human animals. This appears justifiable, since one could point to a number of environmental factors that could influence human life expectancy and gestation period, making them incomparable to those of animals.

c) The correlation is moderately strong. The model explains 72.2% of the variability in gestation period of non-human animals.

d) For every year increase in life expectancy, the model predicts an increase of approximately 15.5 days in gestation period.

e)

$$\widehat{Gest} = -39.5172 + 15.4980(LifEx)$$
$$\widehat{Gest} = -39.5172 + 15.4980(20)$$
$$\widehat{Gest} \approx 270.4428$$

According to the linear model, monkeys with a life expectancy of 20 years are expected to have gestation periods of about 270.5 days. Care should be taken when assessing the accuracy of this prediction. First of all, the residuals plot has not been examined, so the appropriateness of the model is questionable. Second, it is unknown whether or not monkeys were included in the original 17 non-human species studied. Since monkeys and humans are both primates, the monkeys may depart from the overall pattern as well.

27. Elephants and hippos.

a) Hippos are more of a departure from the pattern. Removing that point would make the association appear to be stronger.

b) The slope of the regression line would increase, pivoting away from the hippos point.

c) Anytime data points are removed, there must be a justifiable reason for doing so, and saying, "I removed the point because the correlation was higher without it" is not a justifiable reason.

d) Elephants are an influential point. With the elephants included, the slope of the linear model is 15.4980 days gestation per year of life expectancy. When they are removed, the slope is 11.6 days per year. The decrease is significant.

29. Marriage age 2007, revisited.

a) Modeling decisions may vary, but the important idea is using a subset of the data that allows us to make an accurate prediction for the year in which we are interested. We might model a subset to predict the marriage age in 2015, and model another subset to predict the marriage age in 1911.

In order to predict the average marriage age of American women in 2015, use the data points from the most recent trend only. The data points from 1985 – 2005 look straight enough to apply the linear regression model.

Regression output from a computer program is given below, as well as a residuals plot.

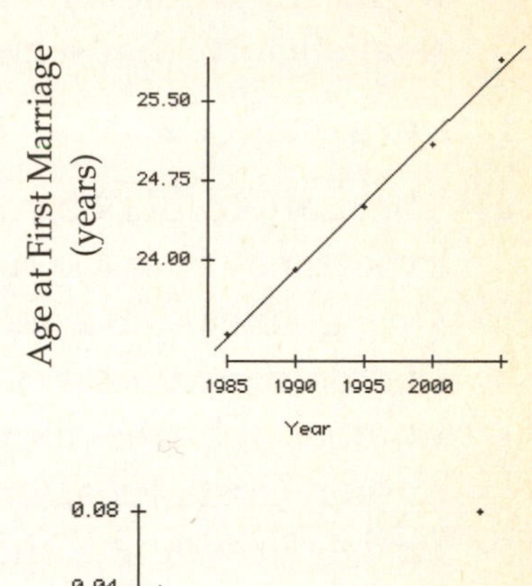


```
Dependent variable is:    Age
No Selector
R squared = 99.6%     R squared (adjusted) = 99.5%
s =  0.07303  with  5 - 2 = 3  degrees of freedom
```

Source	Sum of Squares	df	Mean Square	F-ratio
Regression	4.096	1	4.096	768
Residual	0.016	3	0.00533333	

Variable	Coefficient	s.e. of Coeff	t-ratio	prob
Constant	-230.82	9.215	-25	0.0001
Year	0.128	0.004619	27.7	0.0001

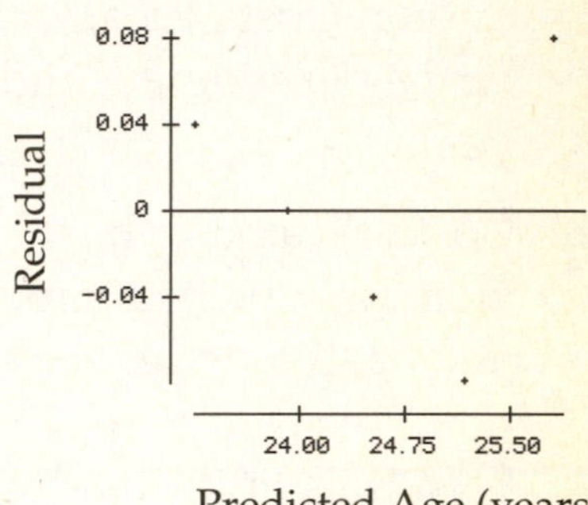

The linear model used to predict average female marriage age from year is: $\widehat{Age} = -230.82 + 0.128(Year)$. The residuals plot shows a pattern (although the number of points used is small, so this may just be random variation), but the residuals are small, and the value of R^2 is high. 99.6% of the variability in average female age at first marriage is accounted for by variability in the year. The model predicts that each year that passes is associated with an increase of 0.161 years in the average female age at first marriage.

$$\widehat{Age} = -230.82 + 0.128(Year)$$
$$\widehat{Age} = -230.82 + 0.128(2015)$$
$$\widehat{Age} \approx 27.1$$

According to the model, the average age at first marriage for women in 2015 will be 27.1 years old. Care should be taken with this prediction, however. It represents an extrapolation of 10 years beyond the highest year, and the residuals plot shows a pattern.

b) This prediction is for a year that is 10 years higher than the highest year for which we have an average female marriage age. Don't place too much faith in this extrapolation.

c) An extrapolation of more than 50 years into the future would be absurd. There is no reason to believe the trend would continue. In fact, given the situation, it is very unlikely that the pattern would continue in this fashion. The model given in part a) predicts that the average marriage age will be 32.86 years in 2040. Realistically, that seems quite high.

31. Life expectancy 2004.

a) The scatterplot of births per woman and life expectancy is at the right. The association is strong, linear, and negative. Countries with higher life expectancies tend to have a lower number of births per woman. There is one outlier, Costa Rica, with an unreasonable 24.9 births per woman, and a life expectancy of 78.7 years.

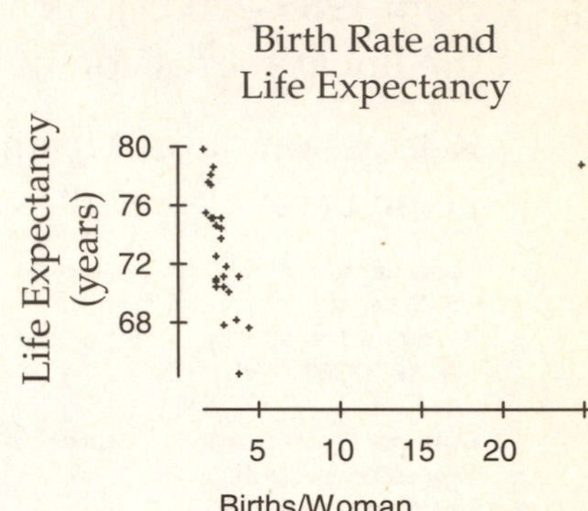

b) Costa Rica has a birth rate of 24.9 births per woman, which common sense would indicate is impossible. There is ample justification to leave that point out of all further calculations. Probably, a decimal point was incorrectly shifted from 2.49 births per woman, but there is no way to be sure. Omit this strange data point!

c)

Dependent variable is: **Life Expectancy**
No Selector
R squared = 63.3%   R squared (adjusted) = 61.7%
s = 2.387 with 25 - 2 = 23 degrees of freedom

Source	Sum of Squares	df	Mean Square	F-ratio
Regression	226.037	1	226.037	39.7
Residual	131.023	23	5.69666	

Variable	Coefficient	s.e. of Coeff	t-ratio	prob
Constant	84.4971	1.902	44.4	< 0.0001
Births/Wo...	-4.43993	0.7048	-6.30	< 0.0001

Without Costa Rica, $R^2 = 63.3\%$, so $r = \sqrt{R^2} = \sqrt{0.633} = -0.796$.
63.3% of the variability in life expectancy is explained by variability in the number of births per year.

(If you assumed that 24.9 births per woman was a data entry error meant to be 2.49 births per woman, the revised value of R^2 is 59.5%, making $r = \sqrt{R^2} = \sqrt{0.595} = -0.771$.)

d) The linear model that predicts a country's life expectancy from the number of births per woman (without Costa Rica) is $\widehat{LifeExp} = 86.8137 - 4.35623(Births)$. (If the number of births per woman in Costa Rica is changed to 2.49, the linear model that predicts a country's life expectancy is $\widehat{LifeExp} = 83.8394 - 4.49366(Births)$

e) The linear model, without Costa Rica, is an appropriate model. The residuals plot shows no pattern. (If the number of births per woman is changed to 2.49, the linear model is also appropriate. The revised residuals plot shows no pattern.)

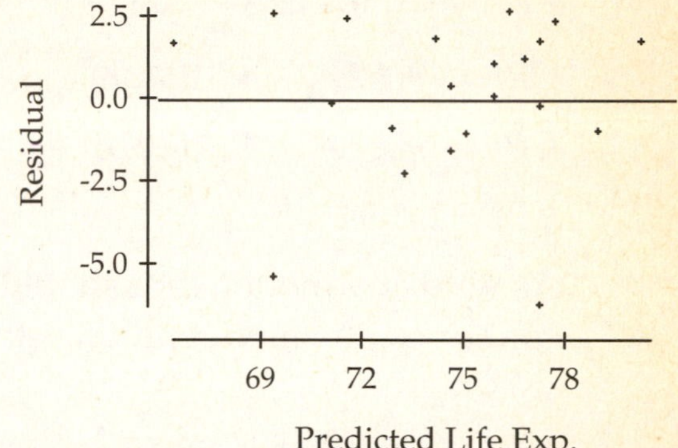

f) According to the model, each additional birth per woman is expected to correspond with a decrease in life expectancy of approximately 4.36 years (or 4.49 years, using the revised model). When the number of births per woman is 0, the model predicts that the life expectancy will be 86.8 years (or 84.8 years, using the revised model). This figure is an extrapolation below the range of the data, and doesn't hold any meaning.

g) The government leaders should not suggest that women have fewer children in order to raise the life expectancy. Although there is evidence of an association between the birth rate and life expectancy, this does not mean that one causes the other. There may be lurking variables involved, such as economic conditions, social factors, or level of health care.

33. Inflation 2006.

a) The trend in Consumer Price Index is strong, non-linear, and positive. Generally, CPI has increased over the years, but the rate of increase has become much greater since approximately 1970. Other characteristics include fluctuations in CPI in the years prior to 1950.

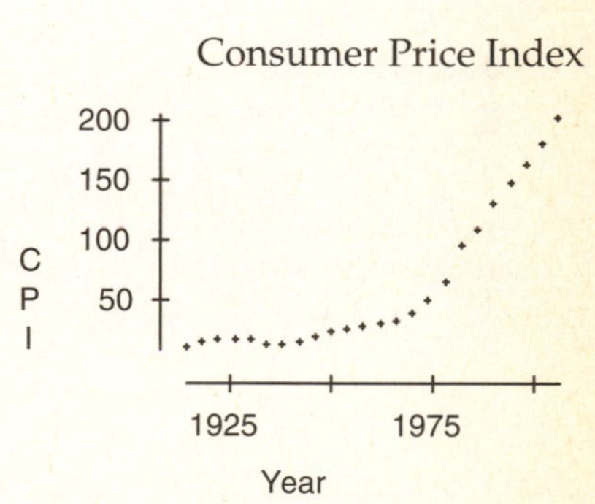

b) In order to effectively predict the CPI in 2016, use only the most recent trend. The trend since 1970 is straight enough to apply the linear model. Prior to 1970, the trend is radically different from that of recent years, and is of no use in predicting CPI for 2016.

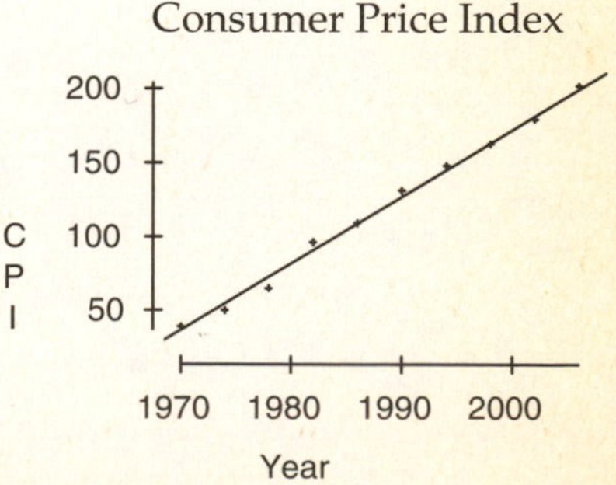

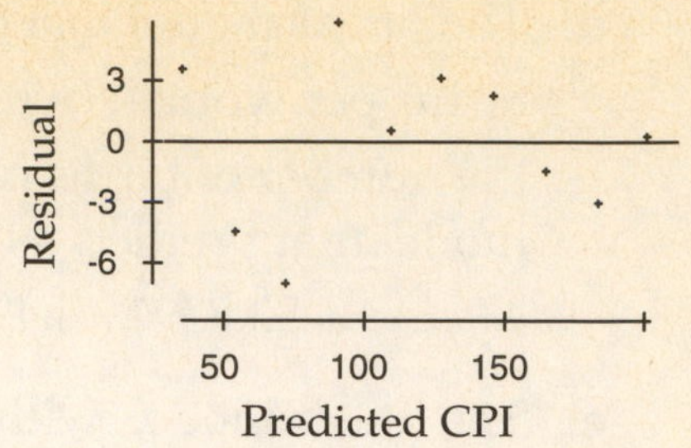

The linear model that predicts CPI from year is $\widehat{CPI} = -9052.42 + 4.61(year)$. $R^2 = 99.5\%$, meaning that the model predicts 99.5% of the variability in CPI. The residuals plot shows no pattern, so the linear model is appropriate. According to the model, the CPI is expected to increase by $4.61 each year, for 1970—2006.

$\widehat{CPI} = -9052.42 + 4.61(year)$

$\widehat{CPI} = -9052.42 + 4.61(2016)$

$\widehat{CPI} = 241.34$

As with any model, care should be taken when extrapolating. If the pattern continues, the model predicts that the CPI in 2006 will be approximately $241.34.

Chapter 10 – Re-expressing Data: Get It Straight!

1. Residuals.

a) The residuals plot shows no pattern. No re-expression is needed.

b) The residuals plot shows a curved pattern. Re-express to straighten the relationship.

c) The residuals plot shows a fan shape. Re-express to equalize spread.

3. Airline passengers revisited.

a) The residuals cycle up and down because there is a yearly pattern in the number of passengers departing Oakland, California.

b) A re-expression should not be tried. A cyclic pattern such as this one cannot be helped by re-expression.

5. Models.

a)
$$\ln \hat{y} = 1.2 + 0.8x$$
$$\ln \hat{y} = 1.2 + 0.8(2)$$
$$\ln \hat{y} = 2.8$$
$$\hat{y} = e^{2.8} = 16.44$$

b)
$$\sqrt{\hat{y}} = 1.2 + 0.8x$$
$$\sqrt{\hat{y}} = 1.2 + 0.8(2)$$
$$\sqrt{\hat{y}} = 2.8$$
$$\hat{y} = 2.8^2 = 7.84$$

c)
$$\frac{1}{\hat{y}} = 1.2 + 0.8x$$
$$\frac{1}{\hat{y}} = 1.2 + 0.8(2)$$
$$\frac{1}{\hat{y}} = 2.8$$
$$\hat{y} = \frac{1}{2.8} = 0.36$$

d)
$$\hat{y} = 1.2 + 0.8 \ln x$$
$$\hat{y} = 1.2 + 0.8 \ln(2)$$
$$\hat{y} = 1.75$$

e)
$$\log \hat{y} = 1.2 + 0.8 \log x$$
$$\log \hat{y} = 1.2 + 0.8 \log(2)$$
$$\log \hat{y} = 1.440823997...$$
$$\hat{y} = 10^{1.4408...}$$
$$\hat{y} = 27.59$$

7. Gas mileage.

a) The association between weight and gas mileage of cars is fairly linear, strong, and negative. Heavier cars tend to have lower gas mileage.

b) For each additional thousand pounds of weight, the linear model predicts a decrease of 7.652 miles per gallon in gas mileage.

c) The linear model is not appropriate. There is a curved pattern in the residuals plot. The model tends to underestimate gas mileage for cars with relatively low and high gas mileages, and overestimates the gas mileage of cars with average gas mileage.

9. Gas mileage revisited.

a) The residuals plot for the re-expressed relationship is much more scattered. This is an indication of an appropriate model.

b) The linear model that predicts the number of gallons per 100 miles in gas mileage from the weight of a car is: $\widehat{Gal}/100 = 0.625 + 1.178(Weight)$.

c) For each additional 1000 pounds of weight, the model predicts that the car will require an additional 1.178 gallons to drive 100 miles.

d)

$$\widehat{Gal}/100 = 0.625 + 1.178(Weight)$$
$$\widehat{Gal}/100 = 0.625 + 1.178(3.5)$$
$$\widehat{Gal}/100 = 4.748$$

According to the model, a car that weighs 3500 pounds (3.5 thousand pounds) is expected to require approximately 4.748 gallons to drive 100 miles, or 0.04748 gallons per mile.

This is $\dfrac{1}{0.04748} \approx 21.06$ miles per gallon.

11. GDP.

a) Although more than 97% of the variation in GDP can be accounted for by the model, the residuals plot should be examined to determine whether or not the model is appropriate.

b) This is not a good model for these data. The residuals plot shows curvature.

13. Better GDP model?

This is a much better model. Although there is still some pattern in the residuals, it is a much weaker pattern than the previous model.

15. Brakes.

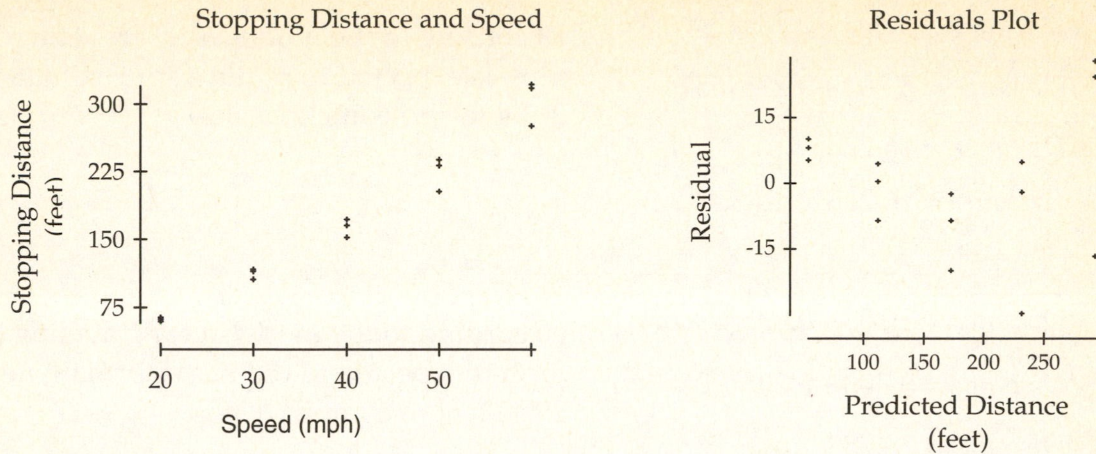

a) The association between speed and stopping distance is strong, positive, and appears straight. Higher speeds are generally associated with greater stopping distances. The linear regression model, with equation $\widehat{Distance} = -65.9 + 5.98(Speed)$, has $R^2 = 96.9\%$, meaning that the model explains 96.9% of the variability in stopping distance. However, the residuals plot has a curved pattern. The linear model is not appropriate. A model using re-expressed variables should be used.

b) Stopping distances appear to be relatively higher for higher speeds. This increase in the rate of change might be able to be straightened by taking the square root of the response variable, stopping distance. The scatterplot of Speed versus $\sqrt{Distance}$ seems like it might be a bit straighter.

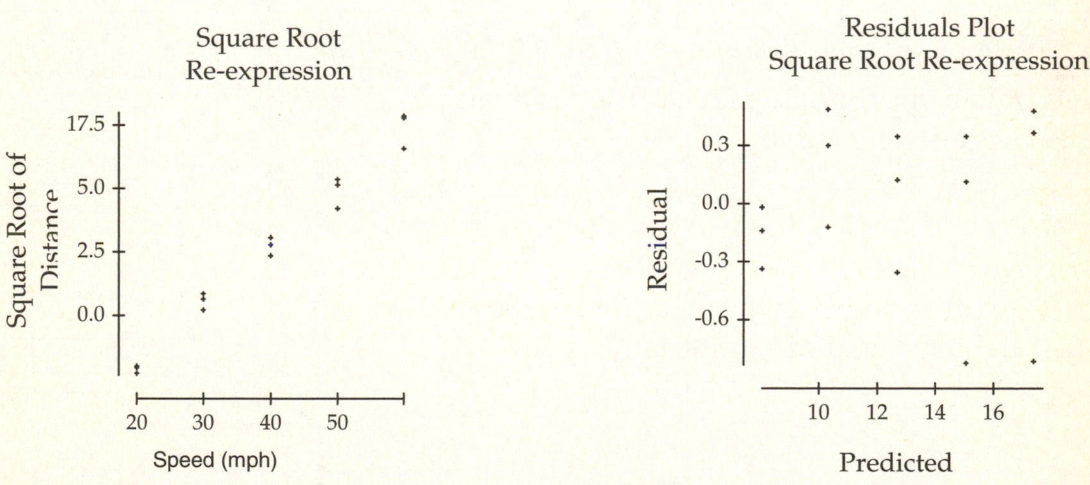

c) The model for the re-expressed data is $\sqrt{\widehat{Distance}} = 3.303 + 0.235(Speed)$. The residuals plot shows no pattern, and $R^2 = 98.4\%$, so 98.4% of the variability in the square root of the stopping distance can be explained by the model.

d)

$$\sqrt{\widehat{Distance}} = 3.303 + 0.235(Speed)$$
$$\sqrt{\widehat{Distance}} = 3.303 + 0.235(55)$$
$$\sqrt{\widehat{Distance}} = 16.228$$
$$\widehat{Distance} = 16.228^2 \approx 263.4$$

According to the model, a car traveling 55 mph is expected to require approximately 263.4 feet to come to a stop.

e)

$$\sqrt{\widehat{Distance}} = 3.303 + 0.235(Speed)$$
$$\sqrt{\widehat{Distance}} = 3.303 + 0.235(70)$$
$$\sqrt{\widehat{Distance}} = 19.753$$
$$\widehat{Distance} = 19.753^2 \approx 390.2$$

According to the model, a car traveling 70 mph is expected to require approximately 390.2 feet to come to a stop.

f) The level of confidence in the predictions should be quite high. R^2 is high, and the residuals plot is scattered. The prediction for 70 mph is a bit of an extrapolation, but should still be reasonably close.

17. Baseball salaries 2009.

a) The scatterplot of year versus highest salary is moderately strong, positive and possibly curved, yet straight enough to try fitting a linear model. The highest salary has generally increased over the years.

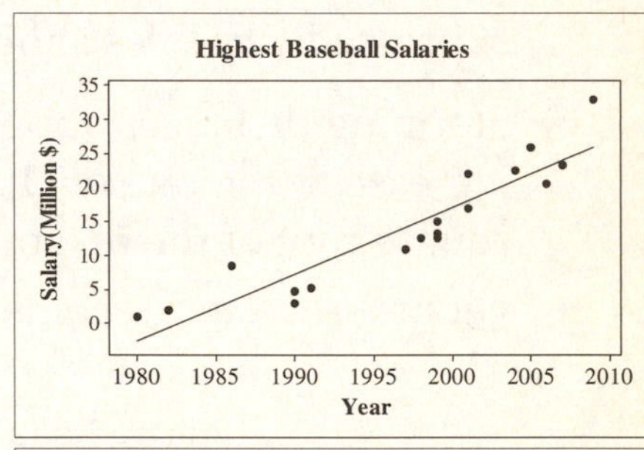

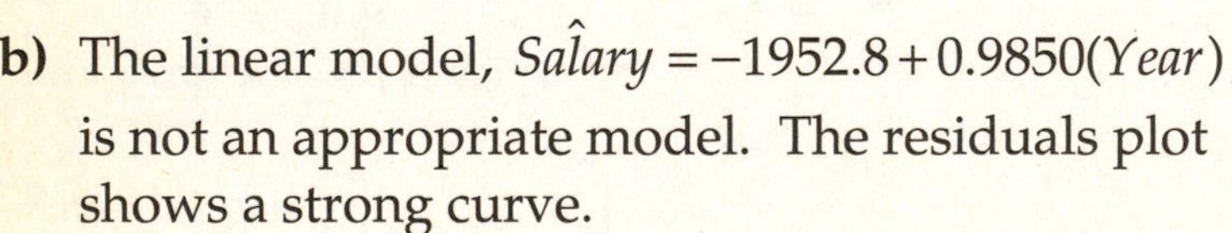

b) The linear model, $\widehat{Salary} = -1952.8 + 0.9850(Year)$, is not an appropriate model. The residuals plot shows a strong curve.

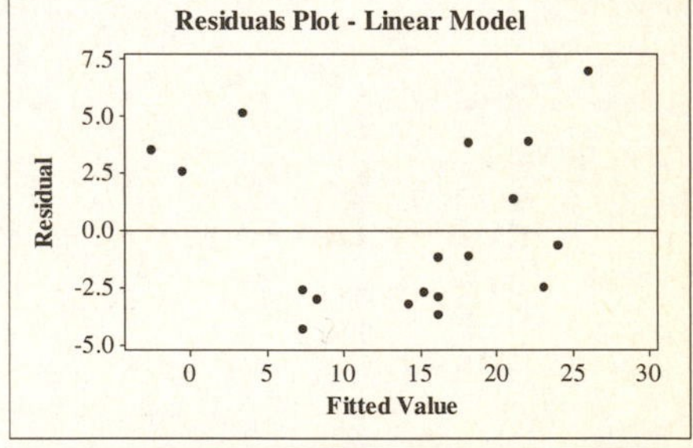

c) Re-expression using the logarithm of the salaries straightens the plot significantly.

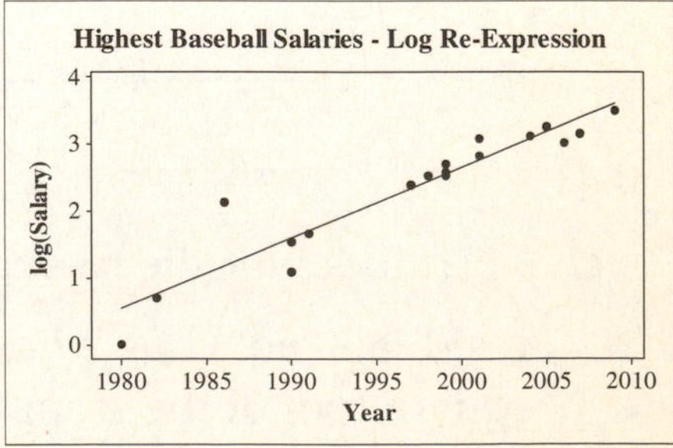

d) $\log(\widehat{Salary}) = -209.24 + 0.105947(Year)$ is a good model for predicting the highest salary from the year. The residuals plot is scattered and R^2 = 88.8%, indicating that 88.8% of the variability in highest salary can be explained by the model.

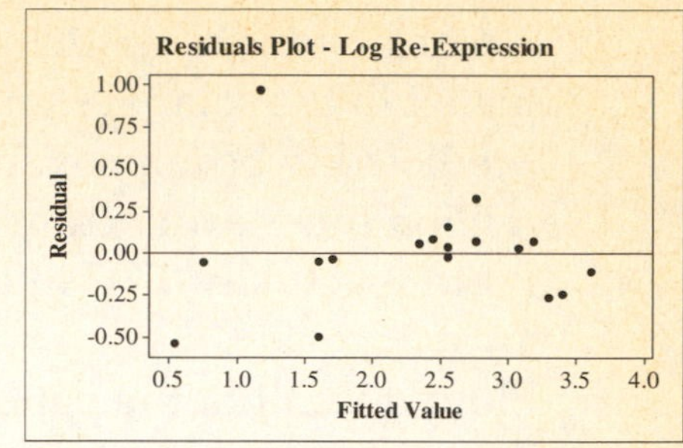

19. Planet distances and order 2006.

a) The association between planetary position and distance from the sun is strong, positive, and curved (below left). A good re-expression of the data is position versus Log(Distance). The scatterplot with regression line (below center) shows the straightened association. The equation of the model is $\log(\widehat{Distance}) = 1.245 + 0.271(Position)$. The residuals plot may have some pattern, but after trying several re-expressions, this is the best that can be done. R^2= 98.2%, so the model explains 98.2% of the variability in the log of the planet's distance from the sun.

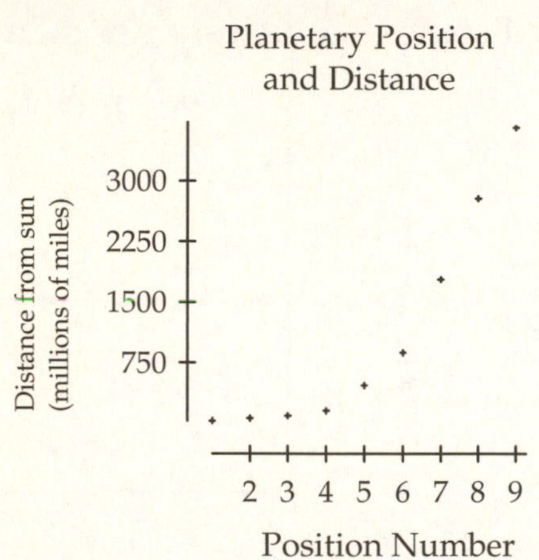

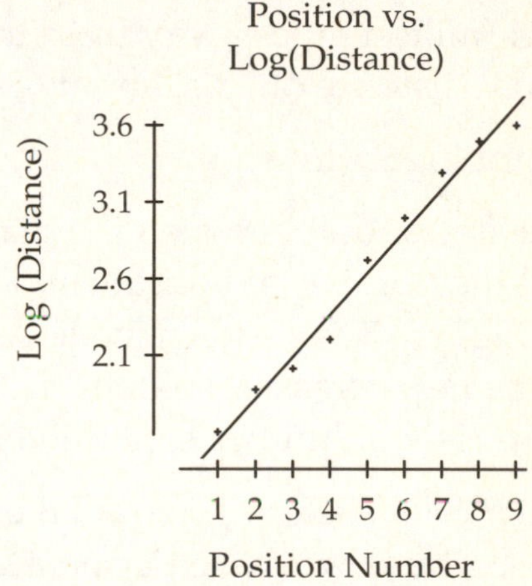

b) At first glance, this model appears to provide little evidence to support the contention of the astronomers. Pluto appears to fit the pattern, although Pluto's distance from the sun is a bit less than expected. A model generated without Pluto does not have a dramatically improved residuals plot, does not have a significantly higher R^2, nor a different slope. Pluto does not appear to be influential.

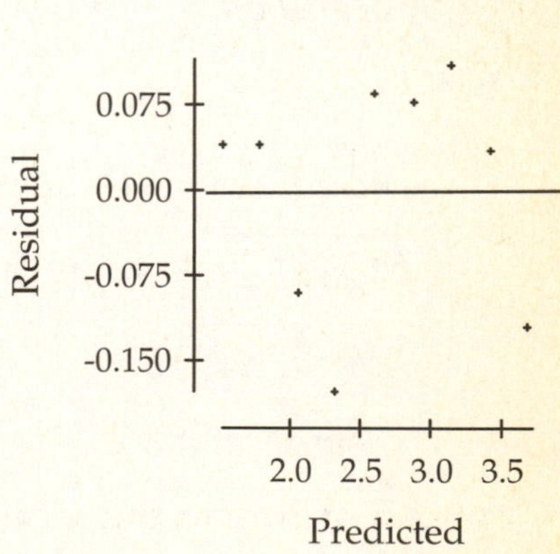

But don't forget that a logarithmic scale is being used for the vertical axis. The higher up the vertical axis you go, the greater the effect of a small change.

According to the model, the 9th planet in the solar system is predicted to be approximately 4844 million miles away from the sun. Pluto is actually 3707 million miles away.

$$\log(\widehat{Distance}) = 1.24418 + 0.271229(Position)$$
$$\log(\widehat{Distance}) = 1.24418 + 0.271229(9)$$
$$\log(\widehat{Distance}) = 3.685241$$
$$\widehat{Distance} = 10^{3.685241} \approx 4844$$

Pluto doesn't fit the pattern for position and distance in the solar system. In fact, the model made with Pluto included isn't a good one, because Pluto influences those predictions. The model without Pluto, $\log(\widehat{Distance}) = 1.20267 + 0.283680(Position)$, works much better. It has a high R^2, and scattered residuals plot. This new model predicts that the 9th planet should be a whopping 5699 million miles away from the sun! There is evidence that the IAU is correct. Pluto doesn't behave like planet in its relation to position and distance.

21. Eris: Planets 2006, part 4.

A planet ninth from the sun was predicted, in a previous exercise, to be about 4844 million miles away from the sun. This distance is much shorter than the actual distance of Eris, about 6300 miles.

23. Logs (not logarithms).

a) The association between the diameter of a log and the number of board feet of lumber is strong, positive, and curved. As the diameter of the log increases, so does the number of board feet of lumber contained in the log.

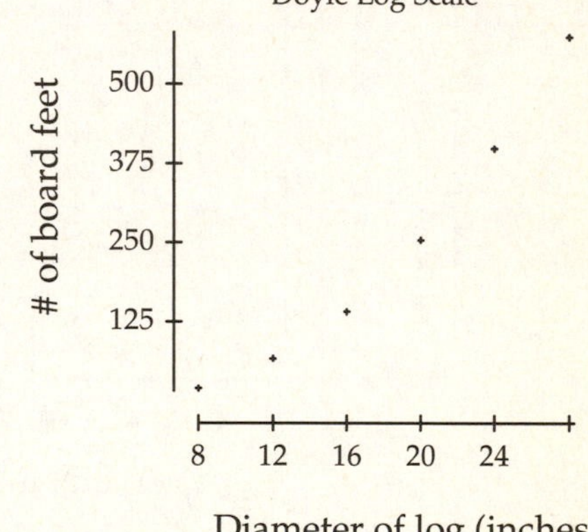

The model used to generate the table used by the log buyers is based upon a square root re-expression. The values in the table correspond exactly to the model $\sqrt{\widehat{BoardFeet}} = -4 + Diameter$.

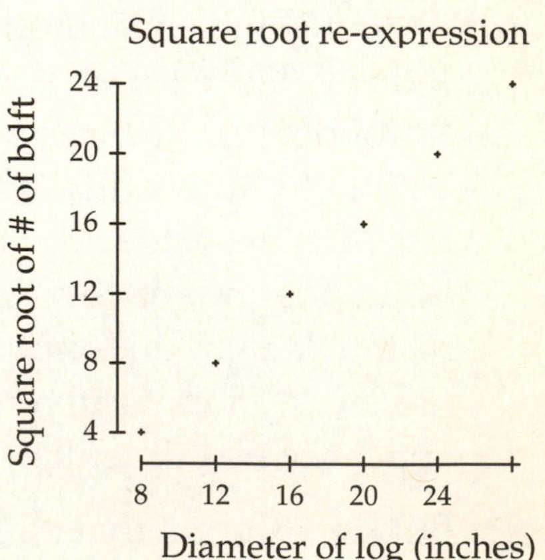

b)

$$\sqrt{\widehat{BoardFeet}} = -4 + Diameter$$
$$\sqrt{\widehat{BoardFeet}} = -4 + (10)$$
$$\sqrt{\widehat{BoardFeet}} = 6$$
$$\widehat{BoardFeet} = 36$$

According to the model, a log 10″ in diameter is expected to contain 36 board feet of lumber.

c)

$$\sqrt{\widehat{BoardFeet}} = -4 + Diameter$$
$$\sqrt{\widehat{BoardFeet}} = -4 + (36)$$
$$\sqrt{\widehat{BoardFeet}} = 32$$
$$\widehat{BoardFeet} = 1024$$

According to the model, a log 36" in diameter is expected to contain 1024 board feet of lumber.

Normally, we would be cautious of this prediction, because it is an extrapolation beyond the given data, but since this is a prediction made from an exact model based on the volume of the log, the prediction will be accurate.

25. Life expectancy.

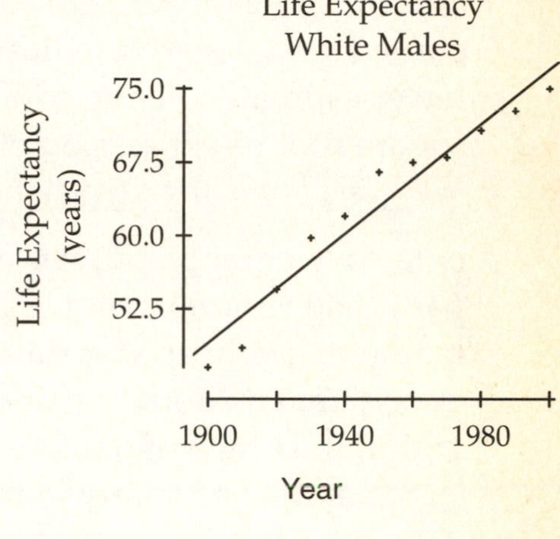

The association between year and life expectancy is strong, curved and positive. As the years passed, life expectancy for white males has increased. The bend (in mathematical language, the change in curvature) in the scatterplot makes it impossible to straighten the association by re-expression.
If all of the data points are to be used, the linear model,
$\widehat{LifeExp} = -454.938 + 0.265697(Year)$

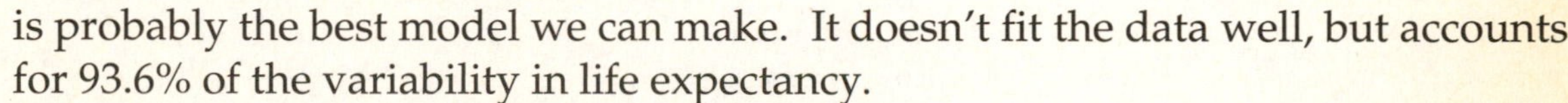

is probably the best model we can make. It doesn't fit the data well, but accounts for 93.6% of the variability in life expectancy.

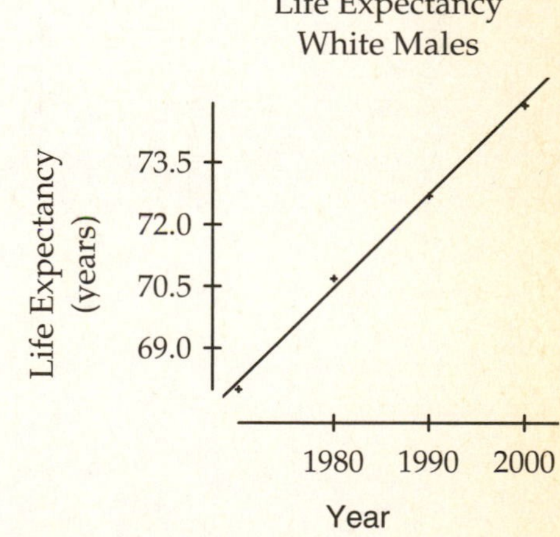

In order to make better future predictions, we might use only a subset of the data points. The scatterplot is very linear from 1970 to 2000, and these recent years are likely to be more indicative of the current trend in life expectancy. Using only these four data points, we might be able to make more accurate future predictions.

The model $\widehat{LifeExp} = -379.020 + 0.227(Year)$ explains 99.6% of the variability in life expectancy, and has a scattered residuals plot. Great care should be taken in using this model for predictions, since it is based upon only 4 points, and, as always, prediction into the future is risky. This is especially true since the response variable is life expectancy, which cannot be expected to increase at the same rate indefinitely.

27. Slower is cheaper?

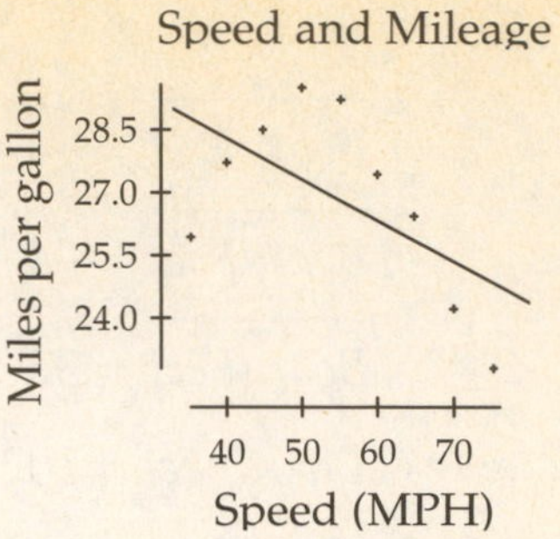

The scatterplot shows the relationship between speed and mileage of the compact car. The association is extremely strong and curved, with mileage generally increasing as speed increases, until around 50 miles per hour, then mileage tends to decrease as speed increases. The linear model is a very poor fit, but the change in direction means that re-expression cannot be used to straighten this association.

29. Years to live 2003.

a) The association between the age and estimated additional years of life for black males is strong, curved, and negative. Older men generally have fewer estimated years of life remaining. The square root re-expression of the data, $\sqrt{Years\hat{L}eft} = 8.465 - 0.06926(Age)$, straightens the data considerably, but has an extremely patterned residuals plot. The model is not a mathematically appropriate model, but fits so closely that it should be fine for predictions within the range of data. The model explains 99.8% of the variability in the estimated number of additional years.

Number of Years Left
Black Males

Years left
60
45
30
15
0 20 40 60 80
Age

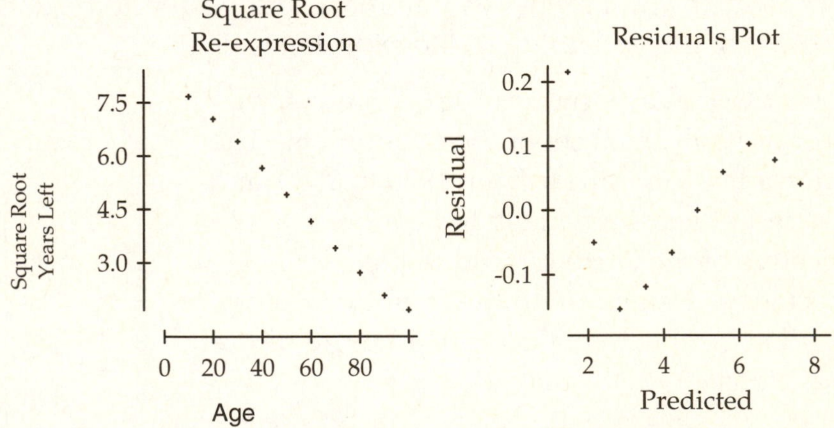

b)

$\sqrt{\widehat{YearsLeft}} = 8.465 - 0.06926(Age)$

$\sqrt{\widehat{YearsLeft}} = 8.465 - 0.06926(18)$

$\sqrt{\widehat{YearsLeft}} = 7.21832$

$\widehat{YearsLeft} = 7.21832^2 \approx 52.1$

According to the model, an 18-year-old black male is expected to live and additional 52.1 years, for a total age of 70.1 years.

c) The residuals plot is extremely patterned, so the model is not appropriate. However, the residuals are very small, making for a tight fit. Since 18 years is within the range of the data, the prediction should be at least reasonable.

Review of Part II – Exploring Relationships Between Variables

1. College.

% over 50: $r = 0.69$ The only moderate, positive correlation in the list.
% under 20: $r = -0.71$ Moderate, negative correlation (–0.98 is too strong)
% Full-time Fac.: $r = 0.09$ No correlation.
% Gr. on time: $r = -0.51$ Moderate, negative correlation (not as strong as %under 20)

3. Vineyards.

a) There does not appear to be an association between ages of vineyards and the cost of products. $r = \sqrt{R^2} = \sqrt{0.027} = 0.164$, indicating a very weak association, at best. The model only explains 2.7% of the variability in case price. Furthermore, the regression equation appears to be influenced by two outliers, products from vineyards over 30 years old, with relatively high case prices.

b) This analysis tells us nothing about vineyards worldwide. There is no reason to believe that the results for the Finger Lakes region are representative of the vineyards of the world.

c) The linear equation used to predict case price from age of the vineyard is:
$\widehat{CasePrice} = 92.765 + 0.567284(Years)$

d) This model is not useful because only 2.7% of the variability in case price is accounted for by the ages of the vineyards. Furthermore, the slope of the regression line seems influenced by the presence of two outliers, products from vineyards over 30 years old, with relatively high case prices.

5. More twins 2006?

a) The association between year and the number of twin births is strong, positive, and appears non-linear. Generally, the number of twin births has increased over the years. The linear model that predicts the number of twin births from the year is: $\widehat{Twins} = -5235191 + 2676.40(Year)$

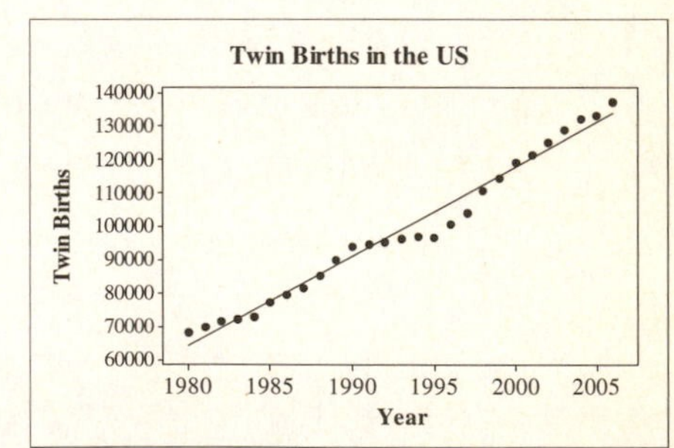

b) For each year that passes, the model predicts that the number of twins born will increase by approximately 2676.40 twin births on average.

c)

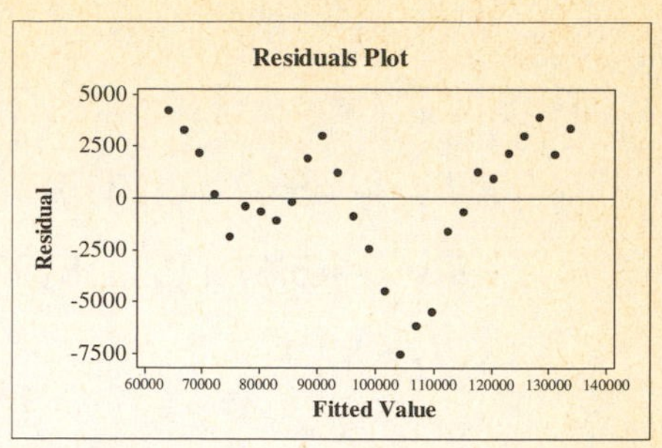

$$\widehat{Twins} = -5235191 + 2676.40(Year)$$

$$\widehat{Twins} = -5235191 + 2676.40(2010)$$

$$\widehat{Twins} = 144,373$$

According to the model, there are expected to be 143,373 twin births in the US in 2010. However, the scatterplot appears non-linear, and there is no reason to believe the number of twin births will keep increasing at the same rate for 4 years beyond the last recorded year. Faith in this prediction is very low.

d) The residuals plot shows a definite curved pattern. The association is not linear, so the linear model is not appropriate.

7. Acid rain.

a) $r = \sqrt{R^2} = \sqrt{0.27} = -0.5196$. The association between pH and BCI appears negative in the scatterplot, so use the negative value of the square root.

b) The association between pH and BCI is negative, moderate, and linear. Generally, higher pH is associated with lower BCI. Additionally, BCI appears more variable for higher values of pH.

c) In a stream with average pH, the BCI would be expected to be average, as well.

d) In a stream where the pH is 3 standard deviations above average, the BCI is expected to be 1.56 standard deviations below the mean level of BCI. (r (3) = —0.5196(3) = —1.56)

9. A manatee model 2005.

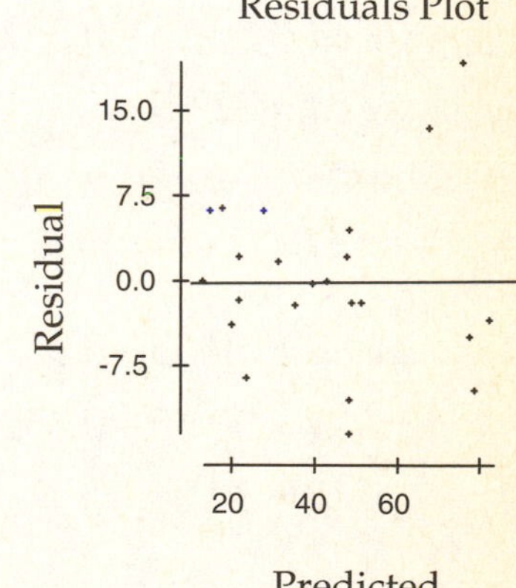

a) The association between the number of powerboat registrations and the number of manatees killed is straight enough to try a linear model. $\widehat{Kills} = -45.67 + 0.1315(ThousandBoats)$ is the best fitting model. The residuals plot is scattered, so the linear model is appropriate.

b) For every additional 10,000 powerboats registered, the model predicts that an additional 1.315 manatees will be killed.

c) The model predicts that if no powerboats were registered, the number of manatee deaths would be approximately —45.67. This is an extrapolation beyond the scope of the data, and doesn't have much contextual meaning.

d)

$$\widehat{Kills} = -45.67 + 0.1315(ThousandBoats)$$
$$\widehat{Kills} = -45.67 + 0.1315(974)$$
$$\widehat{Kills} \approx 82.41$$

The model predicted 82.41 manatee kills in 2005, when the number of powerboat registrations was 974,000. The actual number of kills was 79. The model over-predicted the number of kills by 3.4.

e) Negative residuals are better for the manatees. A negative residual suggests that the actual number of kills was below the number of kills predicted by the model.

f) Over time, the number of powerboat registrations has increased and the number of manatees killed has increased. The trend may continue, resulting in a greater number of manatee deaths in the future.

11. Traffic.

a)

$$b_1 = \frac{rs_y}{s_x}$$
$$-0.352 = \frac{r(9.68)}{27.07}$$
$$r = -0.984$$

The correlation between traffic density and speed is $r = -0.984$

b) $R^2 = (-0.984)^2 = 0.969$.
The variation in the traffic density accounts for 96.9% of the variation in speed.

c)

$$\widehat{speed} = 50.55 - 0.352cars$$
$$\widehat{speed} = 50.55 - 0.352(50)$$
$$\widehat{speed} = 32.95$$

According to the linear model, when traffic density is 50 cars per mile, the average speed of traffic on a moderately large city thoroughfare is expected to be 32.95 miles per hour.

d)

$$\widehat{speed} = 50.55 - 0.352cars$$
$$\widehat{speed} = 50.55 - 0.352(56)$$
$$\widehat{speed} = 30.84$$

According to the linear model, when traffic density is 56 cars per mile, the average speed of traffic on a moderately large city thoroughfare is expected to be 30.84 miles per hour. If traffic is actually moving at 32.5 mph, the residual is 32.5—30.84 = 1.66 miles per hour.

e)

$\hat{speed} = 50.55 - 0.352cars$

$\hat{speed} = 50.55 - 0.352(125)$

$\hat{speed} = 6.55$

According to the linear model, when traffic density is 125 cars per mile, the average speed of traffic on a moderately large city thoroughfare is expected to be 6.55 miles per hour. The point with traffic density 125 cars per minute and average speed 55 miles per hour is considerably higher than the model would predict. If this point were included in the analysis, the slope would increase.

f) The correlation between traffic density and average speed would become weaker. The influential point (125, 55) is a departure from the pattern established by the other data points.

g) The correlation would not change if kilometers were used instead of miles in the calculations. Correlation is a "unitless" measure of the degree of linear association based on *z*-scores, and is not affected by changes in scale. The correlation would remain the same, $r = -0.984$.

13. Correlations.

a) Weight, with a correlation of –0.903, seems to be most strongly associated with fuel economy, since the correlation has the largest magnitude (distance from zero). However, without looking at a scatterplot, we can't be sure that the relationship is linear. Correlation might not be an appropriate measure of the strength of the association if the association is non-linear.

b) The negative correlation between weight and fuel economy indicates that, generally, cars with higher weights tend to have lower mileages than cars with lower weights. Once again, this is only correct if the association between weight and fuel economy is linear.

c) $R^2 = (-0.903)^2 = 0.815$. The variation in weight accounts for 81.5% of the variation in mileage. Once again, this is only correct if the association between weight and fuel economy is linear.

15. Cars, one more time!

a) The linear model that predicts the horsepower of an engine from the weight of the car is: $\hat{Horsepower} = 3.49834 + 34.3144(Weight)$.

b) The weight is measured in thousands of pounds. The slope of the model predicts an increase of about 34.3 horsepower for each additional unit of weight. 34.3 horsepower for each additional thousand pounds makes more sense than 34.3 horsepower for each additional pound.

c) Since the residuals plot shows no pattern, the linear model is appropriate for predicting horsepower from weight.

d)

$$\widehat{Horsepower} = 3.49843 + 34.3144(Weight)$$
$$\widehat{Horsepower} = 3.49843 + 34.3144(2.595)$$
$$\widehat{Horsepower} \approx 92.544$$

According to the model, a car weighing 2595 pounds is expected to have 92.543 horsepower. The actual horsepower of the car is: 92.544 + 22.5 ≈ 115.0 horsepower.

17. Old Faithful.

a) The association between the duration of eruption and the interval between eruptions of Old Faithful is fairly strong, linear, and positive. Long eruptions are generally associated with long intervals between eruptions. There are also two distinct clusters of data, one with many short eruptions followed by short intervals, the other with many long eruptions followed by long intervals, with only a few medium eruptions and intervals in between.

b) The linear model used to predict the interval between eruptions is: $\widehat{Interval} = 33.9668 + 10.3582(Duration)$.

c) As the duration of the previous eruption increases by one minute, the model predicts an increase of about 10.4 minutes in the interval between eruptions.

d) $R^2 = 77.0\%$, so the model accounts for 77% of the variability in the interval between eruptions. The predictions should be fairly accurate, but not precise. Also, the association appears linear, but we should look at the residuals plot to be sure that the model is appropriate before placing too much faith in any prediction.

e)

$$\widehat{Interval} = 33.9668 + 10.3582(Duration)$$
$$\widehat{Interval} = 33.9668 + 10.3582(4)$$
$$\widehat{Interval} \approx 75.4$$

According to the model, if an eruption lasts 4 minutes, the next eruption is expected to occur in approximately 75.4 minutes.

f) The actual eruption at 79 minutes is 3.6 minutes later than predicted by the model. The residual is 79 – 75.4 = 3.6 minutes. In other words, the model under-predicted the interval.

19. How old is that tree?

a) The correlation between tree diameter and tree age is $r = 0.888$. Although the correlation is moderately high, this does not suggest that the linear model is appropriate. We must look at a scatterplot in order to verify that the relationship is straight enough to try the linear model. After finding the linear model, the residuals plot must be checked. If the residuals plot shows no pattern, the linear model can be deemed appropriate.

Tree Age and Diameter

Age (years)

Diameter (inches)

b) The association between diameter and age of these trees is fairly strong, somewhat linear, and positive. Trees with larger diameters are generally older.

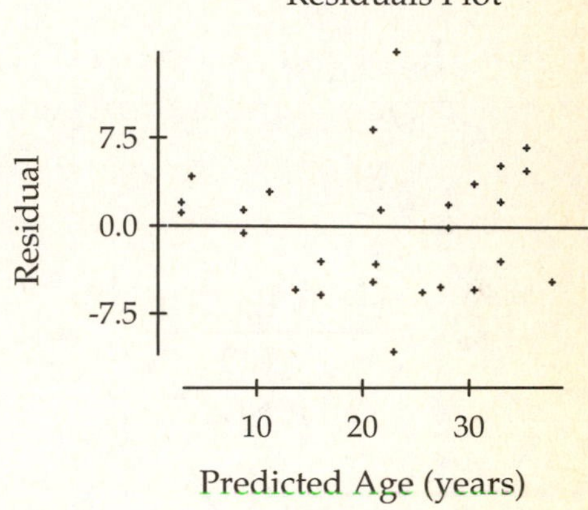

c) The linear model that predicts age from diameter of trees is:

$\widehat{Age} = -0.974424 + 2.20552(Diameter)$. This model explains 78.9% of variability in age.

d) The residuals plot shows a curved pattern, so the linear model is not appropriate. Additionally, there are several trees with large residuals.

e) The largest trees are generally above the regression line, indicating a positive residual. The model is likely to underestimate these values.

21. New homes.

New homes might have areas somewhere between 1000 and 5000 square feet. Doubling the area of a house from 1000 square feet to 2000 square feet would be predicted to increase the price by either 0.008(1000) = \$8 thousand, 0.08(1000) = \$80 thousand, 0.8(1000) = \$800 thousand, or 8(1000) = \$8000 thousand. The only reasonable answer is \$80 thousand, so the slope must be 0.08.

23. No smoking?

a) The model from Exercise 22 is for predicting the percent of expectant mothers who smoked during their pregnancies from the year, not the year from the percent.

b) The model that predicts the year from the percent of expectant mothers who smoked during pregnancy is: $\widehat{Year} = 2027.91 - 2.027(\%)$. This model predicts that 0% of mothers will smoke during pregnancy in $2027.91 - 2.027(0) \approx 2028$.

c) The lowest data point corresponds to 12.7% of expectant mothers smoking during pregnancy in 2003. The prediction for 0% is an extrapolation outside the scope of the data. There is no reason to believe that the model will be accurate at that point.

25. US Cities.

There is a strong, roughly linear, negative association between mean January temperature and latitude. U.S. cities with higher latitudes generally have lower mean January temperatures. There are two outliers, cities with higher mean January temperatures than the pattern would suggest.

27. Winter in the city.

a) $R^2 = (-0.848)^2 \approx 0.719$. The variation in latitude explains 71.9% of the variability in average January temperature.

b) The negative correlation indicates that the as latitude increases, the average January temperature generally decreases.

c)

$$b_1 = \frac{rs_y}{s_x} \qquad \hat{y} = b_0 + b_1 x$$

$$b_1 = \frac{(-0.848)(13.49)}{5.42} \qquad \bar{y} = b_0 + b_1\bar{x}$$

$$26.44 = b_0 - 2.1106125(39.02)$$

$$b_1 = -2.1106125 \qquad b_0 = 108.79610$$

The equation of the linear model for predicting January temperature from latitude is: $\widehat{JanTemp} = 108.796 - 2.111(Latitude)$

d) For each additional degree of latitude, the model predicts a decrease of approximately 2.1°F in average January temperature.

e) The model predicts that the mean January temperature will be approximately 108.8°F when the latitude is 0°. This is an extrapolation, and may not be meaningful.

f)

$$\widehat{JanTemp} = 108.796 - 2.111(Latitude)$$

$$\widehat{JanTemp} = 108.796 - 2.111(40)$$

$$\widehat{JanTemp} \approx 24.4$$

According to the model, the mean January temperature in Denver is expected to be 24.4°F.

g) In this context, a positive residual means that the actual average temperature in the city was higher than the temperature predicted by the model. In other words, the model underestimated the average January temperature.

29. Jumps 2008.

a) The association between Olympic long jump distances and high jump heights is strong, linear, and positive. Years with longer long jumps tended to have higher high jumps. There is one departure from the pattern. The year in which the Olympic gold medal long jump was the longest had a shorter gold medal high jump than we might have predicted.

b) There is an association between long jump and high jump performance, but it is likely that training and technique have improved over time and affected both jump performances.

c) The correlation would be the same, 0.920. Correlation is a measure of the degree of linear association between two quantitative variables and is unaffected by changes in units.

d) In a year when the high jumper jumped one standard deviation better than the average jump, the long jumper would be predicted to jump $r = 0.920$ standard deviations above the average long jump.

31. French.

a) Most of the students would have similar weights. Regardless of their individual French vocabularies, the correlation would be near 0.

b) There are two possibilities. If the school offers French at all grade levels, then the correlation would be positive and strong. Older students, who typically weigh more, would have higher scores on the test, since they would have learned more French vocabulary. If French is not offered, the correlation between weight and test score would be near 0. Regardless of weight, most students would have horrible scores.

c) The correlation would be near 0. Most of the students would have similar weights and vocabulary test scores. Weight would not be a predictor of score.

d) The correlation would be positive and strong. Older students, who typically weigh more, would have higher test scores, since they would have learned more French vocabulary.

33. Lunchtime.

The association between time spent at the table and number of calories consumed by toddlers is moderate, roughly linear, and negative. Generally, toddlers who spent a longer time at the table consumed fewer calories than toddlers who left the table quickly. The scatterplot between time at the table and calories consumed is straight enough to justify the use of the linear model. The linear model that predicts the time number of calories consumed by a toddler from the time spent at the table is $\widehat{Calories} = 560.7 - 3.08(Time)$. For each additional minute spent at the table, the model predicts that the number of calories consumed will be approximately 3.08 fewer. Only 42.1% of the variability in the number of calories consumed can be accounted for by the variability in time spent at the table. The residuals plot shows no pattern, so the linear model is appropriate, if not terribly useful for prediction.

35. Tobacco and alcohol.

The first concern about these data is that they consist of averages for regions in Great Britain, not individual households. Any conclusions reached can only be about the regions, not the individual households living there. The second concern is the data point for Northern Ireland. This point has high leverage, since it has the highest household tobacco spending and the lowest household alcohol spending. With this point included, there appears to be only a weak positive association between tobacco and alcohol spending. Without the point, the association is much stronger. In Great Britain, with the exception of Northern Ireland, higher levels of household spending on tobacco are associated with higher levels of household spending on tobacco. It is not necessary to make the linear model, since we have the household averages for the regions in Great Britain, and the model wouldn't be useful for predicting in other countries or for individual households in Great Britain.

Household Spending on Tobacco and Alcohol in Great Britain

37. Models.

a) $\hat{y} = 2 + 0.8\ln x$

$\hat{y} = 2 + 0.8\ln(10)$

$\hat{y} \approx 3.842$

b) $\log \hat{y} = 5 - 0.23x$

$\log \hat{y} = 5 - 0.23(10)$

$\log \hat{y} = 2.7$

$\hat{y} = 10^{2.7} \approx 501.187$

c) $\frac{1}{\sqrt{\hat{y}}} = 17.1 - 1.66x$

$\frac{1}{\sqrt{\hat{y}}} = 17.1 - 1.66(10) = 0.5$

$\hat{y} = \frac{1}{0.5^2} = 4$

39. Vehicle weights.

a)

$\widehat{Wt} = 10.85 + 0.64 scale$

$\widehat{Wt} = 10.85 + 0.64(31.2)$

$\widehat{Wt} = 30.818$

According to the model, a truck with a scale weight of 31,200 pounds is expected to weigh 30,818 pounds.

b) If the actual weight of the truck is 32,120 pounds, the residual is 32,120 – 30,818 = 1302 pounds. The model underestimated the weight.

c)

$\hat{Wt} = 10.85 + 0.64 scale$

$\hat{Wt} = 10.85 + 0.64(35.590)$

$\hat{Wt} = 33.6276$ thousand pounds

The predicted weight of the truck is 33,627.6 pounds. If the residual is –2440 pounds, the actual weight of the truck is 33,627.6 – 2440 = 31,187.6 pounds.

d) $R^2 = 93\%$, so the model explains 93% of the variability in weight, but some of the residuals are 1000 pounds or more. If we need to be more accurate than that, then this model will not work well.

e) Negative residuals will be more of a problem. Police would be issuing tickets to trucks whose weights had been overestimated by the model. The U.S. justice system is based upon the principle of innocence until guilt is proven. These truckers would be unfairly ticketed, and that is worse than allowing overweight trucks to pass.

41. Down the drain.

The association between diameter of the drain plug and drain time of this water tank is strong, curved, and negative. Tanks with larger drain plugs have lower drain times. The linear model is not appropriate for the curved association, so several re-expressions of the data were tried. The best one was the reciprocal square root re-expression, resulting in the equation

$$\frac{1}{\sqrt{\widehat{DrainTime}}} = 0.00243 + 0.219(Diameter).$$

Drain Diameter and Drain Time

Drain Time (min)

120

90

60

30

0.4 0.8 1.2 1.6

Diameter (inches)

The re-expressed data is nearly linear, and although the residuals plot might still indicate some pattern and has one large residual, this is the best of the models examined. The model explains 99.7% of the variability in drain time.

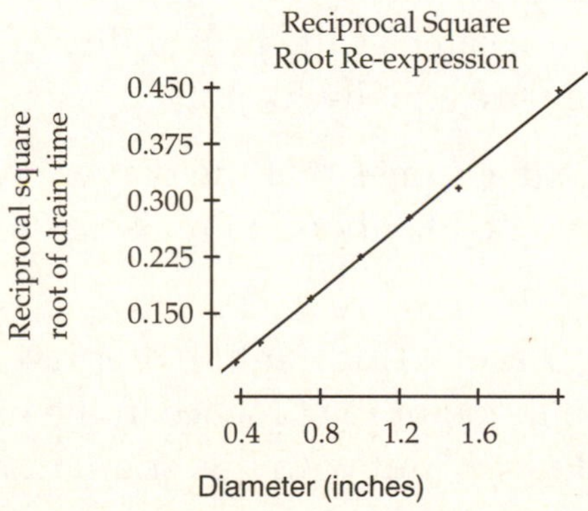

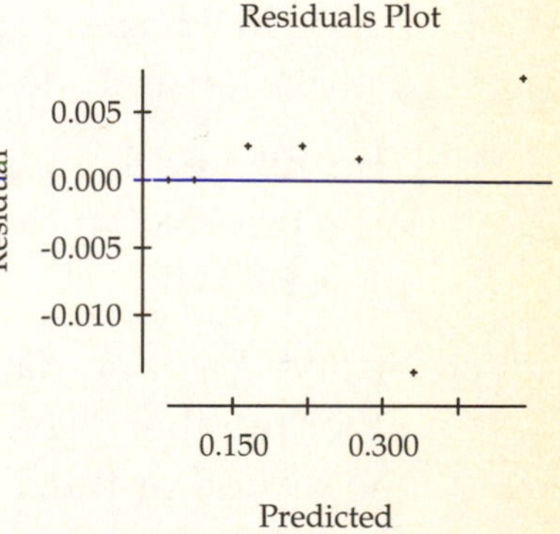

Chapter 11 – Understanding Randomness

1. Coin toss.

A coin flip is random, because the outcome cannot be predicted beforehand.

3. The lottery.

In state lotteries, a machine pops up numbered balls. If the lottery were truly random, the outcome could not be predicted and the outcomes would be equally likely. It is random only if the balls generate numbers in equal frequencies.

5. Birth defects.

Answers may vary. Generate two-digit random numbers, 00-99. Let 00-02 represent a defect. Let 03-99 represent no defect.

7. Geography.

a) Looking at pairs of digits, the first state number is 45, Vermont. The next set is ignored since there is no 92nd state. The next state number is 10, Georgia.

b) Continuing along, the next state number is 17, Kentucky. The next state number, 10, is ignored, since Georgia was already assigned. The final state number is 22, Michigan.

9. Play the lottery.

If the lottery is random, it doesn't matter if you play the same favorite "lucky" numbers or if you play different numbers each time. All numbers are equally likely (or, rather, UNLIKELY) to win.

11. Bad simulations

a) The outcomes are not equally likely. For example, the probability of getting 5 heads in 9 tosses is not the same as the probability of getting 0 heads, but the simulation assumes they are equally likely.

b) The even-odd assignment assumes that the player is equally likely to score or miss the shot. In reality, the likelihood of making the shot depends on the player's skill.

c) Suppose a hand has four aces. This might be represented by 1,1,1,1, and any other number. The likelihood of the first ace in the hand is not the same as for the second or third or fourth. But with this simulation, the likelihood is the same for each.

13. Wrong conclusion.

The conclusion should indicate that the simulation **suggests** that the average length of the line would be 3.2 people. Future results might not match the simulated results exactly.

15. Election.

a) Answers will vary. A component is one voter voting. An outcome is a vote for our candidate. Using two random digits, 00-99, let 01-55 represent a vote for your candidate, and let 55-99 and 00 represent a vote for the underdog.

b) A trial is 100 votes. Examine 100 two-digit random numbers and count how many simulated votes are cast for each candidate. Whoever gets the majority of the votes wins the trial.

c) The response variable is whether the underdog wins or not. To calculate the experimental probability, divide the number of trials in which the simulated underdog wins by the total number of trials.

17. Cereal.

Answers will vary. A component is the simulation of the picture in one box of cereal. One possible way to model this component is to generate random digits 0-9. Let 0 and 1 represent LeBron James, 2-4 represent David Beckham, and 5-9 represent Serena Williams. Each trial will consist of 5 random digits, and the response variable will be whether or not a complete set of pictures is simulated. Trials in which at least one of each picture is simulated will be a success. The total number of successes divided by the total number of trials will be the simulated probability of ending up with a complete set of pictures. According to the simulation, the probability of getting a complete set of pictures is expected to be about 51.5%.

19. Multiple choice

Answers will vary. A component is one multiple-choice question. One possible way to model this component is to generate random digits 0-9. Let digits 0-7 represent a correct answer, and let digits 8 and 9 represent an incorrect answer. Each trial will consist of 6 random digits. The response variable is whether or not all 6 simulated questions are answered correctly (all 6 digits are 0-7). The total number of successes divided by the total number of trials will be the simulated probability of getting all 6 questions right. According to the simulation, the probability of getting all 6 multiple-choice questions correct is expected to be about 26%.

21. Beat the lottery.

a) Answers based on your simulation will vary, but you should win about 10% of the time.

b) You should win at the same rate with any number.

23. It evens out in the end.

Answers based on your simulation will vary, but you should win about 10% of the time. Playing lottery numbers that have turned up the least in recent lottery drawers offers no advantage. Each new drawing is independent of recent drawings.

25. Driving test.

Answers will vary. A component is one drivers test, but this component will be modeled differently, depending on whether or not it is the first test taken. One possible way to model this component is to generate pairs of random digits 00-99. Let 01-34 represent passing the first test and let 35-99 and 00 represent failing the first test. Let 01-72 represent passing a retest, and let 73-99 and 00 represent failing a retest. To simulate one trial, generate pairs of random numbers until a pair is generated that represents passing a test. Begin each trial using the "first test" representation, and switch to the "retest" representation if failure is indicated on the first simulated test. The response variable is the number of simulated tests required to achieve the first passing test. The total number of simulated tests taken divided by the total number of trials is the simulated average number of tests required to pass. According to the simulation, the number of driving tests required to pass is expected to be about 1.9.

27. Basketball strategy.

Answers will vary. A component is one foul shot. One way to model this component would be to generate pairs of random digits 00-99. Let 01-72 represent a made shot, and let 73-99 and 00 represent a missed shot. The response variable is the number of shots made in a "one and one" situation. If the first shot simulated represents a made shot, simulate a second shot. If the first shot simulated represents a miss, the trial is over. The simulated average number of points is the total number of simulated points divided by the number of trials. According to the simulation, the player is expected to score about 1.24 points.

29. Free groceries.

Answers will vary. A component is the selection of one card with the prize indicated. One possible way to model the prize is to generate pairs of random digits 00-99. Let 01-10 represent $200, let 11-20 represent $100, let 21-40 represent $50, and let 41-99 and 00 represent $20. Repeated pairs of digits must be ignored. (For this reason, a simulation in which random digits 0-9 are generated with 0 representing $200, 1 representing $100, etc., is NOT acceptable. Each card must be individually represented.) A trial continues until the total simulated prize is greater than $500. The response variable is the number of simulated customers until the payoff is greater than $500. The simulated average number of customers is the total number of simulated customers divided by the number of trials. According to the simulation, about 10.2 winners are expected each week.

31. The family.

Answers will vary. Each child is a component. One way to model the component is to generate random digits 0-9. Let 0-4 represent a boy and let 5-9 represent a girl. A trial consists of generating random digits until a child of each gender is simulated. The response variable is the number of children simulated until this happens. The simulated average family size is the number of digits generated in each trial divided by the total number of trials. According to the simulation, the expected number of children in the family is about 3.

33. Dice game.

Answers will very. Each roll of the die is a component. One way of modeling this component is to generate random digits 0-9. The digits 1-6 correspond to the numbers on the faces of the die, and digits 7-9 and 0 are ignored. A trial consists of generating random numbers until the sum of the numbers is exactly 10. If the sum exceeds 10, the last roll must be ignored and simulated again, but still counted as a roll. The response variable is the number of rolls until the sum is exactly 10. The simulated average number of rolls until this happens is the total number of rolls simulated divided by the number of trials. According to the simulation, expect to roll the die about 7.5 times.

35. The hot hand.

Answers may vary. Each shot is a component. One way of modeling this component is to generate pairs of random digits 00-99. Let 01-65 represent a made shot, and let 66-99 and 00 represent a missed shot. A trial consists of 20 simulated shots. The response variable is whether or not the 20 simulated shots contained a run of 6 or more made shots. To find the simulated percentage of games in which the player is expected to have a run of 6 or more made shots, divide the total number of successes by the total number of trials. According to the simulation, the player is expected to make 6 or more shots in a row in about 40% of games. This isn't unusual. The announcer was wrong to characterize her performance as extraordinary.

37. Teammates.

Answers will vary. Each player chosen is a component. One way to model this component is to generate random numbers 0-9. Let 1 and 2 represent the first couple, 3 and 4 the second couple, 5 and 6 the third couple, and 7 and 8 the fourth couple. Ignore 9 and 0. A trial consists of generating random digits, ignoring repeats, and organizing them into pairs, until pairs representing the first three teams are generated. (The final team is assigned by default.) The response variable is whether or not each of the simulated teams is a pairing other than 1-2, 3-4, 5-6, or 7-8. The simulated percentage of the time this is expected to happen is the total number of successes (times that the pairings are *different* than the couples) divided by the total number of trials. According to the simulation, all players are expected to be paired with someone other than the person with whom he or she came to the party about 37.5% of the time.

39. Job discrimination?

Answers may vary. Each person hired is a component. One way of modeling this component is to generate pairs of random digits 00-99. Let 01-10 represent each of the 10 women, and let 11-22 represent each of the 12 men. Ignore 23-99 and 00. A trial consists of 3 usable pairs of numbers. Ignore repeated pairs of digits, since the same man or woman cannot be hired more than once. The response variable is whether or not all 3 simulated hires are women. The simulated percentage of the time that 3 women are expected to be hired is the number of successes divided by the number of trials. According to the simulation, the 3 people hired will all be women about 7.8% of time. This seems a bit strange, but not quite strange enough to be evidence of job discrimination.

Chapter 12 – Sample Surveys

1. **Roper.**

   a) Roper is not using a simple random sample. The samples are designed to get 500 males and 500 females. This would be very unlikely to happen in a simple random sample.

   b) They are using stratified sample, with two strata, males and females.

3. **Emoticons.**

   a) This is a voluntary response sample.

   b) We have absolutely no confidence is estimates made from voluntary response samples.

5. **Gallup.**

   a) The population of interest is all adults in the United States aged 18 and older.

   b) The sampling frame is U.S. adults with land-line telephones.

   c) Some members of the population (e.g. many college students) don't have land-line telephones, so they could never be chosen in the sample. This may create a bias.

7. **Population** – all U.S. adults
   Parameter – proportion who have used and benefited from alternative medical treatments.
   Sampling Frame – all Consumers Union subscribers
   Sample – those subscribers who responded
   Method – not specified, but probably a questionnaire mailed to all subscribers
   Bias – nonresponse bias, specifically voluntary response bias. Those who respond may have strong feelings one way or another.

9. **Population** – adults
   Parameter – proportion who think drinking and driving is a serious problem
   Sampling Frame – bar patrons
   Sample – every 10th person leaving the bar
   Method – systematic sampling
   Bias – undercoverage. Those interviewed had just left a bar, and may have opinions about drinking and driving that differ from the opinions of the population in general.

11. **Population** – soil around a former waste dump
Parameter – proportion with elevated levels of harmful substances
Sampling Frame – accessible soil around the dump
Sample – 16 soil samples
Method – not clear. There is no indication of randomness.
Bias – possible convenience sample. Since there is no indication of randomization, the samples may have been taken from easily accessible areas. Soil in these areas may be more or less polluted than the soil in general.

13. **Population** – snack food bags
Parameter – proportion passing inspection
Sampling Frame – all bags produced each day
Sample – 10 bags, one from each of 10 randomly selected cases
Method – multistage sampling. Presumably, they take a simple random sample of 10 cases, followed by a simple random sample of one bag from each case.
Bias – no indication of bias

15. Mistaken poll.

The station's faulty prediction is more likely to be the result of bias. Only people watching the news were able to respond, and their opinions were likely to be different from those of other voters. The sampling method may have systematically produced samples that did not represent the population.

17. Parent opinion, part 1.

a) This is a voluntary response sample. Only those who see the ad, feel strongly about the issue, and have web access will respond.

b) This is cluster sampling, but probably not a good idea. The opinions of parents in one school may not be typical of the opinions of all parents.

c) This is an attempt at a census, and will probably suffer from nonresponse bias.

d) This is stratified sampling. If the follow-up is carried out carefully, the sample should be unbiased.

19. Churches.

a) This is a multistage design, with a cluster sample at the first stage and a simple random sample for each cluster.

b) If any of the three churches you pick at random are not representative of all churches, then your sample will reflect the makeup of that church, not all churches. Also, choosing 100 members at random from each church could introduce bias. The views of the members of smaller churches chosen in the sample will be weighted heavier in your sample than the views of members of larger churches, especially if the views of the members of that small church differ from the views of churchgoers at large. The hope is that random sampling will equalize these sources of variability in the long run.

21. Roller coasters.

a) This is a systematic sample.

b) This sample is likely to be representative of those waiting in line for the roller coaster, especially if those people at the front of the line (after their long wait) respond differently from those at the end of the line.

c) The sampling frame is patrons willing to wait in line for the roller coaster. The sample should be representative of the people in line, but not of all the people at the park.

23. Wording the survey.

a) Responses to these questions will differ. Question 1 will probably get "no" answers, and Question 2 will probably get "yes" answers. This is response bias, based on the wording of the questions.

b) A question with neutral wording might be: "Do you think standardized tests are appropriate for deciding whether a student should be promoted to the next grade?"

25. Survey questions.

a) The question is biased toward "yes" answers because of the word "pollute". A better question might be: "Should companies be responsible for any costs of environmental clean up?"

b) The question is biased toward "no" because of the preamble "18-year-olds are old enough to serve in the military. A better question might be: "Do you think the drinking age should be lowered from 21?"

27. Phone surveys.

a) A simple random sample is difficult in this case because there is a problem with undercoverage. People with unlisted phone numbers and those without phones are not in the sampling frame. People who are at work, or otherwise away from home, are included in the sampling frame. These people could never be in the sample itself.

b) One possibility is to generate random phone numbers and call at random times, although obviously not in the middle of the night! This would take care of the undercoverage of people at work during the day, as well as people with unlisted phone numbers, although there is still a problem avoiding undercoverage of people without phones.

c) Under the original plan, those families in which one person stays home are more likely to be included. Under the second plan, many more are included. People without phones are still excluded.

d) Follow-up of this type greatly improves the chance that a selected household is included, increasing the reliability of the survey.

e) Random dialers allow people with unlisted phone numbers to be selected, although they may not be the most willing participants. There is a reason that the phone number is unlisted. Time of day will still be an issue, as will people without phones.

29. Arm length.

a) Answers will vary. My arm length is 3 hand widths and 2 finger widths.

b) The parameter estimated by 10 measurements is the true length of your arm. The population is all possible measurements of your arm length.

c) The population is now the arm lengths of your friends. The average now estimates the mean of the arm lengths of your friends.

d) These 10 arm lengths are unlikely to be representative of the community, or the country. Your friends are likely to be of the same age, and not very diverse.

31. Accounting.

a) Assign numbers 001-120 to each order. Generate 10 random numbers 001-120, and select those orders to recheck.

b) The supervisor should perform a stratified sample, randomly checking a certain percentage of each type of sales, retail and wholesale.

33. Quality control.

a) Select three cases at random, then select one jar randomly from each case.

b) Generate three random numbers between 61-80, with no repeats, to select three cases. Then assign each of the jars in the case a number 01-12, and generate one random number for each case to select the three jars, one from each case.

c) This is not a simple random sample, since there are groups of three jars that cannot be the sample. For example, it is impossible for three jars in the same case to be the sample. This would be possible if the sample were a simple random sample.

35. Sampling methods.

a) This method would probably result in undercoverage of those doctors that are not listed in the Yellow Pages. Using the "line listings" seems fair, as long as all doctors are listed, but using the advertisements would not be a typical list of doctors.

b) This method is not appropriate. This cluster sample will probably contain listings for only one or two types of businesses, not a representative cross-section of businesses.

Chapter 13 – Experiments and Observational Studies

1. **Standardized test scores.**

   a) No, this is not an experiment. There are no imposed treatments. This is a retrospective observational study.

   b) We cannot conclude that the differences in score are caused by differences in parental income. There may be lurking variables that are associated with both SAT score and parental income.

3. **MS and vitamin D.**

   a) This is a retrospective observational study.

   b) This is an appropriate choice, since MS is a relatively rare disease.

   c) The subjects were U.S. military personnel, some of whom had developed MS.

   d) The variables were the vitamin D blood levels and whether or not the subject developed MS.

5. **Menopause.**

   a) This was a randomized, comparative, placebo-controlled experiment.

   b) Yes, such an experiment is the right way to determine whether black cohosh is an effective treatment for hot flashes.

   c) The subjects were 351 women, aged 45 to 55 who reported at least two hot flashes a day.

   d) The treatments were black cohosh, a multi-herb supplement, plus advice to consume more soy foods, estrogen, and a placebo. The response was the women's self-reported symptoms, presumably the frequency of hot flashes.

7. a) This is an experiment, since treatments were imposed.
   b) The subjects studied were 30 patients with bipolar disorder.
   c) The experiment has 1 factor (omega-3 fats from fish oil), at 2 levels (high dose of omega-3 fats from fish oil and no omega-3 fats from fish oil).
   d) 1 factor, at 2 levels gives a total of 2 treatments.
   e) The response variable is "improvement", but there is no indication of how the response variable was measured.
   f) There is no information about the design of the experiment.
   g) The experiment is blinded, since the use of a placebo keeps the patients from knowing whether or not they received the omega-3 fats from fish oils. It is not stated whether or not the evaluators of the "improvement" were blind to the treatment, which would make the experiment double-blind.

h) Although it needs to be replicated, the experiment can determine whether or not omega-3 fats from fish oils cause improvements in patients with bipolar disorder, at least over the short term. The experiment design would be stronger is it were double-blind.

9. **a)** This is an observational study. The researchers are simply studying traits that already exist in the subjects, not imposing new treatments.

b) This is a prospective study. The subjects were identified first, then traits were observed.

c) The subjects are roughly 200 men and women with moderately high blood pressure and normal blood pressure. There is no information about the selection method.

d) The parameters of interest are difference in memory and reaction time scores between those with normal blood pressure and moderately high blood pressure.

e) An observational study has no random assignment, so there is no way to know that high blood pressure caused subjects to do worse on memory and reaction time tests. A lurking variable, such as age or overall health, might have been the cause. The most we can say is that there was an association between blood pressure and scores on memory and reaction time tests in this group, and recommend a controlled experiment to attempt to determine whether or not there is a cause-and-effect relationship.

11. **a)** This is an experiment, since treatments were imposed on randomly assigned groups.

b) 24 post-menopausal women were the subjects in this experiment.

c) There is 1 factor (type of drink), at 2 levels (alcoholic and non-alcoholic). (Supplemental estrogen is not a factor in the experiment, but rather a blocking variable. The subjects were not given estrogen supplements as part of the experiment.)

d) 1 factor, with 2 levels, is 2 treatments.

e) The response variable is an increase in estrogen level.

f) This experiment utilizes a blocked design. The subjects were blocked by whether or not they used supplemental estrogen. This design reduces variability in the response variable of estrogen level that may be associated with the use of supplemental estrogen.

g) This experiment does not use blinding.

h) This experiment indicates that drinking alcohol leads to increased estrogen level among those taking estrogen supplements.

13. a) This is an observational study.
b) The study is retrospective. Results were obtained from pre-existing church records.
c) The subjects of the study are women in Finland. The data were collected from church records dating 1640 to 1870, but the selection process is unknown.
d) The parameter of interest is difference in average lifespan between mothers of sons and daughters.
e) For this group, having sons was associated with a decrease in lifespan of an average of 34 weeks per son, while having daughters was associated with an unspecified increase in lifespan. As there is no random assignment, there is no way to know that having sons caused a decrease in lifespan.

15. a) This is an observational study. (Although some might say that the sad movie was "imposed" on the subjects, this was merely a stimulus used to trigger a reaction, not a treatment designed to attempt to influence some response variable. Researchers merely wanted to observe the behavior of two different groups when each was presented with the stimulus.)
b) The study is prospective. Researchers identified subjects, and then observed them after the sad movie.
c) The subjects in this study were people with and without depression. The selection process is not stated.
d) The parameter of interest is the difference in crying response between depressed and nondepressed people exposed to sad situations.
e) There is no apparent difference in crying response to sad movies for the depressed and nondepressed groups.

17. a) This is an experiment. Subjects were randomly assigned to treatments.
b) The subjects were people experiencing migraines.
c) There are 2 factors (pain reliever and water temperature). The pain reliever factor has 2 levels (pain reliever or placebo), and the water temperature factor has 2 levels (ice water and regular water).
d) 2 factors, at 2 levels each, results in 4 treatments.
e) The response variable is the level of pain relief.
f) The experiment is completely randomized.
g) The subjects are blinded to the pain reliever factor through the use of a placebo. The subjects are not blinded to the water factor. They will know whether they are drinking ice water or regular water.
h) The experiment may indicate whether pain reliever alone or in combination with ice water give pain relief, but patients are not blinded to ice water, so the placebo effect may also be the cause of any relief seen due to ice water.

19. a) This is an experiment. Athletes were randomly assigned to one of two exercise programs.

b) The subjects are athletes suffering hamstring injuries.

c) There is one factor (type of exercise), at 2 levels (static stretching, and agility and trunk stabilization).

d) 1 factor, at 2 levels, results in 2 treatments.

e) The response variable is the time before the athletes were able to return to sports.

f) The experiment is completely randomized.

g) The experiment employs no blinding. The subjects know what kind of exercise they do.

h) Assuming that the athletes actually followed the exercise program, this experiment can help determine which of the two exercise programs is more effective at rehabilitating hamstring injuries.

21. Omega-3.

The experimenters need to compare omega-3 results to something. Perhaps bipolarity is seasonal and would have improved during the experiment anyway.

23. Omega-3 revisited.

a) Subjects' responses might be related to other factors, like diet, exercise, or genetics. Randomization should equalize the two groups with respect to unknown factors.

b) More subjects would minimize the impact of individual variability in the responses, but the experiment would become more costly and time-consuming.

25. Omega-3 finis.

The researchers believe that people who engage in regular exercise might respond differently to the omega-3. This additional variability could obscure the effectiveness of the treatment.

27. Tomatoes.

Answers may vary. Number the tomatoes plants 1 to 24. Use a random number generator to randomly select 24 numbers from 1 to 24 without replication. Assign the tomato plants matching the first 8 numbers to the first group, the second 8 numbers to the second group, and the third group of 8 numbers to the third group.

29. Shoes.

a) First, the manufacturers are using athletes who have a vested interest in the success of the shoe by virtue of their sponsorship. They should try to find some volunteers that aren't employed by the company! Second, they should randomize the order of the runs, not run all the races with the new shoes second. They should blind the athletes by disguising the shoes, if possible, so they don't know which is which. The experiment could be double blinded, as well, by making sure that the timers don't know which shoes are being tested at any given time. Finally, they should replicate several times since times will vary under both shoe conditions.

b) First of all, the problems identified in part a would have to be remedied before *any* conclusions can be reached. Even if this is the case, the results cannot be generalized to all runners. This experiment compares effects of the shoes on speed for Olympic class runners, not runners in general.

31. Hamstrings.

a) Allowing the athletes to choose their own treatments could confound the results. Other issues such as severity of injury, diet, age, etc., could also affect time to heal, and randomization should equalize the two treatment groups with respect to any such variables.

b) A control group could have revealed whether either exercise program was better (or worse) than just letting the injury heal without exercise.

c) Although the athletes cannot be blinded, the doctors who approve their return to sports should not know which treatment the subject had engaged in.

d) It's difficult to say with any certainty, since we aren't sure if the distributions of return times are unimodal and roughly symmetric, and contain no outliers. Otherwise, the use of mean and standard deviation as measures of center and spread is questionable. Assuming mean and standard deviation are appropriate measures, the subjects who exercised with agility and trunk stabilization had a mean return time of 22.2 days compared to the static stretching group, with a mean return time of 37.4 days. The agility and trunk stabilization group also had a much more consistent distribution of return times, with a standard deviation of 8.3 days, compared to the standard deviation of 27.6 days for the static stretching group.

33. Mozart.

a) The differences in spatial reasoning scores between the students listening to Mozart and the students sitting quietly were more than would have been expected from ordinary sampling variation.

b)

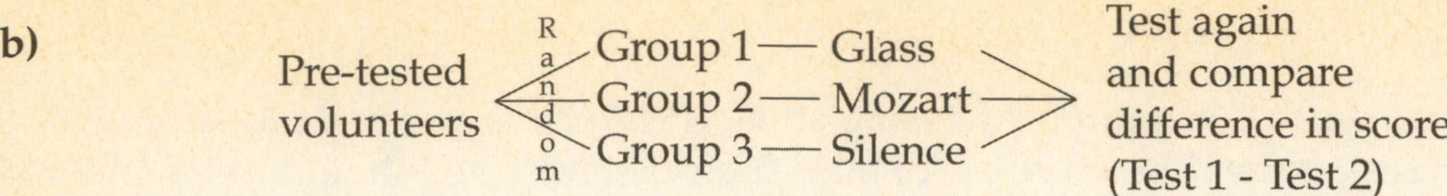

c) The Mozart group seems to have the smallest median difference in spatial reasoning test score and thus the *least* improvement, but there does not appear to be a significant difference.

d) No, the results do not prove that listening to Mozart is beneficial. If anything, there was generally less improvement. The difference does not seem significant compared with the usual variation one would expect between the three groups. Even if type of music has no effect on test score, we would expect some variation between the groups.

35. Wine.

a) This is a prospective observational study. The researchers followed a group of children born at a Copenhagen hospital between 1959 and 1961.

b) The results of the Danish study report a link between high socioeconomic status, education, and wine drinking. Since people with high levels of education and higher socioeconomic status are also more likely to be healthy, the relation between health and wine consumption might be explained by the confounding variables of socioeconomic status and education.

c) Studies such as these prove none of these. While the variables have a relation, there is no indication of a cause-and-effect relationship. The only way to determine causation is through a controlled, randomized, and replicated experiment.

37. Dowsing.

a) Arrange the 20 containers in 20 separate locations. Number the containers 01 – 20, and use a random number generator to identify the 10 containers that should be filled with water.

b) We would expect the dowser to be correct about 50% of the time, just by guessing. A record of 60% (12 out of 20) does not appear to be significantly different than the 10 out of 20 expected.

c) Answers may vary. A high level of success would need to be observed. 90% to 100% success (18 to 20 correct identifications) would be convincing.

39. Reading.

Reading teachers
Random
Group 1 — phonics
Group 2 — whole language
Compare reading scores

Answers may vary. This experiment has 1 factor (reading program), at 2 levels (phonics and whole language), resulting in 2 treatments. The response variable is reading score on an appropriate reading test after a year in the program. After randomly assigning students to teachers, randomly assign half the reading teachers in the district to use each method. There may be variation in reading score based on school within the district, as well as by grade. Blocking by both school and grade will reduce this variation.

41. Weekend deaths.

a) The difference between death rate on the weekend and death rate during the week is greater than would be expected due to natural sampling variation.

b) This was a prospective observational study. The researchers identified hospitals in Ontario, Canada, and tracked admissions to the emergency rooms. This certainly cannot be an experiment. People can't be assigned to become injured on a specific day of the week!

c) Waiting until Monday, if you were ill on Saturday, would be foolish. There are likely to be confounding variables that account for the higher death rate on the weekends. For example, people might be more likely to engage in risky behavior on the weekend.

d) Alcohol use might have something to do with the higher death rate on the weekends. Perhaps more people drink alcohol on weekends, which may lead to more traffic accidents, and higher rates of violence during these days of the week.

43. Beetles.

Answers may vary. This experiment has 1 factor (pesticide), at 3 levels (pesticide A, pesticide B, no pesticide), resulting in 3 treatments. The response variable is the number of beetle larvae found on each plant. Randomly select a third of the plots to be sprayed with pesticide A, a third with pesticide B, and a third to be sprayed with no pesticide (since the researcher also wants to know whether the pesticides even work at all). To control the experiment, the plots of land should be as similar as possible, with regard to amount of sunlight, water, proximity to other plants, etc. If not, plots with similar characteristics should be blocked together. If possible, use some inert substance as a placebo pesticide on the control group, and do not tell the counters of the beetle larvae which plants have been treated with pesticides. After a given period of time, count the number of beetle larvae on each plant and compare the results.

Plots of corn — Random —
Group 1 — pesticide A
Group 2 — pesticide B
Group 3 — no pesticide
→ Count the number of beetle larvae on each plant and compare

45. Safety switch.

Answers may vary. This experiment has 1 factor (hand), at 2 levels (right, left), resulting in 2 treatments. The response variable is the difference in deactivation time between left and right hand. Find a group of volunteers. Using a matched design, we will require each volunteer to deactivate the machine with his or her left hand, as well as with his or her right hand. Randomly assign the left or right hand to be used first. Hopefully, this will equalize any variability in time that may result from experience gained after deactivating the machine the first time. Complete the first attempt for the whole group. Now repeat the experiment with the alternate hand. Check the differences in time for the left and right hands. Since the response variable is difference in times for each hand, workers should be blocked into groups based on their dominant hand. Another way to account for this difference would be to use the absolute value of the difference as the response variable. We are interested in whether or not the difference is significantly different from the zero difference we would expect if the machine were just as easy to operate with either hand.

47. Skydiving, anyone?

a) There is 1 factor, jumping, with 2 levels, with and without a working parachute.

b) You would need some (dim-witted) volunteers skydivers as the subjects.

c) A parachute that looked real, but didn't open, would serve as the placebo.

d) 1 factor at 2 levels is 2 treatments, a good parachute and a placebo parachute.

e) The response variable is whether the skydiver survives the jump (or the extent of injuries).

f) All skydivers should jump from the same altitude, in similar weather conditions, and land on similar surfaces.

g) Make sure that you randomly assign the skydivers to the parachutes.

h) The skydivers (and the distributers of the parachutes) shouldn't know who got a working chute. Additionally, the people evaluating the subjects after the jumps should not be told who had a real chute, either.

Review of Part III – Gathering Data

1. The researchers performed a prospective observational study, since the children were identified at birth and examined at ages 8 and 20. There were indications of behavioral differences between the group of "preemies", and the group of full-term babies. The "preemies" were less likely to engage in risky behaviors, like use of drugs and alcohol, teen pregnancy, and conviction of crimes. This may point to a link between premature birth and behavior, but there may be lurking variables involved. Without a controlled, randomized, and replicated experiment, a cause-and-effect relationship cannot be determined.

3. The researchers at the Purina Pet Institute performed an experiment, matched by gender and weight. The experiment had one factor (diet), at two levels (allowing the dogs to eat as much as they want, or restricted diet), resulting in two treatments. One of each pair of similar puppies was randomly assigned to each treatment. The response variable was length of life. The researchers were able to conclude that, on average, dogs with a lower-calorie diet live longer.

5. It is not apparent whether or not the high folate intake was a randomly imposed treatment. It is likely that the high folate intake was simply an observed trait in the women, so this is a prospective observational study. This study indicates that folate may help in reducing colon cancer for those with family histories of the disease. There may be other variables in the diet and lifestyle of these women that accounted for the decreased risk of colon cancer, so a cause-and-effect relationship cannot be inferred.

7. The fireworks manufacturers are sampling. No information is given about the sampling procedure, so hopefully the tested fireworks are selected randomly. It would probably be a good idea to test a few of each type of firework, so stratification by type seems likely. The population is all fireworks produced each day, and the parameter of interest is the proportion of duds. With a random sample, the manufacturers can make inferences about the proportion of duds in the entire day's production, and use this information to decide whether or not the day's production is suitable for sale.

9. The researchers performed a retrospective observational study. The data were gathered from pre-existing medical records. Living near strong electromagnetic fields may be associated with an increase in leukemia rates. Since this is not a controlled, randomized, and replicated experiment, lurking variables may be involved in this increased risk. For example, the neighborhood around the antennas may consist of people of the same socioeconomic status. Perhaps some variable linked with socioeconomic status may be responsible for the higher leukemia rates.

11. This is an experiment, blocked by sex of the rat. There is one factor (type of hormone), with at least two levels (leptin, insulin). There is a possibility that an additional level of no hormone (or placebo injection) was used as a control, although this is not specifically mentioned. This results in at least two treatments, possibly three, if a control was used. There is no specific mention of random allocation. There are two response variables: amount of food consumed and weight lost. Hormones can help suppress appetite, and lead to weight loss, in laboratory rats. The male rats responded best to insulin, and the female rats responded best to leptin.

13. The researchers performed an experiment. There is one factor (gene therapy), at two levels (gene therapy and no gene therapy), resulting in two treatments. The experiment is completely randomized. The response variable is heart muscle condition. The researchers can conclude that gene therapy is responsible for stabilizing heart muscle in laboratory rats.

15. The orange juice plant depends on sampling to ensure the oranges are suitable for juice. The population is all of the oranges on the truck, and the parameter of interest is the proportion of unsuitable oranges. The procedure used is a random sample, stratified by location in the truck. Using this well-chosen sample, the workers at the plant can estimate the proportion of unsuitable oranges on the truck, and decide whether or not to accept the load.

17. The researchers performed a prospective observational study, since the subjects were identified ahead of time. Physically fit men may have a lower risk of death from cancer than other men.

19. Point spread.

Answers may vary. Perform a simulation to determine the gambler's expected winnings. A component is one game. To model that component, generate random digits 0 to 9. Since the outcome after the point spread is a tossup, let digits 0-4 represent a loss, and let digits 5-9 represent a win. A run consists of 5 games, so generate 5 random digits at a time. The response variable is the profit the gambler makes, after accounting for the $10 bet. If the outcome of the run is 0, 1, or 2 simulated wins, the profit is —$10. If the outcome is 3, 4, or 5 simulated wins, the profit is $0, $10, or $40, respectively. The total profit divided by the number of runs is the average weekly profit. According to the simulation (80 runs were performed), the gambler is expected to break even. His simulated losses equaled his simulated winnings. (In theory, the gambler is expected to lose about $2.19 per game.)

21. Everyday randomness.

Answers will vary. Most of the time, events described as "random" are anything but truly random.

23. Tips.

a) The waiters performed an experiment, since treatments were imposed on randomly assigned groups. This experiment has one factor (candy), at two levels (candy or no candy), resulting in two treatments. The response variable is the percentage of the bill given as a tip.

b) If the decision whether to give candy or not was made before the people were served, the server may have subconsciously introduced bias by treating the customers better. If the decision was made just before the check was delivered, then it is reasonable to conclude that the candy was the cause of the increase in the percentage of the bill given as a tip.

c) "Statistically significant" means that the difference in the percentage of tips between the candy and no candy groups was more than expected due to sampling variability.

25. Cloning.

a) The *USA Weekend* survey suffers from voluntary response bias. Only those who feel strongly will pay for the 900 number phone call.

b) Answers may vary. A strong positive response might be created with the question, "If it would help future generations live longer, healthier lives, would you be in favor of human cloning?"

27. When to stop?

a) Answers may vary. A component in this simulation is rolling 1 die. To simulate this component, generate a random digit 1 to 6. To simulate a run, simulate 4 rolls, stopping if a 6 is rolled. The response variable is the sum of the 4 rolls, or 0 if a 6 is rolled. The average number of points scored is the sum of all rolls divided by the total number of runs. According to the simulation, the average number of points scored will be about 5.8.

b) Answers may vary. A component in this simulation is rolling 1 die. To simulate this component, generate a random digit 1 to 6. To simulate a run, generate random digits until the sum of the digits is at least 12, or until a 6 is rolled. The response variable is the sum of the digits, or 0 if a 6 is rolled. The average number of points scored is the sum of all rolls divided by the total number of runs. According to the simulation, the average number of points scored will be about 5.8, similar to the outcome of the method described in part a).

c) Answers may vary. Be careful when making your decision about the effectiveness of your strategy. If you develop a strategy with a higher simulated average number of points than the other two methods, this is only an indication that you may win in the long run. If the game is played round by round, with the winner of a particular round being declared as the player with the highest roll made during that round, the game is much more variable. For example, if Player B rolls a 12 in a particular game, Player A will always lose that game, provided he or she sticks to the strategy. A better way to get a feel for your chances of winning this type of game might be to simulate several rounds, recording whether each player won or lost the round. Then estimate the percentage of the time that each player is expected to win, according to the simulation.

29. Homecoming.

a) Since telephone numbers were generated randomly, every number that could possibly occur in that community had an equal chance of being selected. This method is "better" than using the phone book, because unlisted numbers are also possible. Those community members who deliberately do not list their phone numbers might not consider this method "better"!

b) Although this method results in a simple random sample of phone numbers, it does not result in a simple random sample of residences. Residences without a phone are excluded, and residences with more than one phone have a greater chance of being included.

c) No, this is not a SRS of local voters. People who respond to the survey may be of the desired age, but not registered to vote. Additionally, some voters who are contacted may choose not to participate.

d) This method does not guarantee an unbiased sample of households. Households in which someone answered the phone may be more likely to have someone at home when the phone call was generated. The attitude about homecoming of these households might not be the same as the attitudes of the community at large.

31. Smoking and Alzheimer's.

a) The studies do not prove that smoking offers any protection from Alzheimer's. The studies merely indicate an association. There may be other variables that can account for this association.

b) Alzheimer's usually shows up late in life. Since smoking is known to be harmful, perhaps smokers have died of other causes before Alzheimer's can be seen.

c) The only way to establish a cause-and-effect relationship between smoking and Alzheimer's is to perform a controlled, randomized, and replicated experiment. This is unlikely to ever happen, since the factor being studied, smoking, has already been proven harmful. It would be unethical to impose this treatment on people for the purposes of this experiment. A prospective observational study could be designed in which groups of smokers and nonsmokers are followed for many years and the incidence of Alzheimer's disease is tracked.

33. Sex and violence.

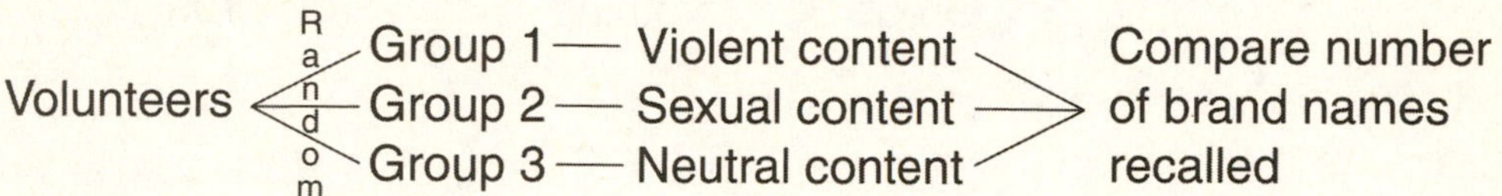

This experiment has one factor (program content), at three levels (violent, sexual, and neutral), resulting in three treatments. The response variable is the number of brand names recalled after watching the program. Numerous subjects will be randomly assigned to see shows with violent, sexual, or neutral content. They will see the same commercials. After the show, they will be interviewed for their recall of brand names in the commercials.

35. Age and party.

a) The number of respondents is roughly the same for each age category. This may indicate a sample stratified by age category, although it may be a simple random sample.

b) 1404 Democrats were surveyed. $\frac{1404}{4002} \approx 35.1\%$ of the people surveyed were Democrats.

c) If this poll is truly representative, this is probably a good estimate of the percentage of voters who are Democrats. However, telephone surveys are likely to systematically exclude those who do not have phones, a group that is likely to be composed of people who cannot afford phones. If socioeconomic status is linked to political affiliation, this might be a problem.

d) The pollsters were probably attempting to determine whether or not political party is associated with age.

37. Save the grapes.

This experiment has one factor (bird control device), at three levels (scarecrow, netting, and no device), resulting in three treatments. Randomly assign different plots in the vineyard to the different treatments, making sure to ensure adequate separation of plots, so that the possible effect of the scarecrow will not be confounded with the other treatments. The response variable to be measured at the end of the season is the proportion of bird-damaged grapes in each plot.

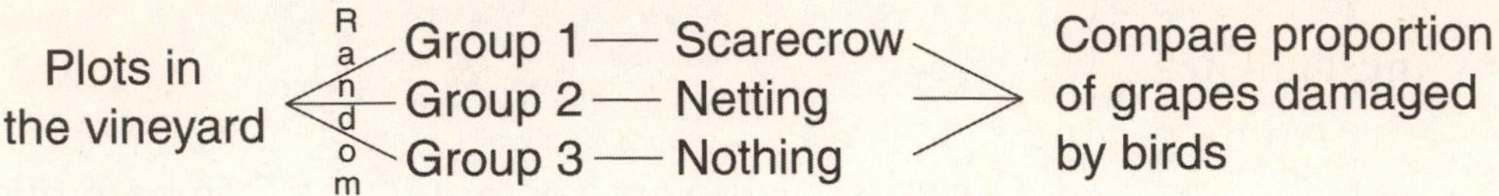

39. Knees.

a) In an experiment, all subjects must be treated as alike as possible. If there were no "placebo surgery", subjects would know that they had not received the operation, and might react differently. Of course, all volunteers for the experiment must be aware of the possibility of being randomly assigned to the placebo group.

b) Experiments always use volunteers. This is not a problem, since experiments are testing response to a treatment, not attempting to determine an unknown population parameter. The randomization in an experiment is random assignment to treatment groups, not random selection from a population. Voluntary response is a problem when sampling, but is not an issue in experimentation..

c) There were differences in the amount of pain relief experienced by the two groups, but these differences were small enough that they could be explained by natural sampling variation, even if surgery were not a factor.

41. Security.

a) To ensure that passengers from first-class, as well as coach, get searched, select 2 passengers from first-class and 12 from coach. Using this stratified random sample, 10% of the first-class passengers are searched, as are 10% of the coach passengers.

b) Answers will vary. Number the passengers alphabetically, with 2-digit numbers. Bergman = 01, Bowman = 02, and so on, ending with Testut = 20. Read the random digits in pairs, ignoring pairs 21 to 99 and 00, and ignoring repeated pairs.

```
65|43|67|11|             27|04|
XX XX XX Fontana         XX Castillo
```

The passengers selected for search from first-class are Fontana and Castillo.

c) Number the passengers alphabetically, with 3 digit numbers, 001 to 120. Use the random number table to generate 3-digit numbers, ignoring numbers 121 to 999 and 000, and ignoring repeated numbers. Search the passengers corresponding to the first 12 valid numbers generated.

43. Par 4.

Answers may vary. A component in this simulation is a shot. Use pairs of random digits 00 to 99 to represent a shot. The way in which this component is simulated depends on the type of shot.

For the first shot, let pairs of digits 01 to 70 represent hitting the fairway, and let pairs of digits 71 to 99, and 00, represent not hitting the fairway.

If the first simulated shot hits the fairway, let 01 to 80 represent landing on the green on the second shot, and let 81 to 99, and 00, represent not landing on the green on the second shot. If the first simulated shot does not hit the fairway, let 01 to 40 represent landing on the green on the second shot, and let 41 to 99, and 00, represent not landing on the green on the second shot.

If the second simulated shot does not land on the green, let 01 to 90 represent landing on the green, and 91 to 99, and 00, represent not landing on the green. Keep simulating shots until the shot lands on the green.

Once on the green, let 01 to 20 represent sinking the putt on the first putt, and let 21 to 99, and 00, represent not sinking the putt on the first putt. If second putts are required, continue simulating putts until a putt goes in, with 01 to 90 representing making the putt, and 91 to 99, and 00, representing not making the putt.

A run consists of following the guidelines above until the final putt is made. The response variable is the number of shots required until the final putt is made.

The simulated average score on the hole is the total number of shots required divided by the total number of runs. According to 40 runs of this simulation, a pretty good golfer can be expected to average about 4.2 strokes per hole. Your simulation results may vary.

Chapter 14 – From Randomness to Probability

1. **Sample spaces.**

   a) S = { HH, HT, TH, TT} All of the outcomes are equally likely to occur.

   b) S = { 0, 1, 2, 3} All outcomes are not equally likely. A family of 3 is more likely to have, for example, 2 boys than 3 boys. There are three equally likely outcomes that result in 2 boys (BBG, BGB, and GBB), and only one that results in 3 boys (BBB).

   c) S = { H, TH, TTH, TTT} All outcomes are not equally likely. For example the probability of getting heads on the first try is $\frac{1}{2}$. The probability of getting three tails is $\left(\frac{1}{2}\right)^3 = \frac{1}{8}$.

   d) S = {1, 2, 3, 4, 5, 6} All outcomes are not equally likely. Since you are recording only the larger number of two dice, 6 will be the larger when the other die reads 1, 2, 3, 4, or 5. The outcome 2 will only occur when the other die shows 1 or 2.

3. **Roulette.**

   If a roulette wheel is to be considered truly random, then each outcome is equally likely to occur, and knowing one outcome will not affect the probability of the next. Additionally, there is an implication that the outcome is not determined through the use of an electronic random number generator.

5. **Winter.**

   Although acknowledging that there is no law of averages, Knox attempts to use the law of averages to predict the severity of the winter. Some winters are harsh and some are mild over the long run, and knowledge of this can help us to develop a long-term probability of having a harsh winter. However, probability does not compensate for odd occurrences in the short term. Suppose that the probability of having a harsh winter is 30%. Even if there are several mild winters in a row, the probability of having a harsh winter is still 30%.

7. **Cold streak.**

   There is no such thing as being "due for a hit". This statement is based on the so-called law of averages, which is a mistaken belief that probability will compensate in the short term for odd occurrences in the past. The batter's chance for a hit does not change based on recent successes or failures.

9. Fire insurance.

a) It would be foolish to insure your neighbor's house for $300. Although you would probably simply collect $300, there is a chance you could end up paying much more than $300. That risk probably is not worth the $300.

b) The insurance company insures many people. The overwhelming majority of customers pay the insurance and never have a claim. The few customers who do have a claim are offset by the many who simply send their premiums without a claim. The relative risk to the insurance company is low.

11. Spinner.

a) This is a legitimate probability assignment. Each outcome has probability between 0 and 1, inclusive, and the sum of the probabilities is 1.

b) This is a legitimate probability assignment. Each outcome has probability between 0 and 1, inclusive, and the sum of the probabilities is 1.

c) This is not a legitimate probability assignment. Each outcome has probability between 0 and 1, inclusive, but the sum of the probabilities is greater than 1.

d) This is a legitimate probability assignment. Each outcome has probability between 0 and 1, inclusive, and the sum of the probabilities is 1. However, this game is not very exciting!

e) This probability assignment is not legitimate. The sum of the probabilities is 0, and there is one probability, -1.5, that is not between 0 and 1, inclusive.

13. Vehicles.

A family may have both a car and an SUV. The events are not disjoint, so the Addition Rule does not apply.

15. Speeders.

When cars are traveling close together, their speeds are not independent. For example, a car following directly behind another can't be going faster than the car ahead. Since the speeds are not independent, the Multiplication Rule does not apply.

17. College admissions.

a) Jorge had multiplied the probabilities.

b) Jorge assumes that being accepted to the colleges are independent events.

c) No. Colleges use similar criteria for acceptance, so the decisions are not independent. Students that meet these criteria are more likely to be accepted at all of the colleges. Since the decisions are not independent, the probabilities cannot be multiplied together.

19. Car repairs.

Since all of the events listed are disjoint, the addition rule can be used.

a) P(no repairs) = 1 – P(some repairs) = 1 – (0.17 + 0.07 + 0.04) = 1 – (0.28) = 0.72

b) P(no more than one repair) = P(no repairs or one repair) = 0.72 + 0.17 = 0.89

c) P(some repairs) = P(one or two or three or more repairs)
= 0.17 + 0.07 + 0.04 = 0.28

21. More repairs.

Assuming that repairs on the two cars are independent from one another, the multiplication rule can be used. Use the probabilities of events from Exercise 19 in the calculations.

a) P(neither will need repair) = (0.72)(0.72) = 0.5184

b) P(both will need repair) = (0.28)(0.28) = 0.0784

c) P(at least one will need repair) = 1 - P (neither will need repair)
= 1 - (0.72)(0.72) = 0.4816

23. Repairs, again.

a) The repair needs for the two cars must be independent of one another.

b) This may not be reasonable. An owner may treat the two cars similarly, taking good (or poor) care of both. This may decrease (or increase) the likelihood that each needs to be repaired.

25. Energy 2007.

a) P(response is "More production") = 342/1005 = 0.340

b) P("Equally important" or "No opinion") = 80/1005 + 30/1005 = 110/1005 = 0.109

27. More energy.

a) P(all three respond "Protect the environment") = $\left(\frac{583}{1005}\right)\left(\frac{583}{1005}\right)\left(\frac{583}{1005}\right) \approx 0.195$

b) P(none respond "Equally important") = $\left(\frac{975}{1005}\right)\left(\frac{975}{1005}\right)\left(\frac{975}{1005}\right) \approx 0.913$

c) In order to compute the probabilities, we must assume that responses are independent.

d) It is reasonable to assume that responses are independent, since the three people were chosen at random.

29. Polling.

a) P(household is contacted and household refuses to cooperate)
$= P(\text{household is contacted})P(\text{household refuses} \mid \text{contacted})$
$= (0.76)(1 - 0.38) = 0.4712$

b) $P(\text{fail to contact household or contacting and not getting interview})$
$= P(\text{fail to contact}) + P(\text{contact household})P(\text{not getting interview} \mid \text{contacted})$
$= (1-0.76)+(0.76)(1-0.38) = 0.7112$

c) The question in part b covers all possible occurrences *except* contacting the house and getting the interview. The probability could also be calculated by subtracting the probability of the complement of the event from 22a.

$P(\text{failing to contact household or contacting and not getting the interview})$
$= 1 - P(\text{contacting the household and getting the interview})$
$= 1-(0.76)(0.38) = 0.7112$

31. M&M's

a) Since all of the events are disjoint (an M&M can't be two colors at once!), use the addition rule where applicable.

1. $P(\text{brown}) = 1 - P(\text{not brown}) = 1 - P(\text{yellow or red or orange or blue or green})$
$= 1 - (0.20 + 0.20 + 0.10 + 0.10 + 0.10) = 0.30$
2. $P(\text{yellow or orange}) = 0.20 + 0.10 = 0.30$
3. $P(\text{not green}) = 1 - P(\text{green}) = 1 - 0.10 = 0.90$
4. $P(\text{striped}) = 0$

b) Since the events are independent (picking out one M&M doesn't affect the outcome of the next pick), the multiplication rule may be used.

1. $P(\text{all three are brown}) = (0.30)(0.30)(0.30) = 0.027$
2. $P(\text{the third one is the first one that is red}) = P(\text{not red and not red and red})$
$= (0.80)(0.80)(0.20) = 0.128$
3. $P(\text{no yellow}) = P(\text{not yellow and not yellow and not yellow})$
$= (0.80)(0.80)(0.80) = 0.512$
4. $P(\text{at least one is green}) = 1 - P(\text{none are green}) = 1 - (0.90)(0.90)(0.90) = 0.271$

33. Disjoint or independent?

a) For one draw, the events of getting a red M&M and getting an orange M&M are disjoint events. Your single draw cannot be both red and orange.

b) For two draws, the events of getting a red M&M on the first draw and a red M&M on the second draw are independent events. Knowing that the first draw is red does not influence the probability of getting a red M&M on the second draw.

c) Disjoint events can never be independent. Once you know that one of a pair of disjoint events has occurred, the other one cannot occur, so its probability has become zero. For example, consider drawing one M&M. If it is red, it cannot possible be orange. Knowing that the M&M is red influences the probability that the M&M is orange. It's zero. The events are not independent.

35. Dice.

a) $P(6) = \frac{1}{6}$, so $P(\text{all 6's}) = \left(\frac{1}{6}\right)\left(\frac{1}{6}\right)\left(\frac{1}{6}\right) \approx 0.005$

b) $P(\text{odd}) = P(1 \text{ or } 3 \text{ or } 5) = \frac{3}{6}$, so $P(\text{all odd}) = \left(\frac{3}{6}\right)\left(\frac{3}{6}\right)\left(\frac{3}{6}\right) = 0.125$

c) $P(\text{not divisible by 3}) = P(1 \text{ or } 2 \text{ or } 4 \text{ or } 5) = \frac{4}{6}$

$P(\text{none divisible by 3}) = \left(\frac{4}{6}\right)\left(\frac{4}{6}\right)\left(\frac{4}{6}\right) \approx 0.296$

d) $P(\text{at least one 5}) = 1 - P(\text{no 5's}) = 1 - \left(\frac{5}{6}\right)\left(\frac{5}{6}\right)\left(\frac{5}{6}\right) \approx 0.421$

e) $P(\text{not all 5's}) = 1 - P(\text{all 5's}) = 1 - \left(\frac{1}{6}\right)\left(\frac{1}{6}\right)\left(\frac{1}{6}\right) \approx 0.995$

37. Champion bowler.

Assuming each frame is independent of others, the multiplication rule may be used.

a) $P(\text{no strikes in 3 frames}) = (0.30)(0.30)(0.30) = 0.027$

b) $P(\text{makes first strike in the third frame}) = (0.30)(0.30)(0.70) = 0.063$

c) $P(\text{at least one strike in first three frames}) = 1 - P(\text{no strikes}) = 1 - (0.30)^3 = 0.973$

d) $P(\text{perfect game}) = (0.70)^{12} \approx 0.014$

39. Voters.

Since you are calling at random, one person's political affiliation is independent of another's. The multiplication rule may be used.

a) $P(\text{all Republicans}) = (0.29)(0.29)(0.29) \approx 0.024$

b) $P(\text{no Democrats}) = (1 - 0.37)(1 - 0.37)(1 - 0.37) \approx 0.25$

c) $P(\text{at least one Ind.}) = 1 - P(\text{no Independents}) = 1 - (0.77)(0.77)(0.77) \approx 0.543$

41. Tires.

Assume that the defective tires are distributed randomly to all tire distributors so that the events can be considered independent. The multiplication rule may be used.

P(at least one of four tires is defective) = 1 – *P*(none are defective)

= 1 – (0.98)(0.98)(0.98)(0.98) ≈ 0.078

43. 9/11?

a) For any date with a valid three-digit date, the chance is 0.001, or 1 in 1000. For many dates in October through December, the probability is 0. For example, there is no way three digits will make 1015, to match October 15.

b) There are 65 days when the chance to match is 0. (October 10 through October 31, November 10 through November 30, and December 10 through December 31.) That leaves 300 days in a year (that is not a leap year) in which a match might occur.
P(no matches in 300 days) = $(0.999)^{300}$ ≈ 0.741.

c) *P*(at least one match in a year) = 1 – *P*(no matches in a year) = 1 – 0.741 ≈ 0.259

d) *P*(at least one match on 9/11 in one of the 50 states)
= 1 – *P*(no matches in 50 states) = 1 – $(0.999)^{50}$ ≈ 0.049

Chapter 15 – Probability Rules!

1. Homes.

a) P(pool or garage)
= P(pool) + P(garage) – P(pool and garage)
= 0.64 + 0.21 – 0.17 = 0.68

Or, from the Venn: 0.47 + 0.17 + 0.04 = 0.68

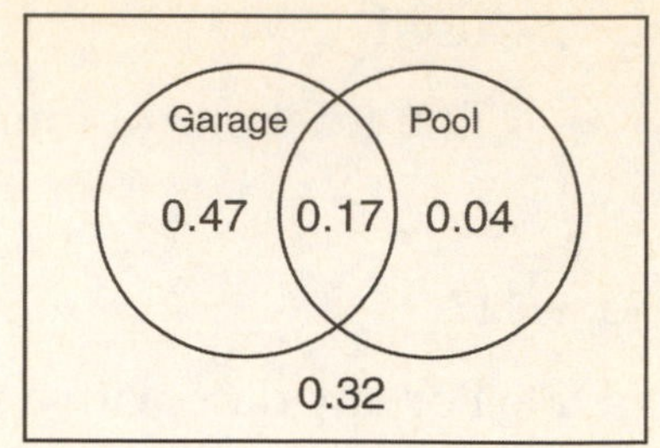

b) P(neither)= 1 – P(pool or garage) = 1 – 0.68 = 0.32

Or, from the Venn: 0.32 (the region outside the circles)

c) P(pool or no garage) = P(pool) – P(pool and garage) = 0.21 – 0.17 = 0.04

Or, from the Venn: 0.04 (the region inside pool circle, yet outside garage circle)

3. Amenities.

a) P(TV and no refrigerator)
= P(TV) – P(TV and refrigerator)
= 0.52 – 0.21 = 0.31

Or, from the Venn: 0.31
(inside the TV circle, yet outside the Fridge circle)

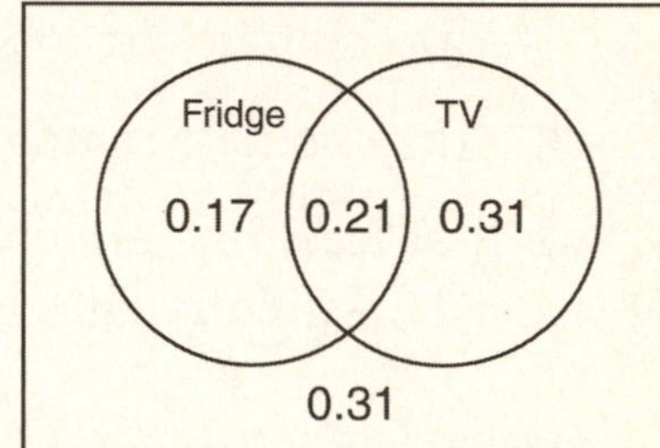

b) P(refrigerator or TV, but not both) =
= [P(refrigerator) – P(refrigerator and TV)] +[P(TV) – P(refrigerator and TV)]
= [0.38 – 0.21] + [0.52 – 0.21] = 0.48

Using the Venn diagram, simply add the probabilities in the two regions for Fridge only and TV only. P(refrigerator or TV, but not both) = 0.17 + 0.31 = 0.48

c) P(neither TV nor refrigerator) = 1 – P(either TV or refrigerator)
= 1 – [P(TV) + P(ref.) – P(TV and ref.)]
= 1 – [0.52 + 0.38 – 0.21]
= 0.31

Or, from the Venn: 0.31 (the region outside the circles)

5. Global survey.

a) $P(USA) = \dfrac{1557}{7690} \approx 0.2025$

b) $P(\text{some high school or primary or less}) = \dfrac{4195}{7690} + \dfrac{1161}{7690} \approx 0.6965$

c)

$$P(\text{France or post-graduate}) = P(\text{France}) + P(\text{post-graduate}) - P(\text{both})$$
$$= \frac{1539}{7690} + \frac{379}{7690} - \frac{69}{7690} \approx 0.2404$$

d) $P(\text{France and primary school or less}) = \frac{309}{7690} \approx 0.0402$

7. Cards.

a) $P(\text{heart} \mid \text{red}) = \frac{P(\text{heart and red})}{P(\text{red})} = \frac{13/52}{26/52} = \frac{1}{2}$

A more intuitive approach is to think about only the red cards. Half of them are hearts.

b) $P(\text{red} \mid \text{heart}) = \frac{P(\text{red and heart})}{P(\text{heart})} = \frac{13/52}{13/52} = 1$

Think about only the hearts. They are all red!

c) $P(\text{ace} \mid \text{red}) = \frac{P(\text{ace and red})}{P(\text{red})} = \frac{2/52}{26/52} = \frac{2}{26} \approx 0.077$

Consider only the red cards. Of those 26 cards, 2 of them are aces.

d) $P(\text{queen} \mid \text{face}) = \frac{P(\text{queen and face})}{P(\text{face})} = \frac{4/52}{12/52} \approx 0.333$

There are 12 faces cards: 4 jacks, 4 queens, and 4 kings. Four of the 12 face cards are queens.

9. Health.

Construct a two-way table of the conditional probabilities, including the marginal probabilities.

Cholesterol	Blood Pressure: High	OK	Total
High	0.11	0.21	0.32
OK	0.16	0.52	0.68
Total	0.27	0.73	1.00

a) $P(\text{both conditions}) = 0.11$

b) $P(\text{high BP}) = 0.11 + 0.16 = 0.27$

c) $P(\text{high chol.} \mid \text{high BP}) = \frac{P(\text{high chol. and high BP})}{P(\text{high BP})} = \frac{0.11}{0.27} \approx 0.407$

Consider only the High Blood Pressure column. Within this column, the probability of having high cholesterol is 0.11 out of a total of 0.27.

d) $P(\text{high BP} \mid \text{high chol.}) = \dfrac{P(\text{high BP and high chol.})}{P(\text{high chol.})} = \dfrac{0.11}{0.32} \approx 0.344$

This time, consider only the high cholesterol row. Within this row, the probability of having high blood pressure is 0.11, out of a total of 0.32.

11. Global survey, take 2.

a) $P(\text{USA and postgraduate work}) = \dfrac{84}{7690} \approx 0.011$

b) $P(\text{USA} \mid \text{post-graduate}) = \dfrac{84}{379} \approx 0.222$

c) $P(\text{post-graduate} \mid \text{USA}) = \dfrac{84}{1557} \approx 0.054$

d) $P(\text{primary} \mid \text{China}) = \dfrac{506}{1502} \approx 0.337$

e) $P(\text{China} \mid \text{primary}) = \dfrac{506}{1161} \approx 0.436$

13. Sick kids.

Having a fever and having a sore throat are not independent events, so:

$P(\text{fever and sore throat}) = P(\text{Fever})\,P(\text{Sore Throat} \mid \text{Fever}) = (0.70)(0.30) = 0.21$

The probability that a kid with a fever has a sore throat is 0.21.

15. Cards.

a)

$$P(\text{first heart drawn is on the third card}) = P(\text{no heart})P(\text{no heart})P(\text{heart})$$
$$= \left(\frac{39}{52}\right)\left(\frac{38}{51}\right)\left(\frac{13}{50}\right) \approx 0.145$$

b)

$$P(\text{all three cards drawn are red}) = P(\text{red})P(\text{red})P(\text{red})$$
$$= \left(\frac{26}{52}\right)\left(\frac{25}{51}\right)\left(\frac{24}{50}\right) \approx 0.118$$

c)

$$P(\text{none of the cards are spades}) = P(\text{no spade})P(\text{no spade})P(\text{no spade})$$
$$= \left(\frac{39}{52}\right)\left(\frac{38}{51}\right)\left(\frac{37}{50}\right) \approx 0.414$$

d)

$$\begin{aligned} P(\text{at least one of the cards is an ace}) &= 1 - P(\text{none of the cards are aces}) \\ &= 1 - \left[P(\text{no ace})P(\text{no ace})P(\text{no ace})\right] \\ &= 1 - \left(\frac{48}{52}\right)\left(\frac{47}{51}\right)\left(\frac{46}{50}\right) \approx 0.217 \end{aligned}$$

17. Batteries.

Since batteries are not being replaced, use conditional probabilities throughout.

a)

$$\begin{aligned} P(\text{the first two batteries are good}) &= P(\text{good})P(\text{good}) \\ &= \left(\frac{7}{12}\right)\left(\frac{6}{11}\right) \approx 0.318 \end{aligned}$$

b)

$$\begin{aligned} &P(\text{at least one of the first three batteries works}) \\ &= 1 - P(\text{none of the first three batt. work}) \\ &= 1 - \left[P(\text{no good})P(\text{no good})P(\text{no good})\right] \\ &= 1 - \left(\frac{5}{12}\right)\left(\frac{4}{11}\right)\left(\frac{3}{10}\right) \approx 0.955 \end{aligned}$$

c)

$$\begin{aligned} P(\text{the first four batteries are good}) &= P(\text{good})P(\text{good})P(\text{good})P(\text{good}) \\ &= \left(\frac{7}{12}\right)\left(\frac{6}{11}\right)\left(\frac{5}{10}\right)\left(\frac{4}{9}\right) \approx 0.071 \end{aligned}$$

d)

$$\begin{aligned} &P(\text{pick five to find one good}) \\ &= P(\text{no good})P(\text{no good})P(\text{no good})P(\text{no good})P(\text{good}) \\ &= \left(\frac{5}{12}\right)\left(\frac{4}{11}\right)\left(\frac{3}{10}\right)\left(\frac{2}{9}\right)\left(\frac{7}{8}\right) \approx 0.009 \end{aligned}$$

19. Eligibility.

a)

$$\begin{aligned} P(\text{eligibility}) &= P(\text{stats}) + P(\text{comp sci}) - P(\text{both}) \\ &= 0.52 + 0.23 - 0.07 \\ &= 0.68 \end{aligned}$$

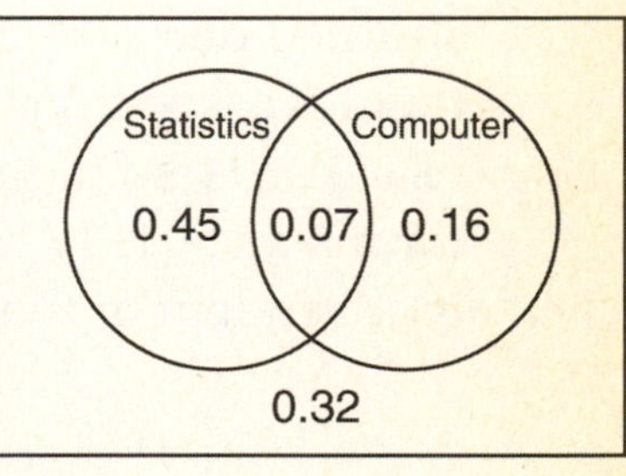

68% of students are eligible for BioResearch, so $100 - 68 = 32\%$ are ineligible.

From the Venn, the region outside the circles represents those students who have taken neither course, and are therefore ineligible for BioResearch.

b)

$$P(\text{computer science} \mid \text{statistics}) = \frac{P(\text{computer science and statistics})}{P(\text{statistics})} = \frac{0.07}{0.52} \approx 0.135$$

From the Venn, consider only the region inside the Statistics circle. The probability of having taken computer science is 0.07 out of a total of 0.52 (the entire Statistics circle).

c) Taking the two courses are not disjoint events, since they have outcomes in common. In fact, 7% of juniors have taken both courses.

d) Taking the two courses are not independent events. The overall probability that a junior has taken a computer science is 0.23. The probability that a junior has taken a computer course given that he or she has taken a statistics course is 0.135. If taking the two courses were independent events, these probabilities would be the same.

21. For sale.

Construct a Venn diagram of the disjoint outcomes.

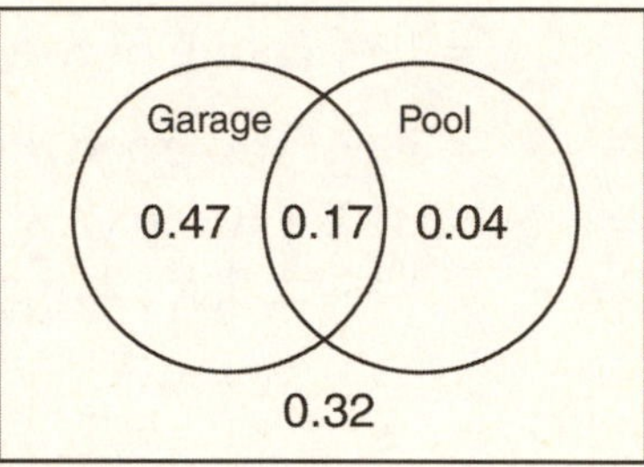

a)

$$P(\text{pool} \mid \text{garage}) = \frac{P(\text{pool and garage})}{P(\text{garage})} = \frac{0.17}{0.64} \approx 0.266$$

From the Venn, consider only the region inside the Garage circle. The probability that the house has a pool is 0.17 out of a total of 0.64 (the entire Garage circle).

b) Having a garage and a pool are not independent events. 26.6% of homes with garages have pools. Overall, 21% of homes have pools. If having a garage and a pool were independent events, these would be the same.

c) No, having a garage and a pool are not disjoint events. 17% of homes have both.

23. Cards.

Yes, getting an ace is independent of the suit when drawing one card from a well shuffled deck. The overall probability of getting an ace is 4/52, or 1/13, since there are 4 aces in the deck. If you consider just one suit, there is only 1 ace out of 13 cards, so the probability of getting an ace given that the card is a diamond, for instance, is 1/13. Since the probabilities are the same, getting an ace is independent of the suit.

25. Unsafe food.

a) Using the Venn diagram, the probability that a tested chicken was not contaminated with either kind of bacteria is 17%.

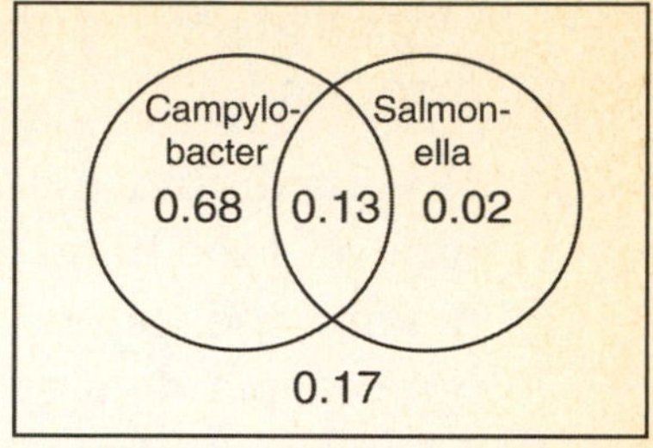

b) Contamination with campylobacter and contamination with salmonella are not disjoint events, since 13% of chicken are contaminated with both.

c) No, contamination with campylobacter and contamination with salmonella are not independent events. The probability that a tested chicken is contaminated with campylobacter is 0.81. The probability that chicken contaminated with salmonella is also contaminated with campylobacter is 0.13/0.15 = 0.87. If chicken is contaminated with salmonella, it is more likely to be contaminated with campylobacter than chicken in general.

27. Men's health, again.

Consider the two-way table from Exercise 9.

Cholesterol	Blood Pressure: High	OK	Total
High	0.11	0.21	0.32
OK	0.16	0.52	0.68
Total	0.27	0.73	1.00

High blood pressure and high cholesterol are not independent events. 28.8% of men with OK blood pressure have high cholesterol, while 40.7% of men with high blood pressurehave high cholesterol. If having high blood pressure and high cholesterol were independent, these percentages would be the same.

29. Phone service.

a) Since 2.8% of U.S. adults have only a cell phone, and 1.6% have no phone at all, polling organizations can reach 100 – 2.8 – 1.6 = 96.5% of U.S. adults.

b) Using the Venn diagram, about 96.5% of U.S. adults have a land line. The probability of a U.S. adults having a land line given that they have a cell phone is 58.2/(58.2+2.8) or about 95.4%. It appears that having a cell phone and having a land line are independent, since the probabilities are roughly the same.

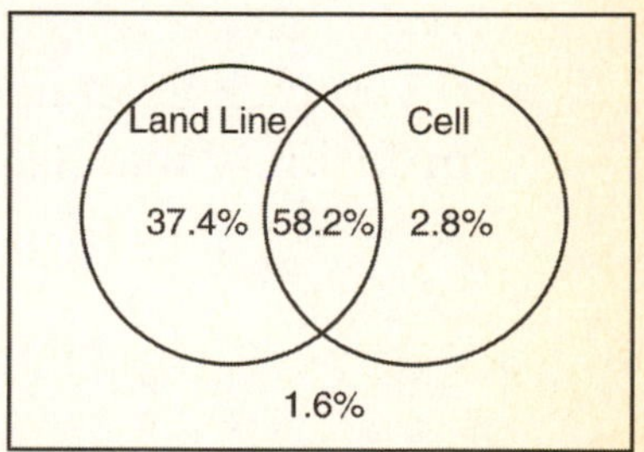

31. Montana.

According to the poll, party affiliation is not independent of sex.
Overall, (36+48)/202 = 41.6% of the respondents were Democrats. Of the men, only 36/105 = 34.3% were Democrats.

33. Luggage.

Organize using a tree diagram.

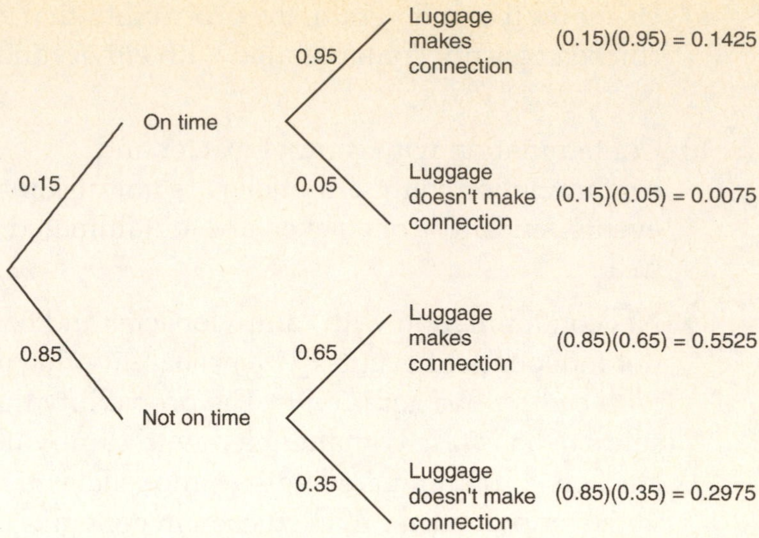

a) No, the flight leaving on time and the luggage making the connection are not independent events. The probability that the luggage makes the connection is dependent on whether or not the flight is on time. The probability is 0.95 if the flight is on time, and only 0.65 if it is not on time.

b)
$$\begin{aligned} P(\text{Luggage}) &= P(\text{On time and Luggage}) + P(\text{Not on time and Luggage}) \\ &= (0.15)(0.95) + (0.85)(0.65) \\ &= 0.695 \end{aligned}$$

35. Late luggage.

Refer to the tree diagram constructed for Exercise 33.

$$\begin{aligned} P(\text{Not on time} \mid \text{No Lug.}) &= \frac{P(\text{Not on time and No Lug.})}{P(\text{No Lug.})} \\ &= \frac{(0.85)(0.35)}{(0.15)(0.05)+(0.85)(0.35)} \approx 0.975 \end{aligned}$$

If you pick Leah up at the Denver airport and her luggage is not there, the probability that her first flight was delayed is 0.975.

37. Absenteeism.

Organize the information in a tree diagram.

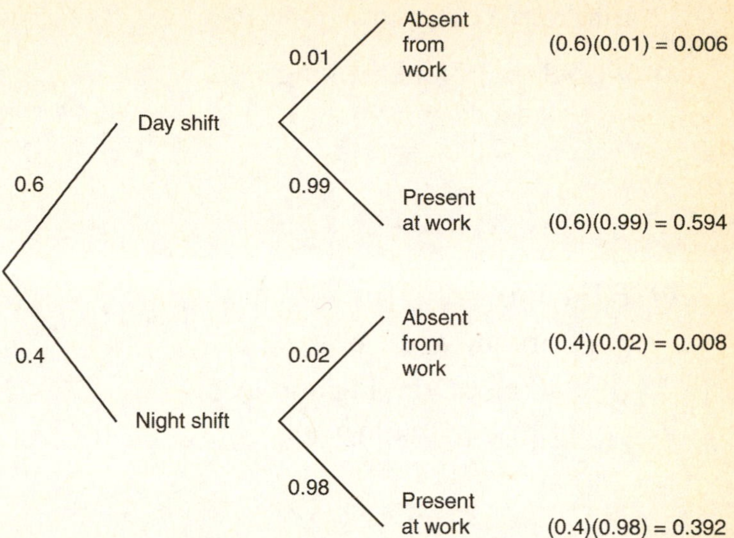

a) No, absenteeism is not independent of shift worked. The rate of absenteeism for the night shift is 2%, while the rate for the day shift is only 1%. If the two were independent, the percentages would be the same.

b)

$P(\text{Absent})=P(\text{Day and Absent})+P(\text{Night and Absent})=(0.6)(0.01)+(0.4)(0.02)=0.014$

The overall rate of absenteeism at this company is 1.4%.

39. Absenteeism, part II.

Refer to the tree diagram constructed for Exercise 37.

$$P(\text{Night} \mid \text{Absent})=\frac{P(\text{Night and Absent})}{P(\text{Absent})}=\frac{(0.4)(0.02)}{(0.6)(0.01)+(0.4)(0.02)}\approx 0.571$$

Approximately 57.1% of the company's absenteeism occurs on the night shift.

41. Drunks.

Organize the information into a tree diagram.

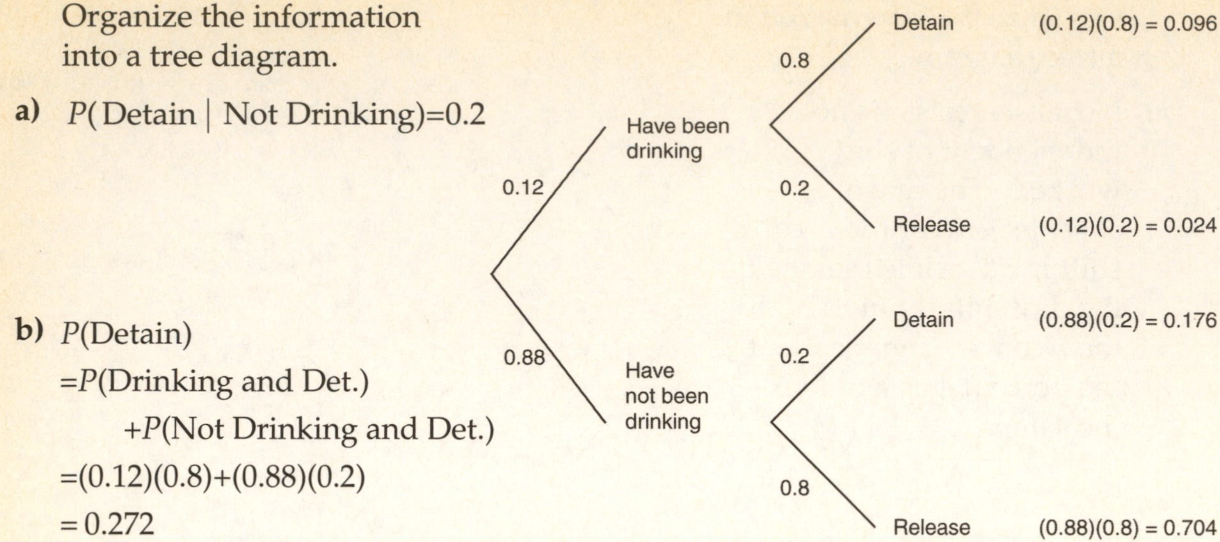

a) $P(\text{Detain} \mid \text{Not Drinking}) = 0.2$

b)
$$\begin{aligned} &P(\text{Detain}) \\ &= P(\text{Drinking and Det.}) \\ &\quad + P(\text{Not Drinking and Det.}) \\ &= (0.12)(0.8) + (0.88)(0.2) \\ &= 0.272 \end{aligned}$$

c)
$$\begin{aligned} P(\text{Drunk} \mid \text{Det.}) &= \frac{P(\text{Drunk and Det.})}{P(\text{Detain})} \\ &= \frac{(0.12)(0.8)}{(0.12)(0.8) + (0.88)(0.2)} \\ &\approx 0.353 \end{aligned}$$

d)
$$\begin{aligned} P(\text{Drunk} \mid \text{Release}) &= \frac{P(\text{Drunk and Release})}{P(\text{Release})} \\ &= \frac{(0.12)(0.2)}{(0.12)(0.2) + (0.88)(0.8)} \\ &\approx 0.033 \end{aligned}$$

43. Dishwashers.

Organize the information in a tree diagram.

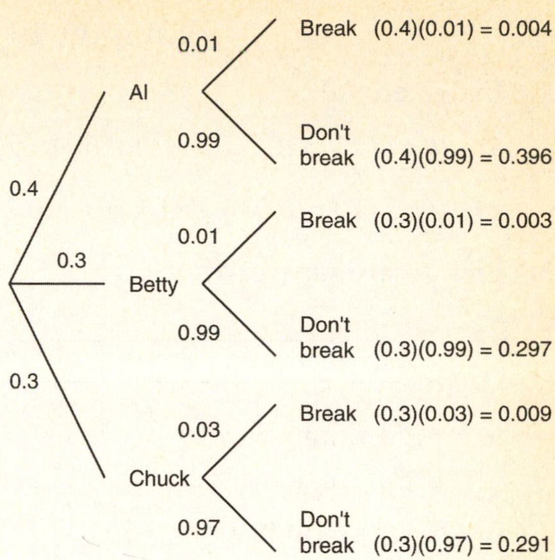

$$
\begin{aligned}
&P(\text{Chuck} \mid \text{Break}) \\
&= \frac{P(\text{Chuck and Break})}{P(\text{Break})} \\
&= \frac{(0.3)(0.03)}{(0.4)(0.01)+(0.3)(0.01)+(0.3)(0.03)} \\
&\approx 0.563
\end{aligned}
$$

If you hear a dish break, the probability that Chuck is on the job is approximately 0.563.

45. HIV Testing.

Organize the information in a tree diagram.

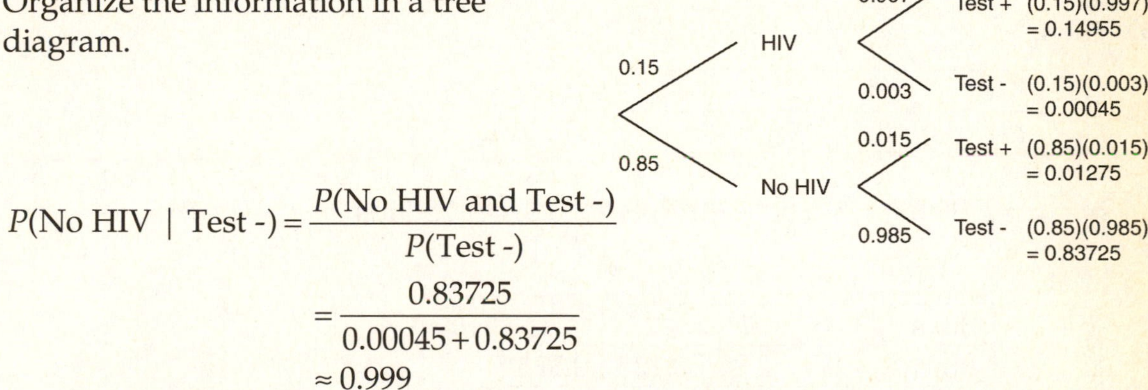

$$
\begin{aligned}
P(\text{No HIV} \mid \text{Test -}) &= \frac{P(\text{No HIV and Test -})}{P(\text{Test -})} \\
&= \frac{0.83725}{0.00045+0.83725} \\
&\approx 0.999
\end{aligned}
$$

The probability that a patient testing negative is truly free of HIV is about 99.9%.

Chapter 16 – Random Variables

1. **Expected value.**

   a) $\mu = E(Y) = 10(0.3) + 20(0.5) + 30(0.2) = 19$

   b) $\mu = E(Y) = 2(0.3) + 4(0.4) + 6(0.2) + 8(0.1) = 4.2$

3. **Pick a card, any card.**

   a)

Win	\$0	\$5	\$10	\$30
P(amount won)	$\frac{26}{52}$	$\frac{13}{52}$	$\frac{12}{52}$	$\frac{1}{52}$

   b) $\mu = E(\text{amount won}) = \$0\left(\frac{26}{52}\right) + \$5\left(\frac{13}{52}\right) + \$10\left(\frac{12}{52}\right) + \$30\left(\frac{1}{52}\right) \approx \$4.13$

   c) Answers may vary. In the long run, the expected payoff of this game is \$4.13 per play. Any amount less than \$4.13 would be a reasonable amount to pay in order to play. Your decision should depend on how long you intend to play. If you are only going to play a few times, you should risk less.

5. **Kids.**

   a)

Kids	1	2	3
P(Kids)	0.5	0.25	0.25

   b) $\mu = E\,(\text{Kids}) = 1(0.5) + 2(0.25) + 3(0.25) = 1.75$ kids

   c)

Boys	0	1	2	3
P(boys)	0.5	0.25	0.125	0.125

   $\mu = E\,(\text{Boys}) = 0(0.5) + 1(0.25) + 2(0.125) + 3(0.125) = 0.875$ boys

7. **Software.**

   Since the contracts are awarded independently, the probability that the company will get both contracts is $(0.3)(0.6) = 0.18$. Organize the disjoint events in a Venn diagram.

Profit	larger only \$50,000	smaller only \$20,000	both \$70,000	neither \$0
P(profit)	0.12	0.42	0.18	0.28

Larger
Smaller
0.12
0.18
0.42
0.28

$$\mu = E(\text{profit}) = \$50{,}000(0.12) + \$20{,}000(0.42) + \$70{,}000(0.18)$$
$$= \$27{,}000$$

9. Variation 1.

a)

$$\sigma^2 = Var(Y) = (10-19)^2(0.3) + (20-19)^2(0.5) + (30-19)^2(0.2) = 49$$
$$\sigma = SD(Y) = \sqrt{Var(Y)} = \sqrt{49} = 7$$

b)

$$\sigma^2 = Var(Y) = (2-4.2)^2(0.3) + (4-4.2)^2(0.4) + (6-4.2)^2(0.2) + (8-4.2)^2(0.1) = 3.56$$
$$\sigma = SD(Y) = \sqrt{Var(Y)} = \sqrt{3.56} \approx 1.89$$

11. Pick another card.

Answers may vary slightly (due to rounding of the mean)

$$\sigma^2 = Var(\text{Won}) = (0-4.13)^2\left(\frac{26}{52}\right) + (5-4.13)^2\left(\frac{13}{52}\right)$$
$$+ (10-4.13)^2\left(\frac{12}{52}\right) + (30-4.13)^2\left(\frac{1}{52}\right) \approx 29.5396$$
$$\sigma = SD(\text{Won}) = \sqrt{Var(\text{Won})} = \sqrt{29.5396} \approx \$5.44$$

13. Kids.

$$\sigma^2 = Var(\text{Kids}) = (1-1.75)^2(0.5) + (2-1.75)^2(0.25) + (3-1.75)^2(0.25) = 0.6875$$
$$\sigma = SD(\text{Kids}) = \sqrt{Var(\text{Kids})} = \sqrt{0.6875} \approx 0.83 \text{ kids}$$

15. Repairs.

a) $\mu = E(\text{Number of Repair Calls}) = 0(0.1) + 1(0.3) + 2(0.4) + 3(0.2) = 1.7 \text{ calls}$

b)

$$\sigma^2 = Var(\text{Calls}) = (0-1.7)^2(0.1) + (1-1.7)^2(0.3) + (2-1.7)^2(0.4) + (3-1.7)^2(0.2) = 0.81$$
$$\sigma = SD(\text{Calls}) = \sqrt{Var(\text{Calls})} = \sqrt{0.81} = 0.9 \text{ calls}$$

17. Defects.

The percentage of cars with *no* defects is 61%.

$$\mu = E(\text{Defects}) = 0(0.61) + 1(0.21) + 2(0.11) + 3(0.07) = 0.64 \text{ defects}$$

$$\sigma^2 = Var(\text{Defects}) = (0-0.64)^2(0.61) + (1-0.64)^2(0.21)$$
$$+ (2-0.64)^2(0.11) + (3-0.64)^2(0.07) \approx 0.8704$$
$$\sigma = SD(\text{Defects}) = \sqrt{Var(\text{Defects})} \approx \sqrt{0.8704} \approx 0.93 \text{ defects}$$

19. Cancelled flights.

a) $\mu = E(\text{gain}) = (-150)(0.20) + 100(0.80) = \$50$

b)

$$\sigma^2 = Var(\text{gain}) = (-150-50)^2(0.20) + (100-50)^2(0.80) = 10{,}000$$

$$\sigma = SD(\text{gain}) = \sqrt{Var(\text{gain})} \approx \sqrt{10{,}000} = \$100$$

21. Contest.

a) The two games are not independent. The probability that you win the second depends on whether or not you win the first.

b)

$$\begin{aligned} P(\text{losing both games}) &= P(\text{losing the first})\,P(\text{losing the second} \mid \text{first was lost}) \\ &= (0.6)(0.7) = 0.42 \end{aligned}$$

c)

$$\begin{aligned} P(\text{winning both games}) &= P(\text{winning the first})\,P(\text{winning the second} \mid \text{first was won}) \\ &= (0.4)(0.2) = 0.08 \end{aligned}$$

d)

X	0	1	2
$P(X = x)$	0.42	0.50	0.08

e)

$$\mu = E(X) = 0(0.42) + 1(0.50) + 2(0.08) = 0.66 \text{ games}$$

$$\sigma^2 = Var(X) = (0-0.66)^2(0.42) + (1-0.66)^2(0.50) + (2-0.66)^2(0.08) = 0.3844$$

$$\sigma = SD(X) = \sqrt{Var(X)} = \sqrt{0.3844} = 0.62 \text{ games}$$

23. Batteries.

a)

# good	0	1	2
P(# good)	$\left(\frac{3}{10}\right)\left(\frac{2}{9}\right) = \frac{6}{90}$	$\left(\frac{3}{10}\right)\left(\frac{7}{9}\right) + \left(\frac{7}{10}\right)\left(\frac{3}{9}\right) = \frac{42}{90}$	$\left(\frac{7}{10}\right)\left(\frac{6}{9}\right) = \frac{42}{90}$

b) $\mu = E(\text{number good}) = 0\left(\frac{6}{90}\right) + 1\left(\frac{42}{90}\right) + 2\left(\frac{42}{90}\right) = 1.4 \text{ batteries}$

c)

$$\sigma^2 = Var(\text{number good}) = (0-1.4)^2\left(\frac{6}{90}\right)+(1-1.4)^2\left(\frac{42}{90}\right)+(2-1.4)^2\left(\frac{42}{90}\right) \approx 0.3733$$

$$\sigma = SD(\text{number good}) = \sqrt{Var(\text{number good})} \approx \sqrt{0.3733} \approx 0.61 \text{ batteries.}$$

25. Random variables.

a)

$$\mu = E(3X) = 3(E(X)) = 3(10) = 30$$
$$\sigma = SD(3X) = 3(SD(X)) = 3(2) = 6$$

b)

$$\mu = E(Y+6) = E(Y)+6 = 20+6 = 26$$
$$\sigma = SD(Y+6) = SD(Y) = 5$$

c)

$$\mu = E(X+Y) = E(X)+E(Y) = 10+20 = 30$$
$$\sigma = SD(X+Y) = \sqrt{Var(X)+Var(Y)} = \sqrt{2^2+5^2} \approx 5.39$$

d)

$$\mu = E(X-Y) = E(X)-E(Y) = 10-20 = -10$$
$$\sigma = SD(X-Y) = \sqrt{Var(X)+Var(Y)} = \sqrt{2^2+5^2} \approx 5.39$$

e)

$$\mu = E(X_1+X_2) = E(X)+E(X) = 10+10 = 20$$
$$\sigma = SD(X_1+X_2) = \sqrt{Var(X)+Var(X)} = \sqrt{2^2+2^2} \approx 2.83$$

27. Random variables.

a)

$$\mu = E(0.8Y) = 0.8(E(Y)) = 0.8(300) = 240$$
$$\sigma = SD(0.8Y) = 0.8(SD(Y)) = 0.8(16) = 12.8$$

b)

$$\mu = E(2X-100) = 2(E(X))-100 = 140$$
$$\sigma = SD(2X-100) = 2(SD(X)) = 2(12) = 24$$

c)

$$\mu = E(X+2Y) = E(X)+2(E(Y)) = 120+2(300) = 720$$
$$\sigma = SD(X+2Y) = \sqrt{Var(X)+2^2Var(Y)} = \sqrt{12^2+2^2(16^2)} \approx 34.18$$

d)

$$\mu = E(3X-Y) = 3(E(X))-E(Y) = 3(120)-300 = 60$$
$$\sigma = SD(3X-Y) = \sqrt{3^2Var(X)+Var(Y)} = \sqrt{3^2(12^2)+16^2} \approx 39.40$$

e)

$$\mu = E(Y_1+Y_2) = E(Y)+E(Y) = 300+300 = 600$$
$$\sigma = SD(Y_1+Y_2) = \sqrt{Var(Y)+Var(Y)} = \sqrt{16^2+16^2} \approx 22.63$$

29. Eggs.

a) $\mu = E(\text{Broken eggs in 3 doz.}) = 3(E(\text{Broken eggs in 1 doz.})) = 3(0.6) = 1.8$ eggs

b) $\sigma = SD(\text{Broken eggs in 3 dozen}) = \sqrt{0.5^2 + 0.5^2 + 0.5^2} \approx 0.87$ eggs

c) The cartons of eggs must be independent of each other.

31. Repair calls.

$\mu = E(\text{calls in 8 hours}) = 8(E(\text{calls in 1 hour}) = 8(1.7) = 13.6$ calls

$\sigma = SD(\text{calls in 8 hours}) = \sqrt{8(Var(\text{calls in 1 hour}))} = \sqrt{8(0.9)^2} \approx 2.55$ calls

This is only valid if the hours are independent of one another.

33. Tickets.

a)

$$\begin{aligned}\mu &= E(\text{tickets for 18 trucks}) \\ &= 18(E(\text{tickets for one truck})) = 18(1.3) = 23.4 \text{ tickets} \\ \sigma &= SD(\text{tickets for 18 trucks}) \\ &= \sqrt{18(Var(\text{tickets for one truck})} = \sqrt{18(0.7)^2} \approx 2.97 \text{ tickets}\end{aligned}$$

b) We are assuming that trucks are ticketed independently.

35. Fire!

a) The standard deviation is large because the profits on insurance are highly variable. Although there will be many small gains, there will occasionally be large losses, when the insurance company has to pay a claim.

b)

$\mu = E(\text{two policies}) = 2(E(\text{one policy})) = 2(150) = \$300$

$\sigma = SD(\text{two policies}) = \sqrt{2(Var(\text{one policy}))} = \sqrt{2(6000^2)} \approx \$8,485.28$

c)

$\mu = E(10,000 \text{ policies}) = 10,000(E(\text{one policy})) = 10,000(150) = \$1,500,000$

$\sigma = SD(10,000 \text{ policies}) = \sqrt{10,000(Var(\text{one policy}))} = \sqrt{10,000(6000^2)} = \$600,000$

d) If the company sells 10,000 policies, they are likely to be successful. A profit of \$0, is 2.5 standard deviations below the expected profit. This is unlikely to happen. However, if the company sells fewer policies, then the likelihood of turning a profit decreases. In an extreme case, where only two policies are sold, a profit of \$0 is more likely, being only a small fraction of a standard deviation below the mean.

e) This analysis depends on each of the policies being independent from each other. This assumption of independence may be violated if there are many fire insurance claims as a result of a forest fire, or other natural disaster.

37. Cereal.

a) $E(\text{large bowl} - \text{small bowl}) = E(\text{large bowl}) - E(\text{small bowl}) = 2.5 - 1.5 = 1$ ounce

b) $\sigma = SD(\text{large} - \text{small}) = \sqrt{Var(\text{large}) + Var(\text{small})} = \sqrt{0.4^2 + 0.3^2} = 0.5$ ounces

c)

$$z = \frac{x - \mu}{\sigma}$$

$$z = \frac{0 - 1}{0.5} = -2$$

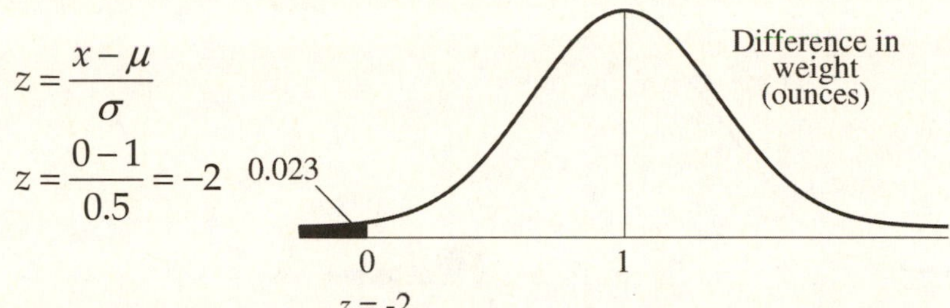

The small bowl will contain more cereal than the large bowl when the difference between the amounts is less than 0. According to the Normal model, the probability of this occurring is approximately 0.023.

d)

$\mu = E(\text{large} + \text{small}) = E(\text{large}) + E(\text{small}) = 2.5 + 1.5 = 4$ ounce

$\sigma = SD(\text{large} + \text{small}) = \sqrt{Var(\text{large}) + Var(\text{small})} = \sqrt{0.4^2 + 0.3^2} = 0.5$ ounces

e)

$$z = \frac{x - \mu}{\sigma}$$

$$z = \frac{4.5 - 4}{0.5} = 1$$

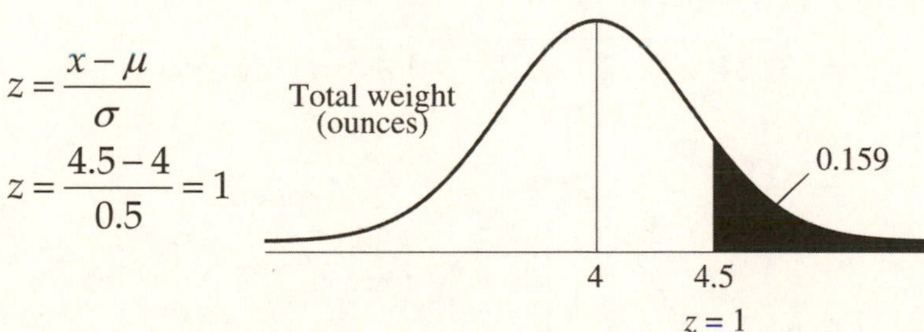

According to the Normal model, the probability that the total weight of cereal in the two bowls is more than 4.5 ounces is approximately 0.159.

f)

$$\mu = E(\text{box} - \text{large} - \text{small}) = E(\text{box}) - E(\text{large}) - E(\text{small})$$
$$= 16.3 - 2.5 - 1.5 = 12.3 \text{ ounces}$$

$$\sigma = SD(\text{box} - \text{large} - \text{small}) = \sqrt{Var(\text{box}) + Var(\text{large}) + Var(\text{small})}$$
$$= \sqrt{0.2^2 + 0.3^2 + 0.4^2} \approx 0.54 \text{ ounces}$$

39. More cereal.

a)

$$\mu = E(\text{box} - \text{large} - \text{small})$$
$$= E(\text{box}) - E(\text{large}) - E(\text{small}) = 16.2 - 2.5 - 1.5 = 12.2 \text{ ounces}$$

b)

$$\sigma = SD(\text{box} - \text{large} - \text{small}) = \sqrt{Var(\text{box}) + Var(\text{large}) + Var(\text{small})}$$
$$= \sqrt{0.1^2 + 0.3^2 + 0.4^2} \approx 0.51 \text{ ounces}$$

c)

$$z = \frac{x - \mu}{\sigma}$$
$$z = \frac{13 - 12.2}{0.51}$$
$$z = 1.57$$

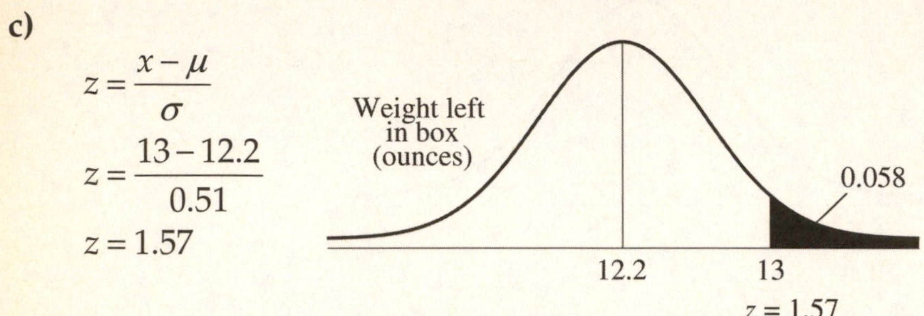

According to the Normal model, the probability that the box contains more than 13 ounces is about 0.058.

41. Medley.

a)

$$\mu = E(\#1 + \#2 + \#3 + \#4) = E(\#1) + E(\#2) + E(\#3) + E(\#4)$$
$$= 50.72 + 55.51 + 49.43 + 44.91 = 200.57 \text{ seconds}$$
$$\sigma = SD(\#1 + \#2 + \#3 + \#4) = \sqrt{Var(\#1) + Var(\#2) + Var(\#3) + Var(\#4)}$$
$$= \sqrt{0.24^2 + 0.22^2 + 0.25^2 + 0.21^2} \approx 0.46 \text{ seconds}$$

b)

$$z = \frac{x - \mu}{\sigma}$$
$$z = \frac{199.48 - 200.57}{0.461}$$
$$z = -2.36$$

Total time (seconds)
0.009
199.48
200.57
z = -2.36

The team is not likely to swim faster than their best time. According to the Normal model, they are only expected to swim that fast or faster about 0.9% of the time.

43. Farmer's market.

a) Let $A =$ price of a pound of apples, and let $P =$ price of a pound of potatoes.

Profit $= 100A + 50P - 2$

b) $\mu = E(100A + 50P - 2) = 100(E(A)) + 50(E(P)) - 2 = 100(0.5) + 50(0.3) - 2 = \$63$

c)

$$\sigma = SD(100A + 50P - 2)$$
$$= \sqrt{100^2(Var(A)) + 50^2(Var(P))} = \sqrt{100^2(0.2^2) + 50^2(0.1^2)} \approx \$20.62$$

d) No assumptions are necessary to compute the mean. To compute the standard deviation, independent market prices must be assumed.

45. Coffee and doughnuts.

a)

$$\mu = E(\text{cups sold in 6 days}) = 6(E(\text{cups sold in 1 day})) = 6(320) = 1920 \text{ cups}$$
$$\sigma = SD(\text{cups sold in 6 days}) = \sqrt{6(Var(\text{cups sold in 1 day})} = \sqrt{6(20)^2} \approx 48.99 \text{ cups}$$

The distribution of total coffee sales for 6 days has distribution $N(1920, 48.99)$.

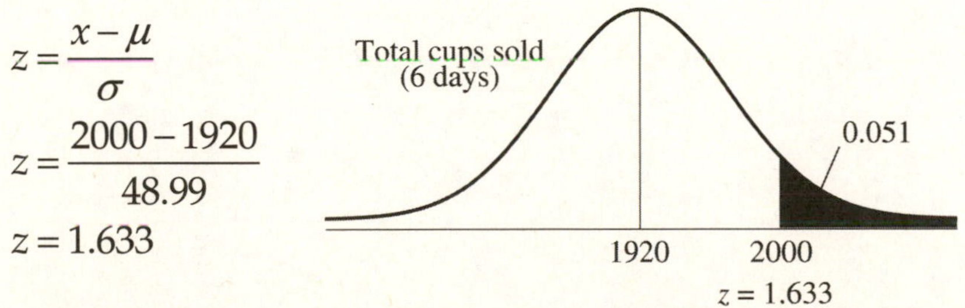

$$z = \frac{x - \mu}{\sigma}$$
$$z = \frac{2000 - 1920}{48.99}$$
$$z = 1.633$$

According to the Normal model, the probability that he will sell more than 2000 cups of coffee in a week is approximately 0.051.

b) Let $C =$ the number of cups of coffee sold. Let $D =$ the number of doughnuts sold.

$$\mu = E(50C + 40D) = 0.50(E(C)) + 0.40(E(D)) = 0.50(320) + 0.40(150) = \$220$$

$$\sigma = SD(0.50C + 0.40D)$$
$$= \sqrt{0.50^2(Var(C)) + 0.40^2(Var(D))} = \sqrt{0.50^2(20^2) + 0.40^2(12^2)} \approx \$11.09$$

The day's profit can be modeled by $N(220, 11.09)$. A day's profit of \$300 is over 7 standard deviations above the mean. This is extremely unlikely. It would not be reasonable for the shop owner to expect the day's profit to exceed \$300.

c) Consider the difference $D-0.5C$. When this difference is greater than zero, the number of doughnuts sold is greater than half the number of cups of coffee sold.

$$\mu = E(D-0.5C) = (E(D)) - 0.5(E(C)) = 150 + 0.5(320) = -\$10$$

$$\sigma = SD(D-0.5C) = \sqrt{(Var(D)) + 0.5(Var(C))} = \sqrt{(12^2) + 0.5^2(20^2)} \approx \$15.62$$

The difference $D-0.5C$ can be modeled by $N(-10, 15.62)$.

According to the Normal model, the probability that the shop owner will sell a doughnut to more than half of the coffee customers is approximately 0.26.

Chapter 17 – Probability Models

1. Bernoulli.

a) These are not Bernoulli trials. The possible outcomes are 1, 2, 3, 4, 5, and 6. There are more than two possible outcomes.

b) These may be considered Bernoulli trials. There are only two possible outcomes, Type A and not Type A. Assuming the 120 donors are representative of the population, the probability of having Type A blood is 43%. The trials are not independent, because the population is finite, but the 120 donors represent less than 10% of all possible donors.

c) These are not Bernoulli trials. The probability of getting a heart changes as cards are dealt without replacement.

d) These are not Bernoulli trials. We are sampling without replacement, so the trials are not independent. Samples without replacement may be considered Bernoulli trials if the sample size is less than 10% of the population, but 500 is more than 10% of 3000.

e) These may be considered Bernoulli trials. There are only two possible outcomes, sealed properly and not sealed properly. The probability that a package is unsealed is constant, at about 10%, as long as the packages checked are a representative sample of all packages. Finally, the trials are not independent, since the total number of packages is finite, but the 24 packages checked probably represent less than 10% of the packages.

3. Simulating the model.

a) Answers will vary. A component is the simulation of the picture in one box of cereal. One possible way to model this component is to generate random digits 0-9. Let 0 and 1 represent LeBron James and 2-9 a picture of another sports star. Each run will consist of generating random numbers until a 0 or 1 is generated. The response variable will be the number of digits generated until the first 0 or 1.

b) Answers will vary.

c) Answers will vary. To construct your simulated probability model, start by calculating the simulated probability that you get a picture of LeBron James in the first box. This is the number of trials in which a 0 or 1 was generated first, divided by the total number of trials. Perform similar calculations for the simulated probability that you have to wait until the second box, the third box, etc.

d) Let X = the number of boxes opened until the first LeBron James picture is found.

X	1	2	3	4	5	6	7	8	≥9
P(X)	0.20	(0.80)(0.20) = 0.16	$(0.80)^2(0.20)$ = 0.128	$(0.80)^3(0.20)$ = 0.1024	0.082	0.066	0.052	0.042	0.168

e) Answers will vary.

5. LeBron again.

a) Answers will vary. A component is the simulation of the picture in one box of cereal. One possible way to model this component is to generate random digits 0-9. Let 0 and 1 represent LeBron James and 2-9 a picture of another sports star. Each run will consist of generating five random numbers. The response variable will be the number of 0s and 1s in the five random numbers.

b) Answers will vary.

c) Answers will vary. To construct your simulated probability model, start by calculating the simulated probability that you get no pictures of LeBron James in the five boxes. This is the number of trials in which neither 0 nor 1 were generated divided by the total number of trials. Perform similar calculations for the simulated probability that you would get one picture, 2 pictures, etc.

d) Let X = the number of LeBron James pictures in 5 boxes.

X	0	1	2	3	4	5
P(X)	$(0.20)^0(0.80)^5$ ≈ 0.33	${}_5C_1(0.20)^1(0.80)^4$ ≈ 0.41	${}_5C_2(0.20)^2(0.80)^3$ ≈ 0.20	${}_5C_3(0.20)^3(0.80)^2$ ≈ 0.05	${}_5C_4(0.20)^4(0.80)^1$ ≈ 0.01	$(0.20)^5(0.80)^0$ ≈ 0.0

e) Answers will vary.

7. On time.

These departures cannot be considered Bernoulli trials. Departures from the same airport during a 2-hour period may not be independent. They all might be affected by weather and delays.

9. Hoops.

The player's shots may be considered Bernoulli trials. There are only two possible outcomes (make or miss), the probability of making a shot is constant (80%), and the shots are independent of one another (making, or missing, a shot does not affect the probability of making the next).

Let X= the number of shots until the first missed shot.
Let Y = the number of shots until the first made shot.

Since these problems deal with shooting until the first miss (or until the first made shot), a geometric model, either *Geom*(0.8) or *Geom*(0.2), is appropriate.

a) Use *Geom*(0.2). $P(X=5)=(0.8)^4(0.2)=0.08192$ (Four shots made, followed by a miss.)

b) Use *Geom*(0.8). $P(Y=4)=(0.2)^3(0.8)=0.0064$ (Three misses, then a made shot.)

c) Use *Geom*(0.8). $P(Y=1)+P(Y=2)+P(Y=3)=(0.8)+(0.2)(0.8)+(0.2)^2(0.8)=0.992$

11. More hoops.

As determined in a previous exercise, the shots can be considered Bernoulli trials, and since the player is shooting until the first miss, *Geom*(0.2) is the appropriate model.

$$E(X)=\frac{1}{p}=\frac{1}{0.2}=5 \text{ shots}$$

The player is expected to take 5 shots until the first miss.

13. Customer center operator.

The calls can be considered Bernoulli trials. There are only two possible outcomes, taking the promotion, and not taking the promotion. The probability of success is constant at 5% (50% of the 10% Platinum cardholders.) The trials are not independent, since there are a finite number of cardholders, but this is a major credit card company, so we can assume we are selecting fewer than 10% of all cardholders. Since we are calling people until the first success, the model *Geom*(0.05) may be used.

$E(\text{calls})=\frac{1}{p}=\frac{1}{0.05}=20$ calls. We expect it to take 20 calls to find the first cardholder to take the double miles promotion.

15. Blood.

These may be considered Bernoulli trials. There are only two possible outcomes, Type AB and not Type AB. Provided that the donors are representative of the population, the probability of having Type AB blood is constant at 4%. The trials are not independent, since the population is finite, but we are selecting fewer than 10% of all potential donors. Since we are selecting people until the first success, the model *Geom*(0.04) may be used.

Let X = the number of donors until the first Type AB donor is found.

a) $E(X) = \frac{1}{p} = \frac{1}{0.04} = 25$ people

We expect the 25th person to be the first Type AB donor.

b)

$$P(\text{a Type AB donor among the first 5 people checked})$$
$$= P(X=1) + P(X=2) + P(X=3) + P(X=4) + P(X=5)$$
$$= (0.04) + (0.96)(0.04) + (0.96)^2(0.04) + (0.96)^3(0.04) + (0.96)^4(0.04) \approx 0.185$$

c)

$$P(\text{a Type AB donor among the first 6 people checked})$$
$$= P(X=1) + P(X=2) + P(X=3) + P(X=4) + P(X=5) + P(X=6)$$
$$= (0.04) + (0.96)(0.04) + (0.96)^2(0.04)$$
$$+ (0.96)^3(0.04) + (0.96)^4(0.04) + (0.96)^5(0.04) \approx 0.217$$

d) $P(\text{no Type AB donor before the 10th person checked}) = P(X > 9) = (0.96)^9 \approx 0.693$

This one is a bit tricky. There is no implication that we actually find a donor on the 10th trial. We only care that nine trials passed with no Type AB donor.

17. Coins and intuition.

a) Intuitively, we expect 50 heads.

b) $E(\text{heads}) = np = 100(0.5) = 50$ heads.

19. Lefties.

These may be considered Bernoulli trials. There are only two possible outcomes, left-handed and not left-handed. Since people are selected at random, the probability of being left-handed is constant at about 13%. The trials are not independent, since the population is finite, but a sample of 5 people is certainly fewer than 10% of all people.

Let $X =$ the number of people checked until the first lefty is discovered.
Let $Y =$ the number of lefties among $n = 5$.

a) Use $Geom(0.13)$.

$$P(\text{first lefty is the fifth person}) = P(X=5) = (0.87)^4(0.13) \approx 0.0745$$

b) Use $Binom(5, 0.13)$.

$$P(\text{some lefties among the 5 people}) = 1 - P(\text{no lefties among the first 5 people})$$
$$= 1 - P(Y=0)$$
$$= 1 - {}_5C_0(0.13)^0(0.87)^5$$
$$\approx 0.502$$

c) Use $Geom(0.13)$.

$P(\text{first lefty is second or third person}) = P(X=2) + P(X=3)$

$= (0.87)(0.13) + (0.87)^2(0.13) \approx 0.211$

d) Use $Binom(5, 0.13)$.

$P(\text{exactly 3 lefties in the group}) = P(Y=3) = {}_5C_3(0.13)^3(0.87)^2 \approx 0.0166$

e) Use $Binom(5, 0.13)$.

$P(\text{at least 3 lefties in the group})$

$= P(Y=3) + P(Y=4) + P(Y=5)$

$= {}_5C_3(0.13)^3(0.87)^2 + {}_5C_4(0.13)^4(0.87)^1 + {}_5C_5(0.13)^5(0.87)^0$

≈ 0.0179

f) Use $Binom(5, 0.13)$.

$P(\text{at most 3 lefties in the group}) = P(Y=0) + P(Y=1) + P(Y=2) + P(Y=3)$

$= {}_5C_0(0.13)^0(0.87)^5 + {}_5C_1(0.13)^1(0.87)^4$

$+ {}_5C_2(0.13)^2(0.87)^3 + {}_5C_3(0.13)^3(0.87)^2$

≈ 0.9987

21. Lefties redux.

a) In a previous exercise, we determined that the selection of lefties could be considered Bernoulli trials. Since our group consists of 5 people, use $Binom(5, 0.13)$.

Let $Y =$ the number of lefties among $n = 5$.

$E(Y) = np = 5(0.13) = 0.65$ lefties

b) $SD(Y) = \sqrt{npq} = \sqrt{5(0.13)(0.87)} \approx 0.75$ lefties

c) Use $Geom(0.13)$. Let $X =$ the number of people checked until the first lefty is discovered.

$E(X) = \frac{1}{p} = \frac{1}{0.13} \approx 7.69$ people

23. Still more lefties.

a) In a previous exercise, we determined that the selection of lefties (and also righties) could be considered Bernoulli trials. Since our group consists of 12 people, and now we are considering the righties, use $Binom(12, 0.87)$.

Let $Y =$ the number of righties among $n = 12$.

$E(Y) = np = 12(0.87) = 10.44$ righties

$SD(Y) = \sqrt{npq} = \sqrt{12(0.87)(0.13)} \approx 1.16$ righties

b)

$$\begin{aligned} P(\text{not all righties}) &= 1 - P(\text{all righties}) \\ &= 1 - P(Y = 12) \\ &= 1 - {}_{12}C_{12}(0.87)^{12}(0.13)^{0} \\ &\approx 0.812 \end{aligned}$$

c)

$$\begin{aligned} P(\text{no more than 10 righties}) &= P(Y \le 10) \\ &= P(Y = 0) + P(Y = 1) + P(Y = 2) + \ldots + P(Y = 10) \\ &= {}_{12}C_{0}(0.87)^{0}(0.13)^{12} + \ldots + {}_{12}C_{10}(0.87)^{10}(0.13)^{2} \\ &\approx 0.475 \end{aligned}$$

d)

$$\begin{aligned} P(\text{exactly six of each}) &= P(Y = 6) \\ &= {}_{12}C_{6}(0.87)^{6}(0.13)^{6} \\ &\approx 0.00193 \end{aligned}$$

e)

$$\begin{aligned} P(\text{majority righties}) &= P(Y \ge 7) \\ &= P(Y = 7) + P(Y = 8) + P(Y = 9) + \ldots + P(Y = 12) \\ &= {}_{12}C_{7}(0.87)^{7}(0.13)^{5} + \ldots + {}_{12}C_{12}(0.87)^{12}(0.13)^{0} \\ &\approx 0.998 \end{aligned}$$

25. Vision.

The vision tests can be considered Bernoulli trials. There are only two possible outcomes, nearsighted or not. The probability of any child being nearsighted is given as $p = 0.12$. Finally, since the population of children is finite, the trials are not independent. However, 169 is certainly less than 10% of all children, and we will assume that the children in this district are representative of all children in relation to nearsightedness. Use *Binom*(169,0.12).

$\mu = E(\text{nearsighted}) = np = 169(0.12) = 20.28$ children.

$\sigma = SD(\text{nearsighted}) = \sqrt{npq} = \sqrt{169(0.12)(0.88)} \approx 4.22$ children.

27. Tennis, anyone?

The first serves can be considered Bernoulli trials. There are only two possible outcomes, successful and unsuccessful. The probability of any first serve being good is given as $p = 0.70$. Finally, we are assuming that each serve is independent of the others. Since she is serving 6 times, use $Binom(6, 0.70)$.

Let $X =$ the number of successful serves in $n = 6$ first serves.

a)

$$\begin{aligned} P(\text{six serves in}) &= P(X=6) \\ &= {}_6C_6(0.70)^6(0.30)^0 \\ &\approx 0.118 \end{aligned}$$

b)

$$\begin{aligned} &P(\text{exactly four serves in}) \\ &= P(X=4) \\ &= {}_6C_4(0.70)^4(0.30)^2 \\ &\approx 0.324 \end{aligned}$$

c)

$$\begin{aligned} P(\text{at least four serves in}) &= P(X=4)+P(X=5)+P(X=6) \\ &= {}_6C_4(0.70)^4(0.30)^2 + {}_6C_5(0.70)^5(0.30)^1 + {}_6C_6(0.70)^6(0.30)^0 \\ &\approx 0.744 \end{aligned}$$

d)

$$\begin{aligned} &P(\text{no more than four serves in}) \\ &= P(X=0)+P(X=1)+P(X=2)+P(X=3)+P(X=4) \\ &= {}_6C_0(0.70)^0(0.30)^6 + {}_6C_1(0.70)^1(0.30)^5 + {}_6C_2(0.70)^2(0.30)^4 \\ &\quad + {}_6C_3(0.70)^3(0.30)^3 + {}_6C_4(0.70)^4(0.30)^2 \\ &\approx 0.580 \end{aligned}$$

29. And more tennis.

The first serves can be considered Bernoulli trials. There are only two possible outcomes, successful and unsuccessful. The probability of any first serve being good is given as $p = 0.70$. Finally, we are assuming that each serve is independent of the others. Since she is serving 80 times, use $Binom(80, 0.70)$.

Let $X =$ the number of successful serves in $n = 80$ first serves.

a) $E(X) = np = 80(0.70) = 56$ first serves in.

$SD(X) = \sqrt{npq} = \sqrt{80(0.70)(0.30)} \approx 4.10$ first serves in.

b) Since $np = 56$ and $nq = 24$ are both greater than 10, $Binom(80, 0.70)$ may be approximated by the Normal model, $N(56, 4.10)$.

c) According to the Normal model, in matches with 80 serves, she is expected to make between 51.9 and 60.1 first serves approximately 68% of the time, between 47.8 and 64.2 first serves approximately 95% of the time, and between 43.7 and 68.3 first serves approximately 99.7% of the time.

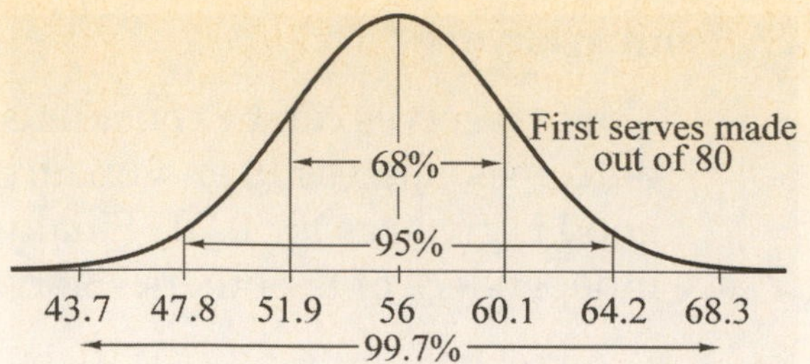

d) **Using *Binom*(80, 0.70):**

$$
\begin{aligned}
P(\text{at least 65 first serves}) &= P(X \geq 65) \\
&= P(X = 65) + P(X = 66) + \ldots + P(X = 80) \\
&= {}_{80}C_{65}(0.70)^{65}(0.30)^{15} + \ldots + {}_{80}C_{80}(0.70)^{80}(0.30)^{0} \\
&\approx 0.0161
\end{aligned}
$$

According to the Binomial model, the probability that she makes at least 65 first serves out of 80 is approximately 0.0161.

Using *N*(56, 4.10):

$$z = \frac{x-\mu}{\sigma}$$

$$z = \frac{65-56}{4.10}$$

$$z \approx 2.195$$

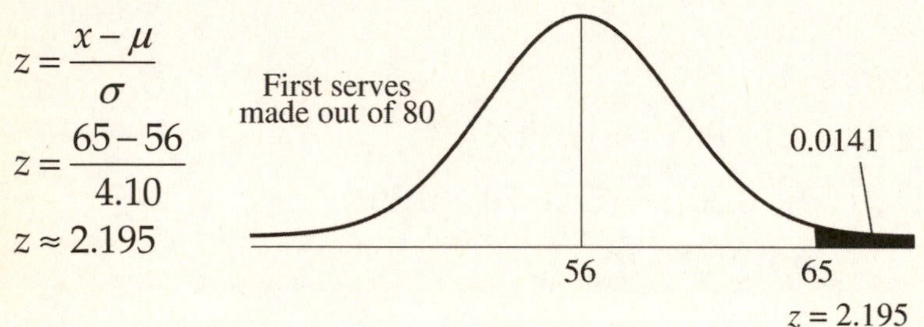

According to the Normal model, the probability that she makes at least 65 first serves out of 80 is approximately 0.0141.

$P(X \geq 65) \approx P(z > 2.195) \approx 0.0141$

31. Apples.

a) A binomial model and a normal model are both appropriate for modeling the number of cider apples that may come from the tree.

Let X = the number of cider apples found in the n = 300 apples from the tree.

The quality of the apples may be considered Bernoulli trials. There are only two possible outcomes, cider apple or not a cider apple. The probability that an apple must be used for a cider apple is constant, given as $p = 0.06$. The trials are not independent, since the population of apples is finite, but the apples on the tree are undoubtedly less than 10% of all the apples that the farmer has ever produced, so model with *Binom*(300, 0.06).

$E(X) = np = 300(0.06) = 18$ cider apples.

$SD(X) = \sqrt{npq} = \sqrt{300(0.06)(0.94)} \approx 4.11$ cider apples.

Since $np = 18$ and $nq = 282$ are both greater than 10, *Binom*(300,0.06) may be approximated by the Normal model, $N(18, 4.11)$.

b) Using *Binom*(300, 0.06):

$$\begin{aligned} P(\text{at most 12 cider apples}) &= P(X \le 12) \\ &= P(X=0) + \ldots + P(X=12) \\ &= {}_{300}C_0(0.06)^0(0.94)^{300} + \ldots + {}_{300}C_{12}(0.06)^{12}(0.94)^{282} \\ &\approx 0.085 \end{aligned}$$

According to the Binomial model, the probability that no more than 12 cider apples come from the tree is approximately 0.085.

Using *N*(18, 4.11):

$$z = \frac{x-\mu}{\sigma}$$

$$z = \frac{12-18}{4.11}$$

$$z = -1.460$$

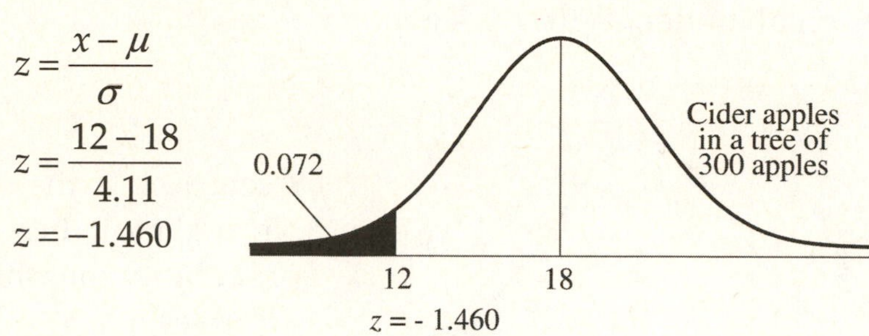

According to the Normal mol, the probability that no more than 12 apples out of 300 are cider apples is approximately 0.072.

$P(X \le 12) \approx P(z < -1.460) \approx 0.072$

c) It is extremely unlikely that the tree will bear more than 50 cider apples. Using the Normal model, $N(18, 4.11)$, 50 cider apples is approximately 7.8 standard deviations above the mean.

33. Lefties again.

Let $X =$ the number of righties among a class of $n = 188$ students.

Using *Binom*(188, 0.87):

These may be considered Bernoulli trials. There are only two possible outcomes, right-handed and not right-handed. The probability of being right-handed is assumed to be constant at about 87%. The trials are not independent, since the population is finite, but a sample of 188 students is certainly fewer than 10% of all people. Therefore, the number of righties in a class of 188 students may be modeled by *Binom*(188, 0.87).

If there are 171 or more righties in the class, some righties have to use a left-handed desk.

$$\begin{aligned} P(\text{at least 171 righties}) &= P(X \geq 171) \\ &= P(X=171)+\ldots+P(X=188) \\ &= {}_{188}C_{171}(0.87)^{171}(0.13)^{17}+\ldots+{}_{188}C_{188}(0.87)^{188}(0.13)^{0} \\ &\approx 0.061 \end{aligned}$$

According to the binomial model, the probability that a right-handed student has to use a left-handed desk is approximately 0.061.

Using $N(163.56, 4.61)$:

$E(X) = np = 188(0.87) = 163.56$ righties.

$SD(X) = \sqrt{npq} = \sqrt{188(0.87)(0.13)} \approx 4.61$ righties.

Since $np = 163.56$ and $nq = 24.44$ are both greater than 10, *Binom*(188,0.87) may be approximated by the Normal model, *N*(163.56, 4.61).

$$z = \frac{x-\mu}{\sigma}$$

$$z = \frac{171-163.56}{4.61}$$

$$z \approx 1.614$$

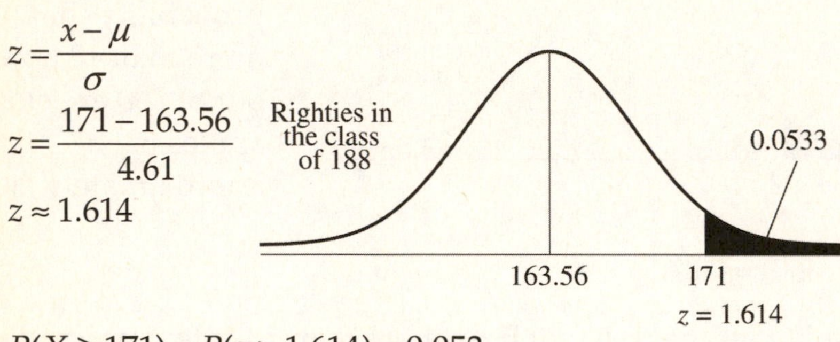

According to the Normal mol, the probability that there are at least 171 righties in the class of 188 is approximately 0.0533.

$P(X \geq 171) \approx P(z > 1.614) \approx 0.053$

35. Annoying phone calls.

Let $X =$ the number of sales made after making $n = 200$ calls.

Using *Binom*(200, 0.12):
These may be considered Bernoulli trials. There are only two possible outcomes, making a sale and not making a sale. The telemarketer was told that the probability of making a sale is constant at about $p = 0.12$. The trials are not independent, since the population is finite, but 200 calls is fewer than 10% of all calls. Therefore, the number of sales made after making 200 calls may be modeled by *Binom*(200, 0.12).

$$\begin{aligned} P(\text{at most } 10) &= P(X \leq 10) \\ &= P(X=0)+\ldots+P(X=10) \\ &= {}_{200}C_{0}(0.12)^{0}(0.88)^{200}+\ldots+{}_{200}C_{1900}(0.12)^{10}(0.88)^{190} \\ &\approx 0.0006 \end{aligned}$$

According to the Binomial model, the probability that the telemarketer would make at most 10 sales is approximately 0.0006.

Using $N(24, 4.60)$:

$E(X) = np = 200(0.12) = 24$ sales.

$SD(X) = \sqrt{npq} = \sqrt{200(0.12)(0.88)} \approx 4.60$ sales.

Since $np = 24$ and $nq = 176$ are both greater than 10, $Binom(200, 0.12)$ may be approximated by the Normal model, $N(24, 4.60)$.

$$z = \frac{x-\mu}{\sigma}$$
$$z = \frac{10-24}{4.60}$$
$$z \approx -3.043$$

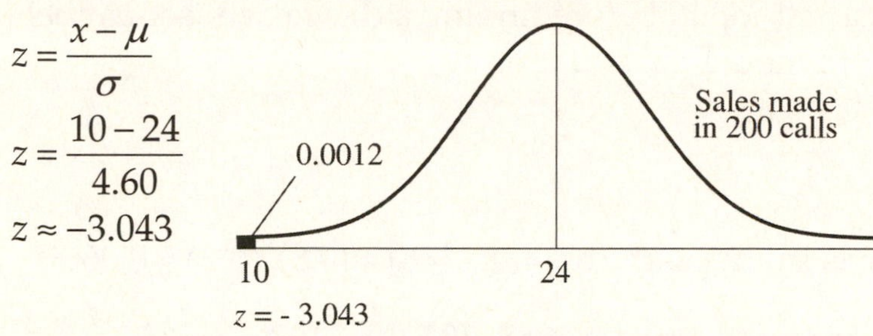

According to the Normal mol, the probability that the telemarketer would make at most 10 sales is approximately 0.0012.

$P(X \le 10) \approx P(z < -3.043) \approx 0.0012$

Since the probability that the telemarketer made 10 sales, given that the 12% of calls result in sales is so low, it is likely that he was misled about the true success rate.

37. Hurricanes, redux.

a)

$E(X) = \lambda = 2.35$

$$P(\text{no hurricanes next year}) = \frac{e^{-2.35}(2.35)^0}{0!} \approx 0.09537$$

b)

$P(\text{exactly one hurricane in next 2 years})$

$= P(\text{hurr. 1}^{\text{st}}\text{ yr})P(\text{no hurr. 2}^{\text{nd}}\text{ yr}) + P(\text{no hurr. 1}^{\text{st}}\text{ yr})P(\text{hurr. 2}^{\text{nd}}\text{ yr})$

$$= \left(\frac{e^{-2.35}(2.35)^1}{1!}\right)\left(\frac{e^{-2.35}(2.35)^0}{0!}\right) + \left(\frac{e^{-2.35}(2.35)^0}{0!}\right)\left(\frac{e^{-2.35}(2.35)^1}{1!}\right)$$

≈ 0.0427

39. TB again.

a) $E(X) = \lambda = np = 8000(0.0005) = 4$ cases.

b)

$P(\text{at least one new case}) = 1 - P(\text{no new cases})$

$$= 1 - \frac{e^{-4}(4)^0}{0!} \approx 0.9817$$

41. Seatbelts II.

These stops may be considered Bernoulli trials. There are only two possible outcomes, belted or not belted. Police estimate that the probability that a driver is buckled is 80%. (The probability of not being buckled is therefore 20%.) Provided the drivers stopped are representative of all drivers, we can consider the probability constant. The trials are not independent, since the population of drivers is finite, but the police will not stop more than 10% of all drivers.

a) Let X = the number of cars stopped before finding a driver whose seat belt is not buckled. Use *Geom*(0.2) to model the situation.

$$E(X) = \frac{1}{p} = \frac{1}{0.2} = 5 \text{ cars.}$$

b) $P(\text{First unbelted driver is in the sixth car}) = P(X = 6) = (0.8)^5(0.2) \approx 0.066$

c) $P(\text{The first ten drivers are wearing seatbelts}) = (0.8)^{10} \approx .107$

d) Let Y = the number of drivers wearing their seatbelts in 30 cars. Use *Binom*(30, 0.8).

$E(Y) = np = 30(0.8) = 24$ drivers.

$SD(Y) = \sqrt{npq} = \sqrt{30(0.8)(0.2)} \approx 2.19$ drivers.

e) Let W = the number of drivers not wearing their seatbelts in 120 cars.

Using *Binom*(120, 0.2):

$$\begin{aligned} P(\text{at least } 20) &= P(W \geq 20) \\ &= P(W = 20) + \ldots + P(W = 120) \\ &= {}_{120}C_{20}(0.2)^{20}(0.8)^{100} + \ldots + {}_{120}C1_{20}(0.2)^{120}(0.8)^{0} \\ &\approx 0.848 \end{aligned}$$

According to the Binomial model, the probability that at least 20 out of 120 drivers are not wearing their seatbelts is approximately 0.848.

Using *N*(24, 4.38):

$E(W) = np = 120(0.2) = 24$ drivers.

$SD(W) = \sqrt{npq} = \sqrt{120(0.2)(0.8)} \approx 4.38$ drivers.

Since $np = 24$ and $nq = 96$ are both greater than 10, *Binom*(120,0.2) may be approximated by the Normal model, *N*(24, 4.38).

$$z = \frac{w - \mu}{\sigma}$$

$$z = \frac{20 - 24}{4.38}$$

$$z \approx -0.913$$

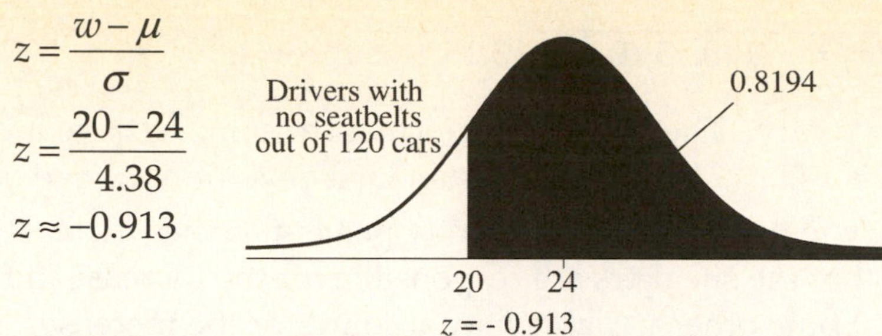

$$P(W \geq 120) \approx P(z > -0.913) \approx 0.8194$$

According to the Normal model, the probability that at least 20 out of 120 drivers stopped are not wearing their seatbelts is approximately 0.8194.

43. ESP.

Choosing symbols may be considered Bernoulli trials. There are only two possible outcomes, correct or incorrect. Assuming that ESP does not exist, the probability of a correct identification from a randomized deck is constant, at p = 0.20. The trials are independent, as long as the deck is shuffled after each attempt. Since 100 trials will be performed, use *Binom*(100, 0.2).

Let X = the number of symbols identified correctly out of 100 cards.

$E(X) = np = 100(0.2) = 20$ correct identifications.

$SD(X) = \sqrt{npq} = \sqrt{100(0.2)(0.8)} = 4$ correct identifications.

Answers may vary. In order be convincing, the "mind reader" would have to identify at least 32 out of 100 cards correctly, since 32 is three standard deviations above the mean. Identifying fewer cards than 32 could happen too often, simply due to chance.

45. Hot hand.

A streak like this is not unusual. The probability that he makes 4 in a row with a 55% free throw percentage is $(0.55)(0.55)(0.55)(0.55) \approx 0.09$. We can expect this to happen nearly one in ten times for every set of 4 shots that he makes. One out of ten times is not that unusual.

47. Hotter hand.

The shots may be considered Bernoulli trials. There are only two possible outcomes, make or miss. The probability of success is constant at 55%, and the shots are independent of one another. Therefore, we can model this situation with *Binom*(32, 0.55).

Let X = the number of free throws made out of 40.

$E(X) = np = 40(0.55) = 22$ free throws made.

$SD(X) = \sqrt{npq} = \sqrt{40(0.55)(0.45)} \approx 3.15$ free throws.

Answers may vary. The player's performance seems to have increased. 32 made free throws is $(32-22)/3.15 \approx 3.17$ standard deviations above the mean, an extraordinary feat, unless his free throw percentage has increased. This does NOT mean that the sneakers are responsible for the increase in free throw percentage. Some other variable may account for the increase. The player would need to set up a controlled experiment in order to determine what effect, if any, the sneakers had on his free throw percentage.

49. Web visitors.

a) The Poisson model because it is a good model to use when the data consists of counts of occurrences. The events must be independent and the mean number of occurrences stays constant.

b) Probability that in any one minute at least one purchase is made:

$$P(X=1)+P(X=2)+\ldots = \frac{e^{-3}3^1}{1!}+\frac{e^{-3}3^2}{2!}+\ldots = 0.9502$$

c) Probability that no one makes a purchase in the next 2 minutes:

$$P(X=0) = \frac{e^{-6}6^0}{0!} = 0.0025$$

51. Web visitors, part 2.

a) The exponential model can be used to model the time between events especially when the number of arrivals of those events can be modeled by a Poisson model.

b) The mean time between purchases: 3 purchases are made per minute so each purchase is made every 1/3 minute.

c) Probability that the time to the next purchase will be between 1 and 2 minutes:

$$e^{-3*1} - e^{-3*2} = 0.0473$$

Review of Part IV – Randomness and Probability

1. **Quality Control.**

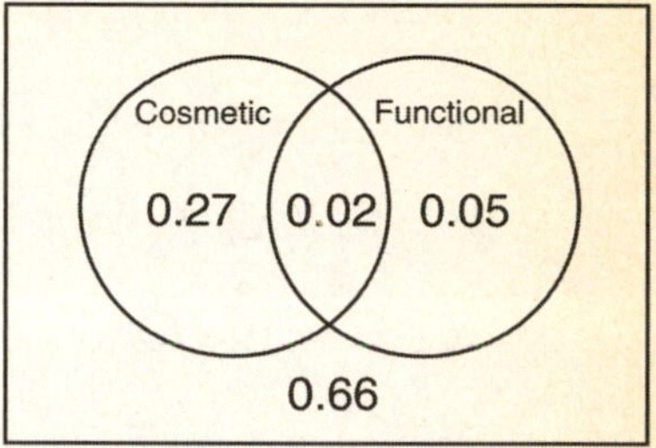

a) $P(\text{defect}) = P(\text{cosm.}) + P(\text{func.}) - P(\text{cosm. and func.})$
$= 0.29 + 0.07 - 0.02 = 0.34$

Or, from the Venn: $0.27 + 0.02 + 0.05 = 0.34$

b) 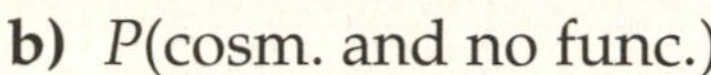 $P(\text{cosm. and no func.})$

$= P(\text{cosm.}) - P(\text{cosm. and func.})$
$= 0.29 - 0.02 = 0.27$

Or, from the Venn: 0.27 (region inside Cosmetic circle, outside Functional circle)

c) $$P(\text{func.} \mid \text{cosm.}) = \frac{P(\text{func. and cosm.})}{P(\text{cosm.})} = \frac{0.02}{0.29} \approx 0.069$$

From the Venn, consider only the region inside the Cosmetic circle. The probability that the car has a functional defect is 0.02 out of a total of 0.29 (the entire Cosmetic circle).

d) The two kinds of defects are not disjoint events, since 2% of cars have both kinds.

e) Approximately 6.9% of cars with cosmetic defects also have functional defects. Overall, the probability that a car has a cosmetic defect is 7%. The probabilities are estimates, so these are probably close enough to say that they two types of defects are independent.

3. **Airfares.**

a) Let C = the price of a ticket to China
Let F = the price of a ticket to France.

Total price of airfare = $3C + 5F$

b) $\mu = E(3C + 5F) = 3E(C) + 5E(F) = 3(1000) + 5(500) = \$5500$

$$\sigma = SD(3C + 5F) = \sqrt{3^2(Var(C)) + 5^2(Var(F))} = \sqrt{3^2(150^2) + 5^2(100^2)} \approx \$672.68$$

c) $\mu = E(C - F) = E(C) - E(F) = 1000 - 500 = \$500$

$$\sigma = SD(C - F) = \sqrt{Var(C) + Var(F)}$$
$$= \sqrt{150^2 + 100^2} \approx \$180.28$$

d) No assumptions are necessary when calculating means. When calculating standard deviations, we must assume that ticket prices are independent of each other for different countries but all tickets to the same country are at the same price.

5. A game.

a) Let X = net amount won

X	\$0	\$2	– \$2
P(X)	0.10	0.40	0.50

$$\mu = E(X) = 0(0.10) + 2(0.40) - 2(0.50) = -\$0.20$$

$$\sigma^2 = Var(X) = (0-(-0.20))^2(0.10) + (2-(-0.20))^2(0.40) + (-2-(-0.20))^2(0.50) = 3.56$$

$$\sigma = SD(X) = \sqrt{Var(X)} = \sqrt{3.56} \approx \$1.89$$

b) $X + X$ = the total winnings for two plays.

$$\mu = E(X+X) = E(X) + E(X) = (-0.20) + (-0.20) = -\$0.40$$

$$\sigma = SD(X+X) = \sqrt{Var(X) + Var(X)}$$

$$= \sqrt{3.56 + 3.56} \approx \$2.67$$

7. Twins.

The selection of these women can be considered Bernoulli trials. There are two possible outcomes, twins or no twins. As long as the women selected are representative of the population of all pregnant women, then $p = 1/90$. (If the women selected are representative of the population of women taking Clomid, then $p = 1/10$.) The trials are not independent since the population of all women is finite, but 10 women are fewer than 10% of the population of women.

Let X = the number of twin births from n = 10 pregnant women.

Let Y = the number of twin births from n = 10 pregnant women taking Clomid.

a) Use *Binom*(10, 1/90)

$$P(\text{at least one has twins}) = 1 - P(\text{none have twins})$$
$$= 1 - P(X = 0)$$
$$= 1 - {}_{10}C_0\left(\frac{1}{90}\right)^0\left(\frac{89}{90}\right)^{10}$$
$$\approx 0.106$$

b) Use *Binom*(10, 1/10)

$$P(\text{at least one has twins}) = 1 - P(\text{none have twins})$$
$$= 1 - P(Y = 0)$$
$$= 1 - {}_{10}C_0\left(\frac{1}{10}\right)^0\left(\frac{9}{10}\right)^{10}$$
$$\approx 0.651$$

c) Use *Binom*(5, 1/90) and *Binom*(5, 1/10).

$P(\text{at least one has twins})$
$= 1 - P(\text{no twins without Clomid})P(\text{no twins with Clomid})$

$$= 1 - \left({}_5C_0\left(\frac{1}{90}\right)^0\left(\frac{89}{90}\right)^5\right)\left({}_5C_0\left(\frac{1}{10}\right)^0\left(\frac{9}{10}\right)^5\right) \approx 0.442$$

9. More twins.

In the previous exercise, it was determined that these were Bernoulli trials. Use *Binom*(5, 0.10).

Let X = the number of twin births from $n = 5$ pregnant women taking Clomid.

a)

$$\begin{aligned} P(\text{none have twins}) &= P(X=0) \\ &= {}_5C_0(0.1)^0(0.9)^5 \\ &\approx 0.590 \end{aligned}$$

b)

$$\begin{aligned} P(\text{one has twins}) &= P(X=1) \\ &= {}_5C_1(0.1)^1(0.9)^4 \\ &\approx 0.328 \end{aligned}$$

c)

$$\begin{aligned} P(\text{at least three will have twins}) &= P(X=3)+P(X=4)+P(X=5) \\ &= {}_5C_3(0.1)^3(0.9)^2 + {}_5C_4(0.1)^4(0.9)^1 + {}_5C_5(0.1)^5(0.9)^0 \\ &= 0.00856 \end{aligned}$$

11. Twins, part III.

In a previous exercise, it was determined that these were Bernoulli trials. Use *Binom*(152, 0.10).

Let X = the number of twin births from $n = 152$ pregnant women taking Clomid.

a) $E(X) = np = 152(0.10) = 15.2$ births.

$SD(X) = \sqrt{npq} = \sqrt{152(0.10)(0.90)} \approx 3.70$ births.

b) Since $np = 15.2$ and $nq = 136.8$ are both greater than 10, the Success/Failure condition is satisfied and *Binom*(152, 0.10) may be approximated by *N*(15.2, 3.70).

c) **Using *Binom*(152, 0.10):**

$$\begin{aligned} P(\text{no more than } 10) &= P(X \le 10) \\ &= P(X=0)+\ldots+P(X=10) \\ &= {}_{152}C_0(0.10)^0(0.90)^{152} + \ldots + {}_{152}C_{10}(0.10)^{10}(0.90)^{142} \\ &\approx 0.097 \end{aligned}$$

According to the Binomial model, the probability that no more than 10 women would have twins is approximately 0.097.

Using *N*(15.2, 3.70):

$$z = \frac{x - \mu}{\sigma}$$

$$z = \frac{10 - 15.2}{3.70}$$

$$z \approx -1.405$$

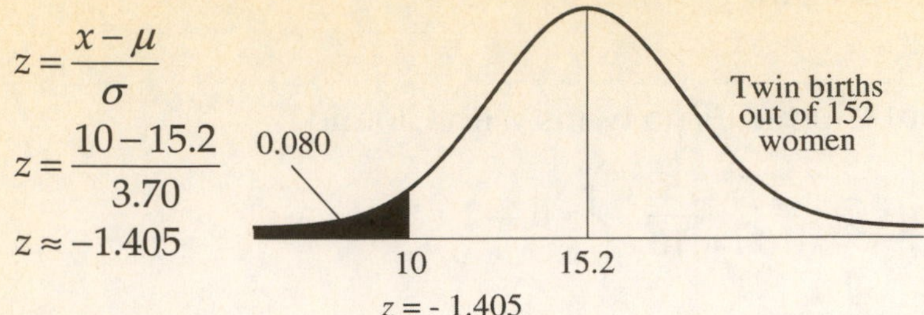

$P(X \le 10) \approx P(z < -1.405) \approx 0.080$

According to the Normal model, the probability that no more than 10 women would have twins is approximately 0.080.

13. Language.

Assuming that the freshman composition class consists of 25 randomly selected people, these may be considered Bernoulli trials. There are only two possible outcomes, having a specified language center or not having the specified language center. The probabilities of the specified language centers are constant at 80%, 10%, or 10%, for right, left, and two-sided language center, respectively. The trials are not independent, since the population of people is finite, but we will select fewer than 10% of all people.

a) Let $L =$ the number of people with left-brain language control from $n = 25$ people.

Use *Binom*(25, 0.80).

$$\begin{aligned} P(\text{no more than } 15) &= P(L \le 15) \\ &= P(L = 0) + \ldots + P(L = 15) \\ &= {}_{25}C_0(0.80)^0(0.20)^{25} + \ldots + {}_{25}C_{15}(0.80)^{15}(0.20)^{10} \\ &\approx 0.0173 \end{aligned}$$

According to the Binomial model, the probability that no more than 15 students in a class of 25 will have left-brain language centers is approximately 0.0173.

b) Let $T =$ the number of people with two-sided language control from $n = 5$ people.

Use *Binom*(5, 0.10).

$$\begin{aligned} P(\text{none have two-sided language control}) &= P(T = 0) \\ &= {}_5C_0(0.10)^0(0.90)^5 \\ &\approx 0.590 \end{aligned}$$

c) Use Binomial models:

$E(\text{left}) = np_L = 1200(0.80) = 960$ people

$E(\text{right}) = np_R = 1200(0.10) = 120$ people

$E(\text{two}-\text{sided}) = np_T = 1200(0.10) = 120$ people

d) Let R = the number of people with right-brain language control.

$E(R) = np_R = 1200(0.10) = 120$ people

$SD(R) = \sqrt{np_R q_R} = \sqrt{1200(0.10)(0.90)} \approx 10.39$ people.

e) Since $np_R = 120$ and $nq_R = 1080$ are both greater than 10, the Normal model, $N(120, 10.39)$, may be used to approximate $Binom(1200, 0.10)$. According to the Normal model, about 68% of randomly selected groups of 1200 people could be expected to have between 109.61 and 130.39 people with right-brain language control. About 95% of randomly selected groups of 1200 people could be expected to have between 99.22 and 140.78 people with right-brain language control. About 99.7% of randomly selected groups of 1200 people could be expected to have between 88.83 and 151.17 people with right-brain language control.

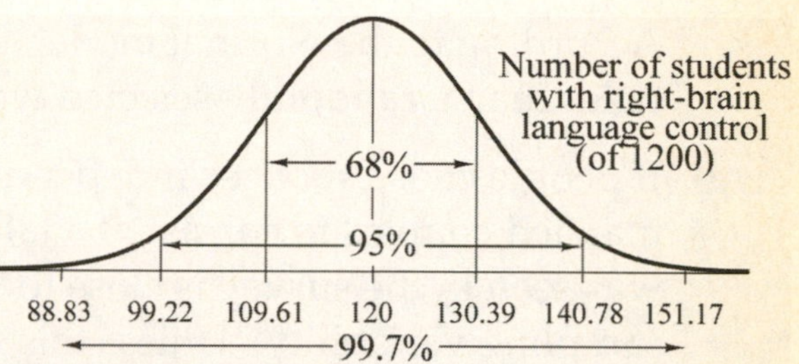

15. Beanstalks.

a) The greater standard deviation for men's heights indicates that men's heights are more variable than women's heights.

b) Admission to a Beanstalk Club is based upon extraordinary height for both men and women, but men are slightly more likely to qualify. The qualifying height for women is about 2.4 SDs above the mean height of women, while the qualifying height for men is about 1.75 SDS above the mean height for men.

c) Let M = the height of a randomly selected man from $N(69.1, 2.8)$.

Let W =the height of a randomly selected woman from $N(64.0, 2.5)$.

$M - W$ = the difference in height of a randomly selected man and woman.

d) $E(M-W) = E(M) - E(W) = 69.1 - 64.0 = 5.1$ inches

e) $SD(M-W) = \sqrt{Var(M) + Var(W)} = \sqrt{2.8^2 + 2.5^2} \approx 3.75$ inches

f) Since each distribution is described by a Normal model, the distribution of the difference in height between a randomly selected man and woman is $N(5.1, 3.75)$.

$$z = \frac{y - \mu}{\sigma}$$
$$z = \frac{0 - 5.1}{\sqrt{2.8^2 + 2.5^2}}$$
$$z \approx -1.359$$

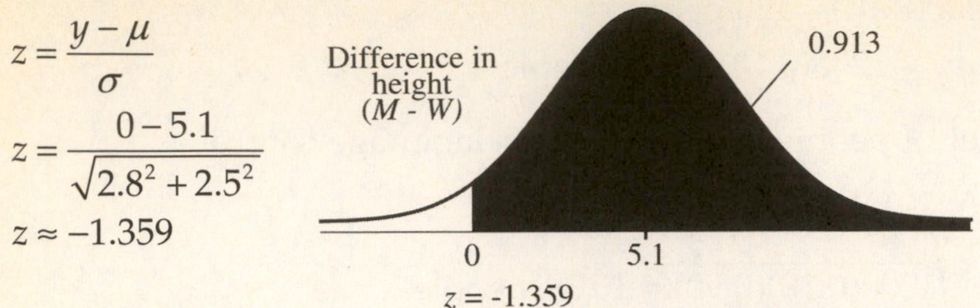

According to the Normal model, the probability that a randomly selected man is taller than a randomly selected woman is approximately 0.913.

g) If people chose spouses independent of height, we would expect 91.3% of married couples to consist of a taller husband and shorter wife. The 92% that was seen in the survey is close to 91.3%, and the difference may be due to natural sampling variability. Unless this survey is very large, there is not sufficient evidence of association between height and choice of partner.

17. Multiple choice.

Guessing at questions can be considered Bernoulli trials. There are only two possible outcomes, correct or incorrect. If you are guessing, the probability of success is $p = 0.25$, and the questions are independent. Use *Binom*(50, 0.25) to model the number of correct guesses on the test.

a) Let $X =$ the number of correct guesses.

$$\begin{aligned} P(\text{at least 30 of 50 correct}) &= P(X \geq 30) \\ &= P(X = 30) + \ldots + P(X = 50) \\ &= {}_{50}C_{30}(0.25)^{30}(0.75)^{20} + \ldots + {}_{50}C_{30}(0.25)^{50}(0.75)^{0} \\ &\approx 0.00000016 \end{aligned}$$

You are **very** unlikely to pass by guessing on every question.

b) Use *Binom*(50, 0.70).

$$\begin{aligned} P(\text{at least 30 of 50 correct}) &= P(X \geq 30) \\ &= P(X = 30) + \ldots + P(X = 50) \\ &= {}_{50}C_{30}(0.70)^{30}(0.30)^{20} + \ldots + {}_{50}C_{50}(0.70)^{50}(0.30)^{0} \\ &\approx 0.952 \end{aligned}$$

According to the Binomial model, your chances of passing are about 95.2%.

c) Use *Geom*(0.70).

$$P(\text{first correct on third question}) = (0.30)^2(0.70) = 0.063$$

19. Insurance.

The company is expected to pay \$100,000 only 2.6% of the time, while always gaining \$520 from every policy sold. When they pay, they actually only pay \$99,480.

$E(\text{profit}) = \$520(0.974) - \$99{,}480(0.026) = -\$2{,}080.$

The expected profit is actually a **loss** of \$2,080 per policy. The company had better raise its premiums if it hopes to stay in business.

21. Passing stats.

Organize the information in a tree diagram.

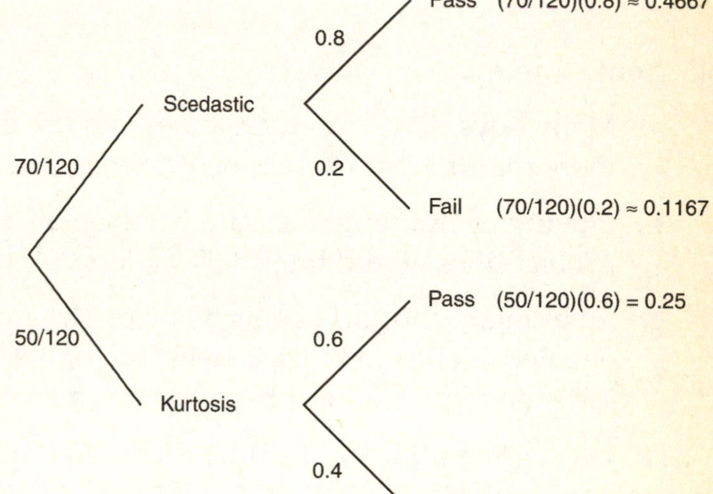

a)

$$\begin{aligned} &P(\text{Passing Statistics}) \\ &\quad = P(\text{Scedastic and Pass}) \\ &\qquad + P(\text{Kurtosis and Pass}) \\ &\quad \approx 0.4667 + 0.25 \\ &\quad \approx 0.717 \end{aligned}$$

b)

$$P(\text{Kurtosis} \mid \text{Fail}) = \frac{P(\text{Kurtosis and Fail})}{P(\text{Fail})} \approx \frac{0.1667}{0.1167 + 0.1667} \approx 0.588$$

23. Random variables.

a)

$$\begin{aligned} \mu &= E(X+50) \\ &= E(X) + 50 = 50 + 50 = 100 \\ \sigma &= SD(X+50) = SD(X) = 8 \end{aligned}$$

b)

$$\begin{aligned} \mu &= E(10Y) = 10E(Y) \\ &= 10(100) = 1000 \\ \sigma &= SD(10Y) = 10SD(Y) = 60 \end{aligned}$$

c)

$$\begin{aligned} \mu &= E(X+0.5Y) = E(X) + 0.5E(Y) \\ &= 50 + 0.5(100) = 100 \\ \sigma &= SD(X+0.5Y) = \sqrt{Var(X) + 0.5^2 Var(Y)} \\ &= \sqrt{8^2 + 0.5^2(6^2)} \approx 8.54 \end{aligned}$$

d)

$$\mu = E(X - Y) = E(X) - E(Y) = 50 - 100 = -50$$

$$\sigma = SD(X - Y) = \sqrt{Var(X) + Var(Y)}$$

$$= \sqrt{8^2 + 6^2} = 10$$

e)

$$\mu = E(X_1 + X_2) = E(X) + E(X) = 50 + 50 = 100$$

$$\sigma = SD(X_1 + X_2) = \sqrt{Var(X) + Var(X)} = \sqrt{8^2 + 8^2} \approx 11.31$$

25. Youth survey.

a) Many boys play computer games and use email, so the probabilities can total more than 100%. There is no evidence that there is a mistake in the report.

b) Playing computer games and using email are not disjoint. If they were, the probabilities would total 100% or less.

c) Emailing friends and being a boy or girl are not independent. 76% of girls emailed friends in the last week, but only 65% of boys emailed. If emailing were independent of being a boy or girl, the probabilities would be the same.

d) Let X = the number of students chosen until the first student is found who does not use the Internet. Use $Geom(0.07)$. $P(X = 5) = (0.93)^4(0.07) \approx 0.0524$.

27. Travel to Kyrgyzstan.

a) Spending an average of 4237 soms per day, you can stay about $\frac{90{,}000}{4237} \approx 21$ days.

b) Assuming that your daily spending is independent, the standard deviation is the square root of the sum of the variances for 21 days.

$\sigma = \sqrt{21(360)^2} \approx 1649.73$ soms

c) The standard deviation in your total expenditures is about 1650 soms, so if you don't think you will exceed your expectation by more than 2 standard deviations, bring an extra 3300 soms. This gives you a cushion of about 157 soms for each of the 21 days.

29. Home sweet home.

Since the homes are randomly selected, these can be considered Bernoulli trials. There are only two possible outcomes, owning the home or not owning the home. The probability of any randomly selected resident home being owned by the current resident is 0.66. The trials are not independent, since the population is finite, but as long as the city has more than 8200 homes, we are not sampling more than 10% of the population. The Binomial model, *Binom*(820, 0.66), can be used to model the number of homeowners among the 820 homes surveyed. Let H = the number of homeowners found in n = 820 homes.

$E(H) = np = 820(0.66) = 541.2$ homes.

$SD(H) = \sqrt{npq} = \sqrt{820(0.66)(0.34)} \approx 13.56$ homes.

The 523 homeowners found in the candidate's survey represent a number of homeowners that is only about 1.34 standard deviations below the expected number of homeowners. It is not particularly unusual to be 1.34 standard deviations below the mean. There is little support for the candidate's claim of a low level of home ownership.

31. Who's the boss?

a) $P(\text{first three owned by women}) = (0.26)^3 \approx 0.018$

b) $P(\text{none of the first four are owned by women}) = (0.74)^4 \approx 0.300$

c) $P(\text{sixth firm called is owned by women} \mid \text{none of the first five were}) = 0.26$

Since the firms are chosen randomly, the fact that the first five firms were owned by men has no bearing on the ownership of the sixth firm.

33. When to stop?

a) Since there are only two outcomes, 6 or not 6, the probability of getting a 6 is 1/6, and the trials are independent, these are Bernoulli trials. Use *Geom*(1/6).

$$\mu = \frac{1}{p} = \frac{1}{1/6} = 6 \text{ rolls}$$

b) If 6's are not allowed, the mean of each die roll is $\dfrac{1+2+3+4+5}{5} = 3$. You would expect to get 15 if you rolled 5 times.

c) $P(\text{5 rolls without a 6}) = \left(\dfrac{5}{6}\right)^5 \approx 0.402$

35. Technology on campus.

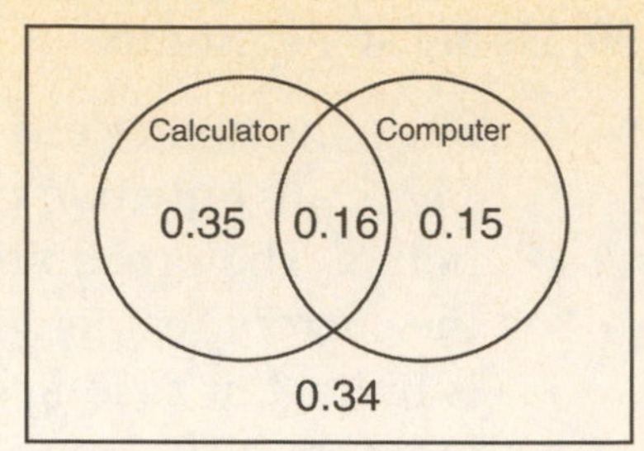

a)

$$\begin{aligned} &P(\text{neither tech.}) \\ &= 1 - P(\text{either tech.}) \\ &= 1 - [P(\text{calculator}) + P(\text{computer}) - P(\text{both})] \\ &= 1 - [0.51 + 0.31 - 0.16] = 0.34 \end{aligned}$$

Or, from the Venn: 0.34 (the region outside both circles) This is MUCH easier.

34% of students use neither type of technology.

b) $P(\text{calc. and no comp.}) = P(\text{calc.}) - P(\text{calc. and comp.}) = 0.51 - 0.16 = 0.35$

Or, from the Venn: 0.35 (region inside the Calculator circle, outside the Computer circle)

35% of students use calculators, but not computers.

c) $P(\text{computer} \mid \text{calculator}) = \dfrac{P(\text{comp. and calc.})}{P(\text{calc.})} = \dfrac{0.16}{0.51} \approx 0.314$

About 31.4% of calculator users have computer assignments.

d) The percentage of computer users overall is 31%, while 31.4% of calculator users were computer users. These are very close. There is no indication of an association between computer use and calculator use.

37. O-rings.

a) A Poisson model would be used.

b) If the probability of failure for one O-ring is 0.01, then the mean number of failures for 10 O-rings is $E(X) = \lambda = np = 10(0.01) = 0.1$ O-ring. We are able to calculate this because the Poisson model scales according to sample size.

c) $P(\text{one failed O-ring}) = \dfrac{e^{-0.1}(0.1)^1}{1!} \approx 0.090$

d)

$$\begin{aligned} P(\text{at least one failed O-ring}) &= 1 - P(\text{no failures}) \\ &= 1 - \frac{e^{-0.1}(0.1)^0}{0!} \approx 0.095 \end{aligned}$$

39. Socks.

Since we are sampling without replacement, use conditional probabilities throughout.

a) $P(2\text{ blue}) = \left(\dfrac{4}{12}\right)\left(\dfrac{3}{11}\right) = \dfrac{12}{132} = \dfrac{1}{11}$

b) $P(\text{no grey}) = \left(\frac{7}{12}\right)\left(\frac{6}{11}\right) = \frac{42}{132} = \frac{7}{22}$

c) $P(\text{at least one black}) = 1 - P(\text{no black}) = 1 - \left(\frac{9}{12}\right)\left(\frac{8}{11}\right) = \frac{60}{132} = \frac{5}{11}$

d) $P(\text{green}) = 0$ (There aren't any green socks in the drawer.)

e) $P(\text{match}) = P(2\text{ blue}) + P(2\text{ grey}) + P(2\text{ black})$

$$= \left(\frac{4}{12}\right)\left(\frac{3}{11}\right) + \left(\frac{5}{12}\right)\left(\frac{4}{11}\right) + \left(\frac{3}{12}\right)\left(\frac{2}{11}\right) = \frac{19}{66}$$

41. The Drake equation.

a) $N \cdot f_p$ represents the number of stars in the Milky Way Galaxy expected to have planets.

b) $N \cdot f_p \cdot n_e \cdot f_i$ represents the number of planets in the Milky Way Galaxy expected to have intelligent life.

c) $f_l \cdot f_i$ is the probability that a planet has a suitable environment and has intelligent life.

d) $f_l = P(\text{life} \mid \text{suitable environment})$. This is the probability that life develops, if a planet has a suitable environment.

$f_i = P(\text{intelligence} \mid \text{life})$. This is the probability that the life develops intelligence, if a planet already has life.

$f_c = P(\text{communication} \mid \text{intelligence})$. This is the probability that radio communication develops, if a planet already has intelligent life.

43. Pregnant?

Organize the information in a tree diagram.

$$P(\text{pregnant} \mid + \text{test}) = \frac{P(\text{preg. and + test})}{P(+\text{ test})} = \frac{0.686}{0.686 + 0.006} \approx 0.991$$

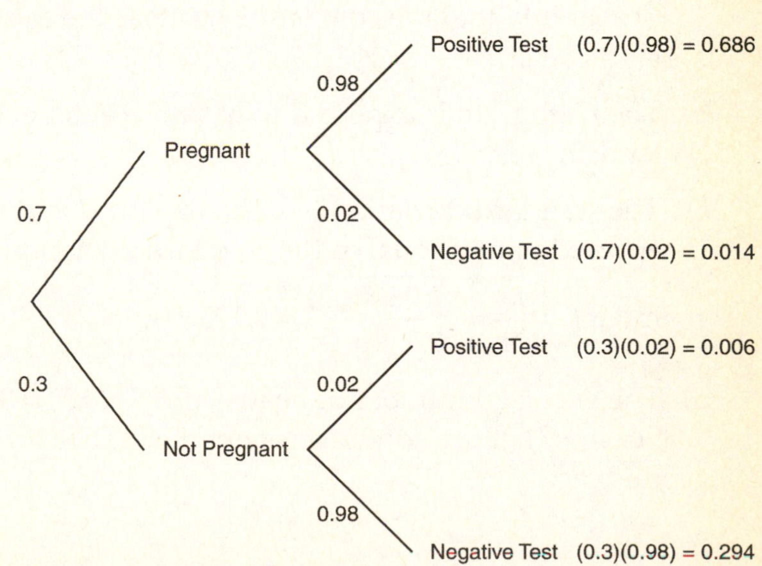

Chapter 18 – Sampling Distribution Models

1. Send money.

All of the histograms are centered around $p = 0.05$. As n gets larger, the shape of the histograms get more unimodal and symmetric, approaching a Normal model, while the variability in the sample proportions decreases.

3. Send money, again.

a)

n	Observed mean	Theoretical mean	Observed st. dev.	Theoretical standard deviation
20	0.0497	0.05	0.0479	$\sqrt{(0.05)(0.95)/20} \approx 0.0487$
50	0.0516	0.05	0.0309	$\sqrt{(0.05)(0.95)/50} \approx 0.0308$
100	0.0497	0.05	0.0215	$\sqrt{(0.05)(0.95)/100} \approx 0.0218$
200	0.0501	0.05	0.0152	$\sqrt{(0.05)(0.95)/200} \approx 0.0154$

b) All of the values seem very close to what we would expect from theory.

c) The histogram for $n = 200$ looks quite unimodal and symmetric. We should be able to use the Normal model.

d) The success/failure condition requires np and nq to both be at least 10, which is not satisfied until $n = 200$ for $p = 0.05$. The theory supports the choice in part c.

5. Coin tosses.

a) The histogram of these proportions is expected to be symmetric, but **not** because of the Central Limit Theorem. The sample of 16 coin flips is not large. The distribution of these proportions is expected to be symmetric because the probability that the coin lands heads is the same as the probability that the coin lands tails.

b) The histogram is expected to have its center at 0.5, the probability that the coin lands heads.

c) The standard deviation of data displayed in this histogram should be approximately equal to the standard deviation of the sampling distribution model, $\sqrt{\frac{pq}{n}} = \sqrt{\frac{(0.5)(0.5)}{16}} = 0.125$.

d) The expected number of heads, $np = 16(0.5) = 8$, which is less than 10. The Success/Failure condition is not met. The Normal model is not appropriate in this case.

7. More coins.

a) $\mu_{\hat{p}} = p = 0.5$ and $SD(\hat{p}) = \sqrt{\frac{pq}{n}} = \sqrt{\frac{(0.5)(0.5)}{25}} = 0.1$

About 68% of the sample proportions are expected to be between 0.4 and 0.6, about 95% are expected to be between 0.3 and 0.7, and about 99.7% are expected to be between 0.2 and 0.8.

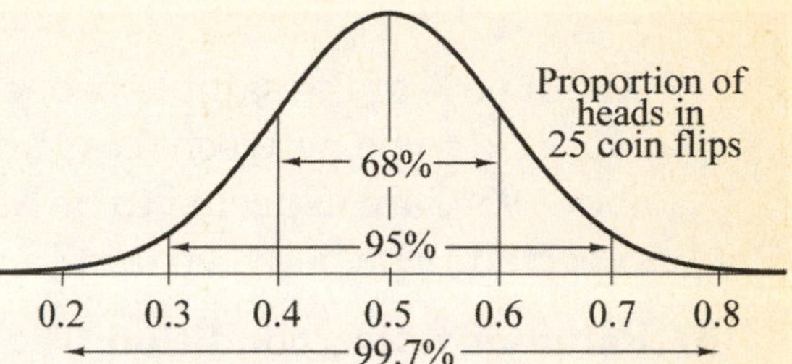

b) Coin flips are independent of one another. There is no need to check the 10% Condition. $np = nq = 12.5$, so both are greater than 10. The Success/Failure condition is met, so the sampling distribution model is $N(0.5, 0.1)$.

c) $\mu_{\hat{p}} = p = 0.5$ and $SD(\hat{p}) = \sqrt{\frac{pq}{n}} = \sqrt{\frac{(0.5)(0.5)}{64}} = 0.0625$

About 68% of the sample proportions are expected to be between 0.4375 and 0.5625, about 95% are expected to be between 0.375 and 0.625, and about 99.7% are expected to be between 0.3125 and 0.6875.

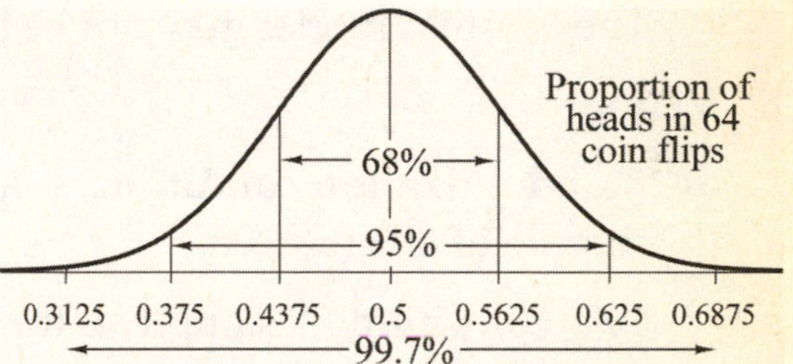

Coin flips are independent of one another, and $np = nq = 32$, so both are greater than 10. The Success/Failure condition is met, so the sampling distribution model is $N(0.5, 0.0625)$.

d) As the number of tosses increases, the sampling distribution model will still be Normal and centered at 0.5, but the standard deviation will decrease. The sampling distribution model will be less spread out.

9. Just (un)lucky.

For 200 flips, the sampling distribution model is Normal with $\mu_{\hat{p}} = p = 0.5$ and $SD(\hat{p}) = \sqrt{\frac{pq}{n}} = \sqrt{\frac{(0.5)(0.5)}{200}} \approx 0.0354$. Her sample proportion of $\hat{p} = 0.42$ is about 2.26 standard deviations below the expected proportion, which is unusual, but not extraordinary. According to the Normal model, we expect sample proportions this low or lower about 1.2% of the time.

11. Speeding.

a) $\mu_{\hat{p}} = p = 0.70$

$$SD(\hat{p}) = \sqrt{\frac{pq}{n}} = \sqrt{\frac{(0.7)(0.3)}{80}} \approx 0.051.$$

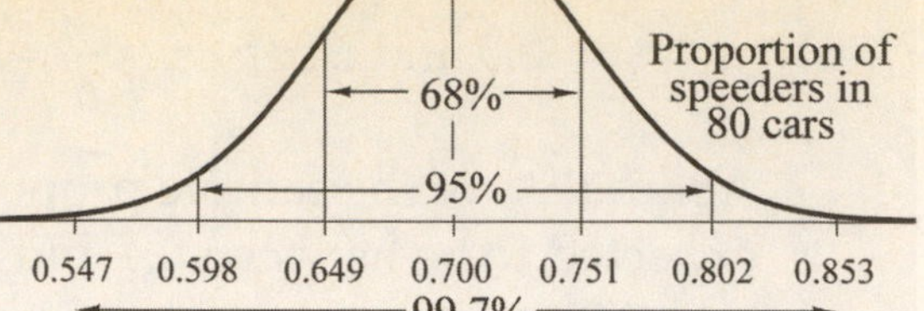

About 68% of the sample proportions are expected to be between 0.649 and 0.751, about 95% are expected to be between 0.598 and 0.802, and about 99.7% are expected to be between 0.547 and 0.853.

b) Randomization condition: The sample may not be representative. If the flow of traffic is very fast, the speed of the other cars around may have some effect on the speed of each driver. Likewise, if traffic is slow, the police may find a smaller proportion of speeders than they expect.

10% condition: 80 cars represent less than 10% of all cars

Success/Failure condition: $np = 56$ and $nq = 24$ are both greater than 10.

The Normal model may not be appropriate. Use caution. (And don't speed!)

13. Vision.

a) Randomization condition: Assume that the 170 children are a representative sample of all children.

10% condition: A sample of this size is less than 10% of all children.

Success/Failure condition: $np = 20.4$ and $nq = 149.6$ are both greater than 10.

Therefore, the sampling distribution model for the proportion of 170 children who are nearsighted is $N(0.12, 0.025)$.

b) The Normal model is to the right.

c) They might expect that the proportion of nearsighted students to be within 2 standard deviations of the mean. According to the Normal model, this means they might expect between 7% and 17% of the students to be nearsighted, or between about 12 and 29 students.

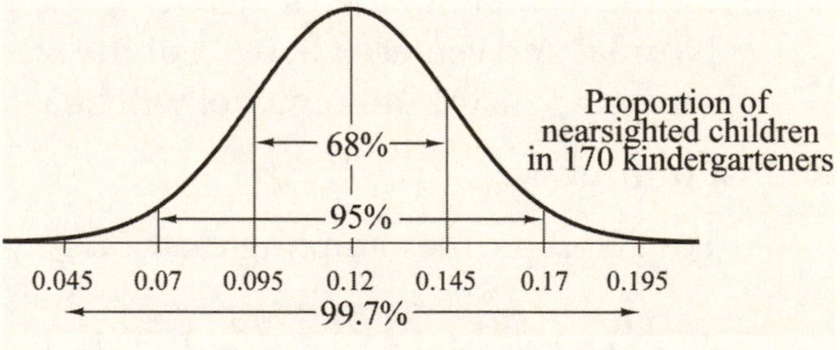

15. Loans.

a) $\mu_{\hat{p}} = p = 7\%$

$$SD(\hat{p}) = \sqrt{\frac{pq}{n}} = \sqrt{\frac{(0.07)(0.93)}{200}} \approx 1.8\%$$

b) **Randomization condition:** Assume that the 200 people are a representative sample of all loan recipients.

10% condition: A sample of this size is less than 10% of all loan recipients.

Success/Failure condition: $np = 14$ and $nq = 186$ are both greater than 10.

Therefore, the sampling distribution model for the proportion of 200 loan recipients who will not make payments on time is $N(0.07, 0.018)$.

c) According to the Normal model, the probability that over 10% of these clients will not make timely payments is approximately 0.048.

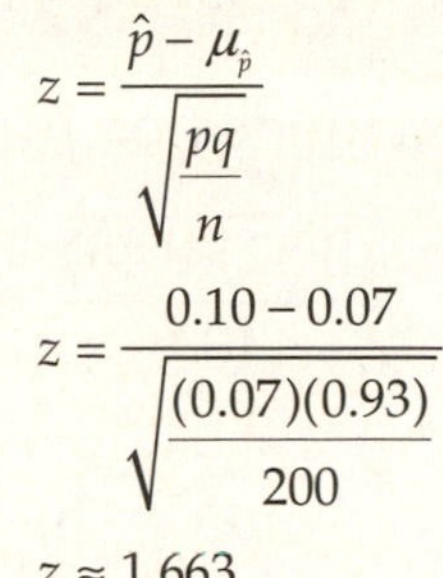

$$z = \frac{\hat{p} - \mu_{\hat{p}}}{\sqrt{\frac{pq}{n}}}$$

$$z = \frac{0.10 - 0.07}{\sqrt{\frac{(0.07)(0.93)}{200}}}$$

$$z \approx 1.663$$

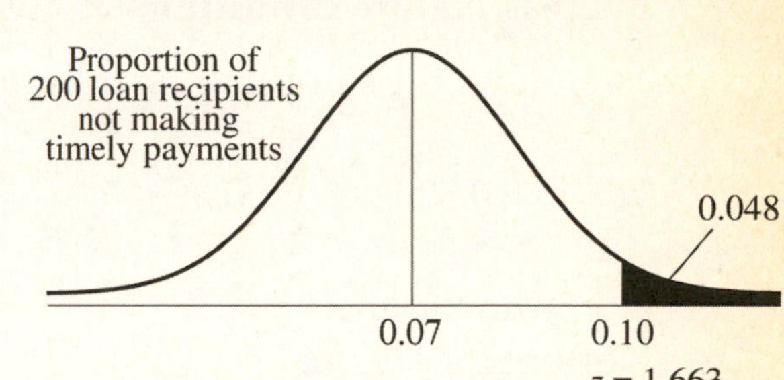

17. Back to school?

Randomization condition: We are considering random samples of 400 students who took the ACT.

10% Condition: 400 students is less than 10% of all college students.

Success/Failure condition: $np = 296$ and $nq = 104$ are both greater than 10.

Therefore, the sampling distribution model for $\hat{p}$ is Normal, with:

$$\mu_{\hat{p}} = p = 0.74$$

$$SD(\hat{p}) = \sqrt{\frac{pq}{n}} = \sqrt{\frac{(0.74)(0.26)}{400}} \approx 0.022$$

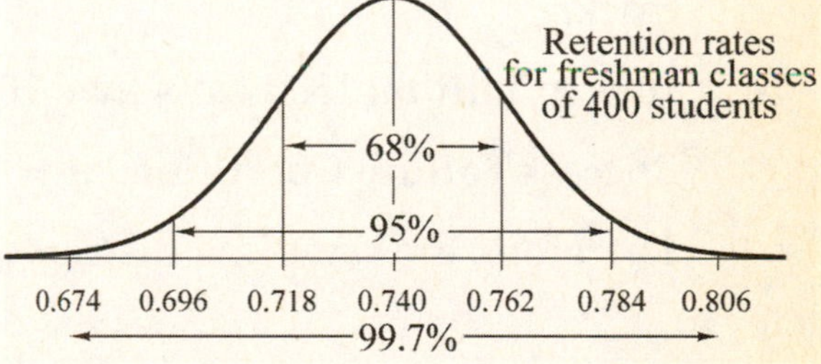

According to the sampling distribution model, about 68% of the colleges are expected to have retention rates between 0.718 and 0.762, about 95% of the colleges are expected to have retention rates between 0.696 and 0.784, and about 99.7% of the colleges are expected to have retention rates between 0.674 and 0.806. However, the conditions for the use of this model may not be met. We should be cautious about making any conclusions based on this model.

19. Back to school, again.

Provided that the students at this college are typical, the sampling distribution model for the retention rate, $\hat{p}$, is Normal with $\mu_{\hat{p}} = p = 0.74$ and standard deviation $SD(\hat{p}) = \sqrt{\frac{pq}{n}} = \sqrt{\frac{(0.74)(0.26)}{603}} \approx 0.018$

This college has a right to brag about their retention rate. 522/603 = 86.6% is over 7 standard deviations above the expected rate of 74%.

21. Polling.

Randomization condition: We must assume that the 400 voters were polled randomly.

10% condition: 400 voters polled represent less than 10% of potential voters.

Success/Failure condition: $np = 208$ and $nq = 192$ are both greater than 10.

Therefore, the sampling distribution model for $\hat{p}$ is Normal, with:

$$\mu_{\hat{p}} = p = 0.52 \qquad SD(\hat{p}) = \sqrt{\frac{pq}{n}} = \sqrt{\frac{(0.52)(0.48)}{400}} \approx 0.025$$

According to the Normal model, the probability that the newspaper's sample will lead them to predict defeat (that is, predict budget support below 50%) is approximately 0.212.

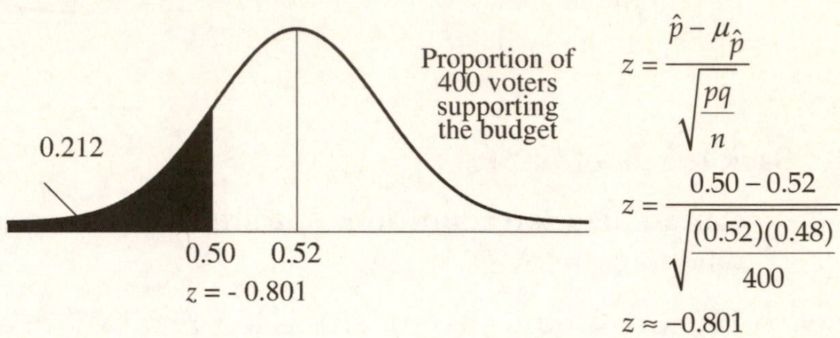

$$z = \frac{\hat{p} - \mu_{\hat{p}}}{\sqrt{\frac{pq}{n}}}$$

$$z = \frac{0.50 - 0.52}{\sqrt{\frac{(0.52)(0.48)}{400}}}$$

$$z \approx -0.801$$

23. Apples.

Randomization condition: A random sample of 150 apples is taken from each truck.

10% condition: 150 is less than 10% of all apples.

Success/Failure Condition: $np = 12$ and $nq = 138$ are both greater than 10.

Therefore, the sampling distribution model for $\hat{p}$ is Normal, with:

$$\mu_{\hat{p}} = p = 0.08 \qquad SD(\hat{p}) = \sqrt{\frac{pq}{n}} = \sqrt{\frac{(0.08)(0.92)}{150}} \approx 0.0222$$

According to the Normal model, the probability that less than 5% of the apples in the sample are unsatisfactory is approximately 0.088.

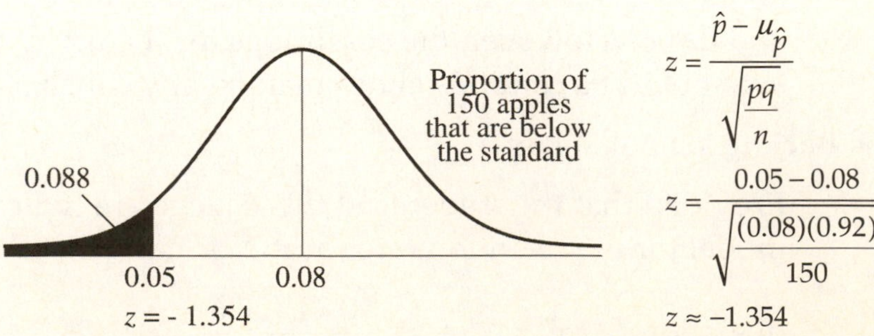

$$z = \frac{\hat{p} - \mu_{\hat{p}}}{\sqrt{\frac{pq}{n}}}$$

$$z = \frac{0.05 - 0.08}{\sqrt{\frac{(0.08)(0.92)}{150}}}$$

$$z \approx -1.354$$

25. Nonsmokers.

Randomization condition: We will assume that the 120 customers (to fill the restaurant to capacity) are representative of all customers.

10% condition: 120 customers represent less than 10% of all potential customers.

Success/Failure condition: $np = 72$ and $nq = 48$ are both greater than 10.

Therefore, the sampling distribution model for $\hat{p}$ is Normal, with:

$$\mu_{\hat{p}} = p = 0.60 \qquad SD(\hat{p}) = \sqrt{\frac{pq}{n}} = \sqrt{\frac{(0.60)(0.40)}{120}} \approx 0.0447$$

Answers may vary. We will use 3 standard deviations above the expected proportion of customers who demand nonsmoking seats to be "very sure".

$$\mu_{\hat{p}} + 3\left(\sqrt{\frac{pq}{n}}\right) \approx 0.60 + 3(0.0447) \approx 0.734$$

Since 120(0.734) = 88.08, the restaurant needs at least 89 seats in the nonsmoking section.

27. Sampling.

a) The sampling distribution model for the sample mean is $N\left(\mu, \frac{\sigma}{\sqrt{n}}\right)$.

b) If we choose a larger sample, the mean of the sampling distribution model will remain the same, but the standard deviation will be smaller.

29. Waist size.

a) The distribution of waist size of 250 men in Utah is unimodal and slightly skewed to the right. A typical waist size is approximately 36 inches, and the standard deviation in waist sizes is approximately 4 inches.

b) All of the histograms show distributions of sample means centered near 36 inches. As n gets larger the histograms approach the Normal model in shape, and the variability in the sample means decreases. The histograms are fairly Normal by the time the sample reaches size 5.

31. Waist size revisited.

a)

n	Observed mean	Theoretical mean	Observed st. dev.	Theoretical standard deviation
2	36.314	36.33	2.855	$4.019/\sqrt{2} \approx 2.842$
5	36.314	36.33	1.805	$4.019/\sqrt{5} \approx 1.797$
10	36.341	36.33	1.276	$4.019/\sqrt{10} \approx 1.271$
20	36.339	36.33	0.895	$4.019/\sqrt{20} \approx 0.897$

b) The observed values are all very close to the theoretical values.

c) For samples as small as 5, the sampling distribution of sample means is unimodal and symmetric. The Normal model would be appropriate.

d) The distribution of the original data is nearly unimodal and symmetric, so it doesn't take a very large sample size for the distribution of sample means to be approximately Normal.

33. GPAs.

Randomization condition: Assume that the students are randomly assigned to seminars.

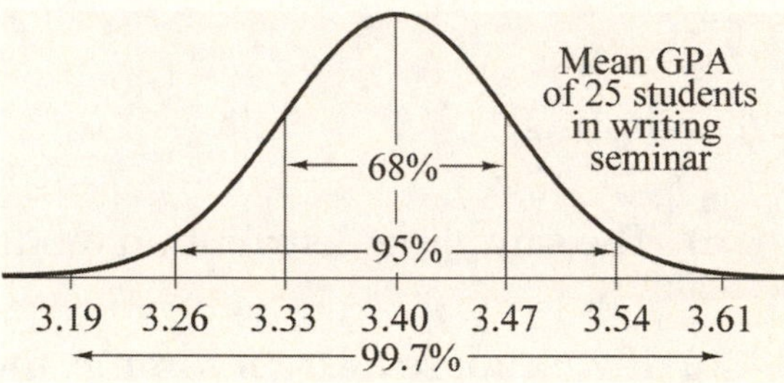

Independence assumption: It is reasonable to think that GPAs for randomly selected students are mutually independent.

10% condition: The 25 students in the seminar certainly represent less than 10% of the population of students.

Large Enough Sample condition: The distribution of GPAs is roughly unimodal and symmetric, so the sample of 25 students is large enough.

The mean GPA for the freshmen was $\mu = 3.4$, with standard deviation $\sigma = 0.35$. Since the conditions are met, the Central Limit Theorem tells us that we can model the sampling distribution of the mean GPA with a Normal model, with $\mu_{\bar{y}} = 3.4$ and standard deviation $SD(\bar{y}) = \frac{0.35}{\sqrt{25}} \approx 0.07$.

The sampling distribution model for the sample mean GPA is approximately $N(3.4, 0.07)$.

35. Lucky spot?

a) Smaller outlets have more variability than the larger outlets, just as the Central Limit Theorem predicts.

b) If the lottery is truly random, all outlets are equally likely to sell winning tickets.

37. Pregnancy.

a)

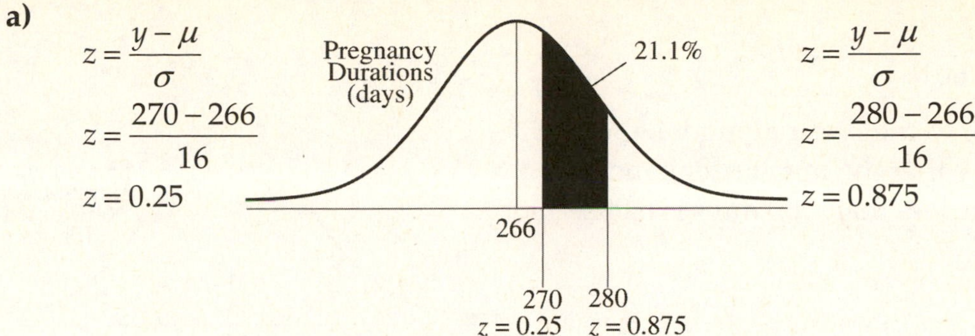

According to the Normal model, approximately 21.1% of all pregnancies are expected to last between 270 and 280 days.

b)

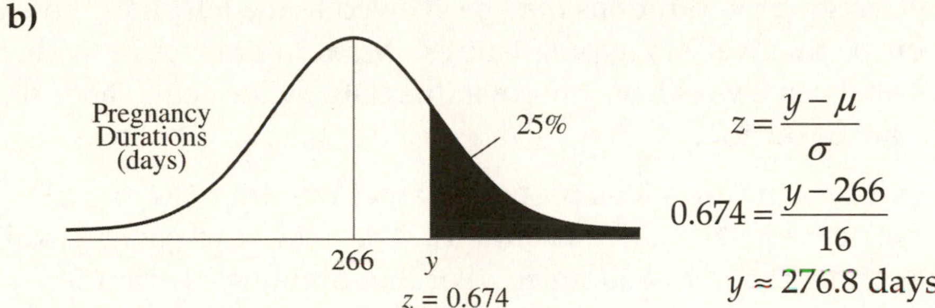

According to the Normal model, the longest 25% of pregnancies are expected to last approximately 276.8 days or more.

c) **Randomization condition:** Assume that the 60 women the doctor is treating can be considered a representative sample of all pregnant women.

Independence assumption: It is reasonable to think that the durations of the patients' pregnancies are mutually independent.

10% condition: The 60 women that the doctor is treating certainly represent less than 10% of the population of all women.

Large Enough Sample condition: The sample of 60 women is large enough. In this case, any sample would be large enough, since the distribution of pregnancies is Normal.

The mean duration of the pregnancies was $\mu = 266$ days, with standard deviation $\sigma = 16$ days. Since the distribution of pregnancy durations is Normal, we can model the sampling distribution of the mean pregnancy duration with a Normal model, with $\mu_{\bar{y}} = 266$ days and standard deviation

$SD(\bar{y}) = \frac{16}{\sqrt{60}} \approx 2.07$ days.

d) According to the Normal model, the probability that the mean pregnancy duration is less than 260 days is 0.002.

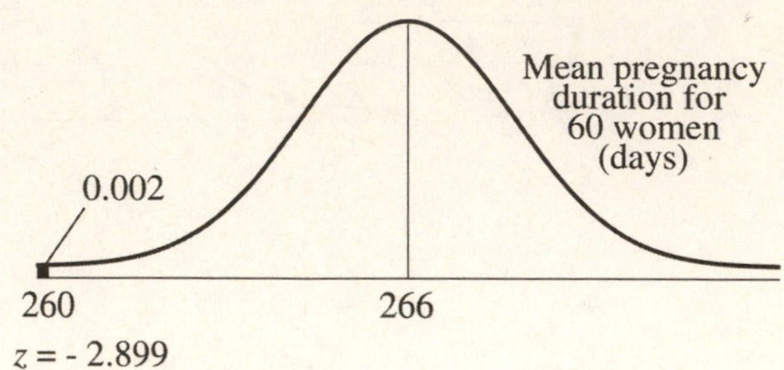

39. Pregnant again.

a) The distribution of pregnancy durations may be skewed to the left since there are more premature births than very long pregnancies. Modern practice of medicine stops pregnancies at about 2 weeks past normal due date by inducing labor or performing a Caesarean section.

b) We can no longer answer the questions posed in parts a and b. The Normal model is not appropriate for skewed distributions. The answer to part c is still valid. The Central Limit Theorem guarantees that the sampling distribution model is Normal when the sample size is large.

41. Dice and dollars.

a) Let X = the number of dollars won in one play.

$$\mu = E(X) = 0\left(\frac{3}{6}\right) + 1\left(\frac{2}{6}\right) + 10\left(\frac{1}{6}\right) = \$2$$

$$\sigma^2 = Var(X) = (0-2)^2\left(\frac{3}{6}\right) + (1-2)^2\left(\frac{2}{6}\right) + (10-2)^2\left(\frac{1}{6}\right) = 13$$

$$\sigma = SD(X) = \sqrt{Var(X)} = \sqrt{13} \approx \$3.61$$

b) $X + X$ = the total winnings for two plays.

$$\mu = E(X+X) = E(X) + E(X) = 2 + 2 = \$4$$

$$\sigma = SD(X+X) = \sqrt{Var(X) + Var(X)}$$

$$= \sqrt{13+13} \approx \$5.10$$

c) In order to win at least \$100 in 40 plays, you must average at least $\frac{100}{40} = \$2.50$ per play.

The expected value of the winnings is $\mu = \$2$, with standard deviation $\sigma = \$3.61$. Rolling a die is random and the outcomes are mutually independent, so the Central Limit Theorem guarantees that the sampling distribution model is Normal with $\mu_{\bar{x}} = \$2$ and standard deviation $SD(\bar{x}) = \frac{\$3.61}{\sqrt{40}} \approx \$0.571$.

According to the Normal model, the probability that you win at least \$100 in 40 plays is approximately 0.191.

(This is equivalent to using $N(80, 22.83)$ to model your total winnings.)

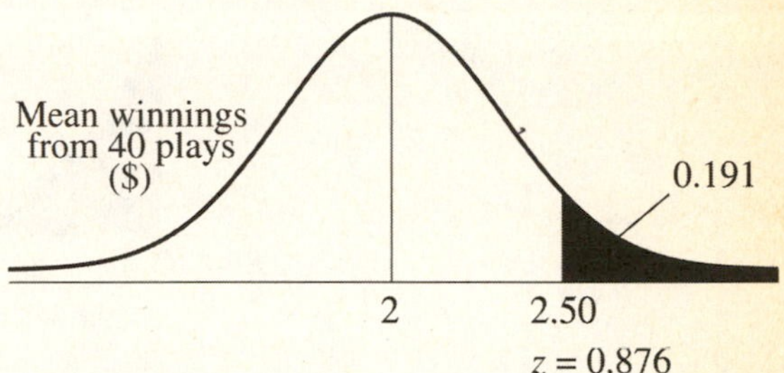

43. AP Stats 2006.

a) $\mu = 5(0.126) + 4(0.222) + 3(0.253) + 2(0.183) + 1(0.216) \approx 2.859$

$$\sigma = \sqrt{\begin{aligned}&(5-2.859)^2(0.126) + (4-2.859)^2(0.222) + (3-2.859)^2(0.253)\\ &+ (2-2.859)^2(0.183) + (1-2.859)^2(0.216)\end{aligned}} \approx 1.324$$

The calculation for standard deviation is based on a rounded mean. Use technology to calculate the mean and standard deviation to avoid inaccuracy.

b) The distribution of scores for 40 randomly selected students would not follow a Normal model. The distribution would resemble the population, mostly uniform for scores 1 – 4, with about half as many 5s.

c) **Randomization condition:** The scores are selected randomly.

Independence assumption: It is reasonable to think that the randomly selected scores are independent of one another.

10% condition: The 40 scores represent less than 10% of all scores.

Large Enough Sample condition: A sample of 40 scores is large enough.

Since the conditions are satisfied, the sampling distribution model for the mean of 40 randomly selected AP Stat scores is Normal, with $\mu_{\bar{y}} = \mu \approx 2.859$ and standard deviation $SD(\bar{y}) = \frac{\sigma}{\sqrt{n}} = \frac{1.324}{\sqrt{40}} \approx 0.2093$.

45. AP Stats 2006, again.

Since the teacher considers his 63 students "typical", and 63 is less than 10% of all students, the sampling distribution model for the mean AP Stat score for 63 students is Normal, with mean $\mu_{\bar{y}} = \mu \approx 2.859$ and standard deviation

$SD(\bar{y}) = \frac{\sigma}{\sqrt{n}} = \frac{1.324}{\sqrt{63}} \approx 0.167$.

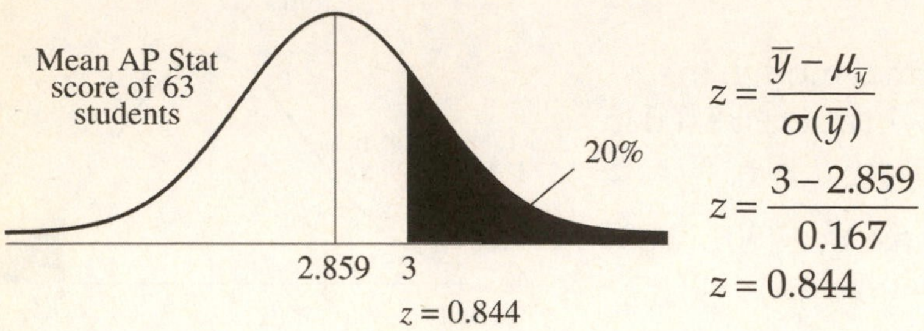

$$z = \frac{\bar{y} - \mu_{\bar{y}}}{\sigma(\bar{y})}$$

$$z = \frac{3 - 2.859}{0.167}$$

$$z = 0.844$$

According to the sampling distribution model, the probability that the class of 63 students achieves an average of 3 on the AP Stat exam is about 20%.

47. Pollution.

a) **Randomization condition:** Assume that the 80 cars can be considered a representative sample of all cars of this type.

Independence assumption: It is reasonable to think that the CO emissions for these cars are mutually independent.

10% condition: The 80 cars in the fleet certainly represent less than 10% of all cars of this type.

Large Enough Sample condition: A sample of 80 cars is large enough.

The mean CO level was $\mu = 2.9$ gm/mi, with standard deviation $\sigma = 0.4$ gm/mi. Since the conditions are met, the CLT allows us to model the sampling distribution of the $\bar{y}$ with a Normal model, with $\mu_{\bar{y}} = 2.9$ gm/mi and standard deviation $SD(\bar{y}) = \frac{0.4}{\sqrt{80}} = 0.045$ gm/mi.

b) According to the Normal model, the probability that $\bar{y}$ is between 3.0 and 3.1 gm/mi is approximately 0.0131.

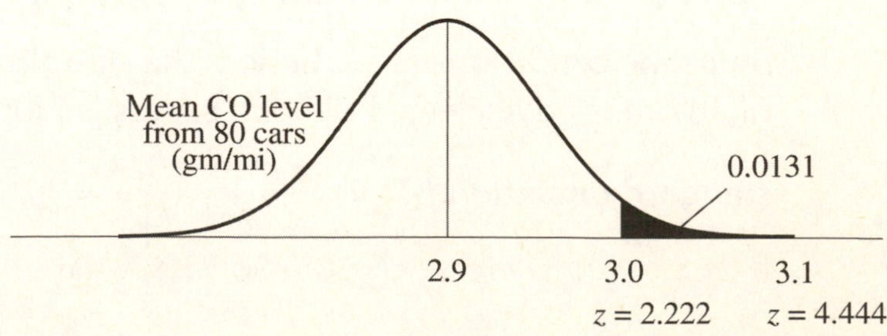

c) According to the Normal model, there is only a 5% chance that the fleet's mean CO level is greater than approximately 2.97 gm/mi.

$$z = \frac{\bar{y} - \mu_{\bar{y}}}{\sigma(\bar{y})}$$

$$1.645 = \frac{\bar{y} - 2.9}{0.045}$$

$$\bar{y} \approx 2.97$$

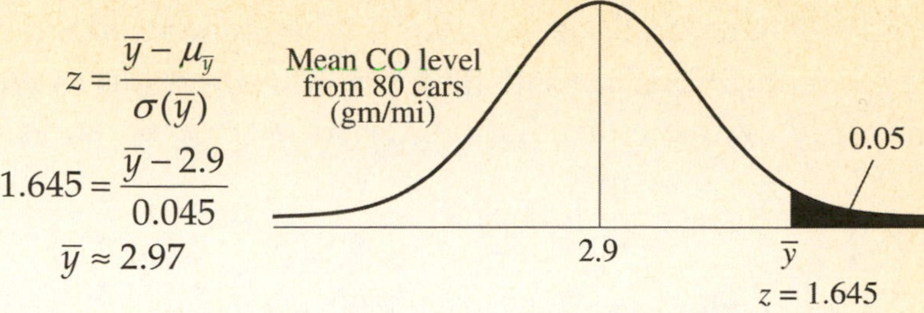

49. Tips.

a) Since the distribution of tips is skewed to the right, we can't use the Normal model to determine the probability that a given party will tip at least $20.

b) No. A sample of 4 parties is probably not a large enough sample for the CLT to allow us to use the Normal model to estimate the distribution of averages.

c) A sample of 10 parties may not be large enough to allow the use of a Normal model to describe the distribution of averages. It would be risky to attempt to estimate the probability that his next 10 parties tip an average of $15. However, since the distribution of tips has $\mu = \$9.60$, with standard deviation $\sigma = \$5.40$, we still know that the mean of the sampling distribution model is $\mu_{\bar{y}} = \$9.60$ with standard deviation $SD(\bar{y}) = \frac{5.40}{\sqrt{10}} \approx \$1.71$.

We don't know the exact shape of the distribution, but we can still assess the likelihood of specific means. A mean tip of $15 is over 3 standard deviations above the expected mean tip for 10 parties. That's not very likely to happen.

According to the Normal model, the probability the mean Sunday purchase of 50 customers is at least $40 is about 0.0023.

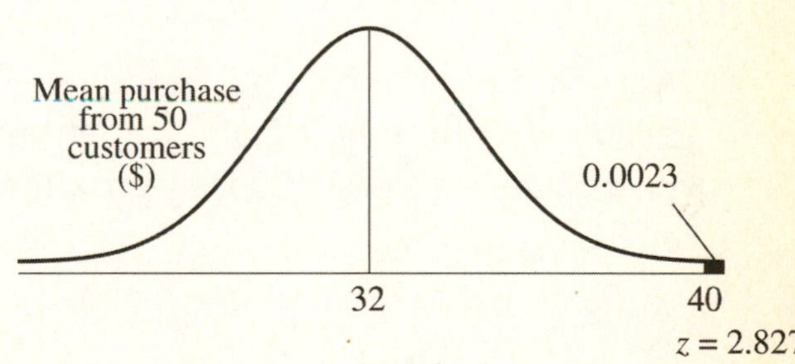

51. More tips.

a) Randomization condition: Assume that the tips from 40 parties can be considered a representative sample of all tips.

Independence assumption: It is reasonable to think that the tips are mutually independent, unless the service is particularly good or bad during this weekend.

10% condition: The tips of 40 parties certainly represent less than 10% of all tips.

Large Enough Sample condition: The sample of 40 parties is large enough.

The mean tip is $\mu = \$9.60$, with standard deviation $\sigma = \$5.40$. Since the conditions are satisfied, the CLT allows us to model the sampling distribution of $\bar{y}$ with a Normal model, with $\mu_{\bar{y}} = \$9.60$ and standard deviation

$$SD(\bar{y}) = \frac{5.40}{\sqrt{40}} \approx \$0.8538.$$

In order to earn at least \$500, the waiter would have to average $\frac{500}{40} = \$12.50$ per party.

According to the Normal model, the probability that the waiter earns at least \$500 in tips in a weekend is approximately 0.0003.

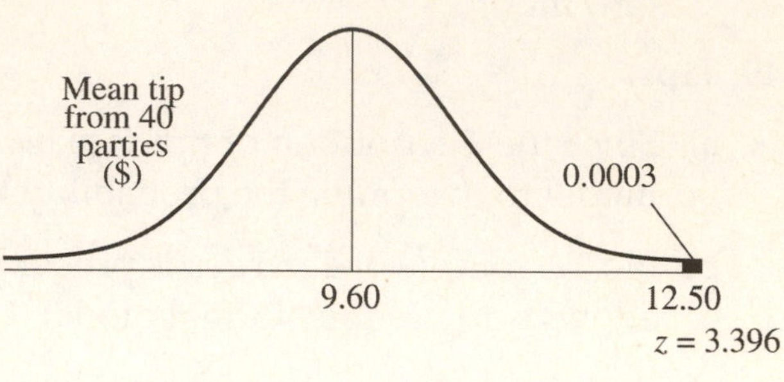

b) According to the Normal model, the waiter can expect to have a mean tip of about \$10.6942, which corresponds to about \$427.77 for 40 parties, in the best 10% of such weekends.

$$z = \frac{\bar{y} - \mu_{\bar{y}}}{\sigma(\bar{y})}$$

$$1.2816 = \frac{\bar{y} - 9.60}{0.8538}$$

$$\bar{y} \approx 10.6942$$

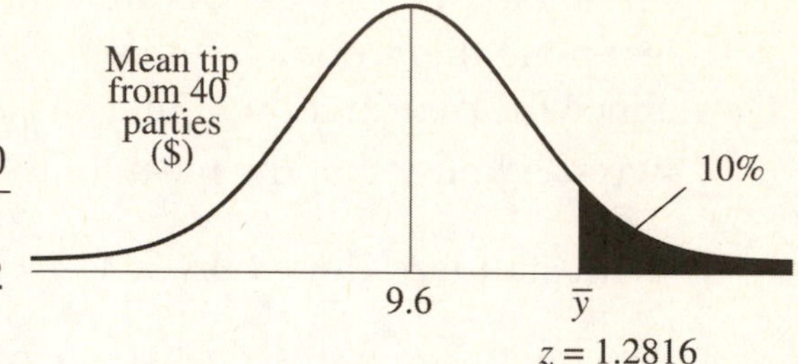

53. IQs.

a) According to the Normal model, the probability that the IQ of a student from East State is at least 125 is approximately 0.734.

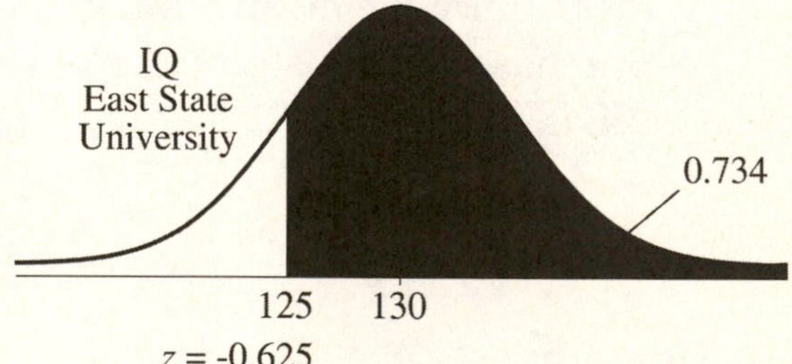

b) First, we will need to generate a model for the difference in IQ between the two schools. Since we are choosing at random, it is reasonable to believe that the students' IQs are independent, which allows us to calculate the standard deviation of the difference.

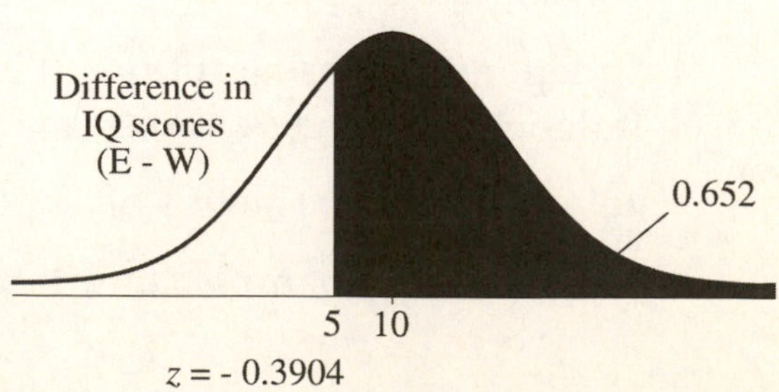

$\mu = E(E - W) = E(E) - E(W) = 130 - 120 = 10$

$\sigma = SD(E - W) = \sqrt{Var(E) + Var(W)}$

$= \sqrt{8^2 + 10^2} \approx 12.806$

Since both distributions are Normal, the distribution of the difference is $N(10, 12.806)$.

According to the Normal model, the probability that the IQ of a student at ESU is at least 5 points higher than a student at WSU is approximately 0.652.

c) **Randomization condition:** Students are randomly sampled from WSU.

Independence assumption: It is reasonable to think that the IQs are mutually independent.

10% condition: The 3 students certainly represent less than 10% of students.

Large Enough Sample condition: The distribution of IQs is Normal, so the distribution of sample means of samples of any size will be Normal, so a sample of 3 students is large enough.

The mean IQ is $\mu_w = 120$, with standard deviation $\sigma_w = 10$. Since the distribution IQs is Normal, we can model the sampling distribution of $\overline{w}$ with a Normal model, with $\mu_{\overline{w}} = 120$ with standard deviation $\sigma(\overline{w}) = \frac{10}{\sqrt{3}} \approx 5.7735.$

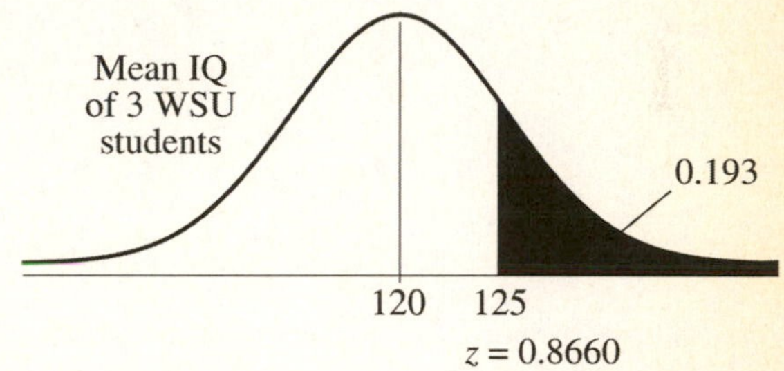

According to the Normal model, the probability that the mean IQ of the 3 WSU students is above 125 is approximately 0.193.

d) As in part c, the sampling distribution of $\overline{e}$, the mean IQ of 3 ESU students, can be modeled with a Normal model, with $\mu_{\overline{e}} = 130$ with standard deviation

$SD(\overline{e}) = \frac{8}{\sqrt{3}} \approx 4.6188.$

The distribution of the difference in mean IQ is Normal, with the following parameters:

$\mu_{\overline{e}-\overline{w}} = E(\overline{e} - \overline{w}) = E(\overline{e}) - E(\overline{w}) = 130 - 120 = 10$

$\sigma_{\overline{e}-\overline{w}} = SD(\overline{e} - \overline{w}) = \sqrt{Var(\overline{e}) + Var(\overline{w})}$

$= \sqrt{\left(\frac{10}{\sqrt{3}}\right)^2 + \left(\frac{8}{\sqrt{3}}\right)^2} \approx 7.3937$

According to the Normal model, the probability that the mean IQ of 3 ESU students is at least 5 points higher than the mean IQ of 3 WSU students is approximately 0.751.

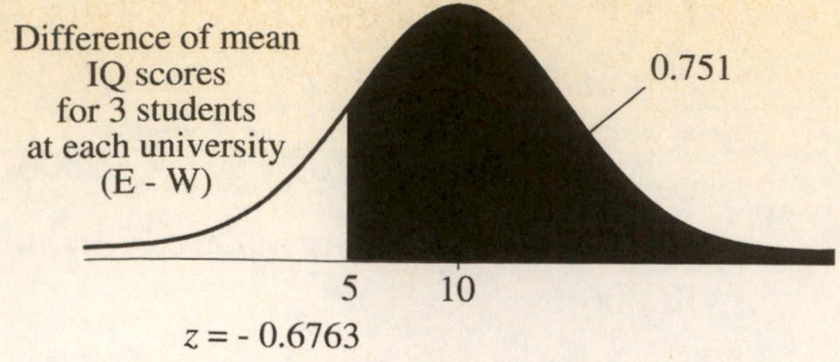

Chapter 19 – Confidence Intervals for Proportions

1. **Margin of error.**

   He believes the true proportion of voters with a certain opinion is within 4% of his estimate, with some degree of confidence, perhaps 95% confidence.

3. **Conditions.**

   a) *Population* – all cars; *sample* – 134 cars actually stopped at the checkpoint; p – proportion of all cars with safety problems; $\hat{p}$ – proportion of cars in the sample that actually have safety problems (10.4%).
   Plausible Independence condition: There is no reason to believe that the safety problems of cars are related to each other.
   Randomization condition: This sample is not random, so hopefully the cars stopped are representative of cars in the area.
   10% condition: The 134 cars stopped represent a small fraction of all cars, certainly less than 10%.
   Success/Failure condition: $n\hat{p} = 14$ and $n\hat{q} = 120$ are both greater than 10, so the sample is large enough.
   A one-proportion z-interval can be created for the proportion of all cars in the area with safety problems.

   b) *Population* – the general public; *sample* – 602 viewers that logged on to the Web site;
   p –proportion of the general public that support prayer in school; $\hat{p}$ - proportion of viewers that logged on to the Web site and voted that support prayer in schools (81.1%). **Randomization condition:** This sample is not random, but biased by voluntary response.
   It would be very unwise to attempt to use this sample to infer anything about the opinion of the general public related to school prayer.

   c) *Population* – parents at the school; *sample* – 380 parents who returned surveys; p – proportion of all parents in favor of uniforms; $\hat{p}$ – proportion of those who responded that are in favor of uniforms (60%).
   Randomization condition: This sample is not random, but rather biased by nonresponse. There may be lurking variables that affect the opinions of parents who return surveys (and the children who deliver them!).

   It would be very unwise to attempt to use this sample to infer anything about the opinion of the parents about uniforms.

d) *Population* – all freshmen enrollees at the college (not just one year); *sample* – 1632 freshmen during the specified year; p – proportion of all students who will graduate on time; $\hat{p}$ – proportion of students from that year who graduate on time (85.05%).
Plausible independence condition: It is reasonable to think that the abilities of students to graduate on time are mutually independent.
Randomization condition: This sample is not random, but this year's freshmen class is probably representative of freshman classes in other years.
10% condition: The 1632 students in that years freshmen class represent less than 10% of all possible students.
Success/Failure condition: $n\hat{p} = 1388$ and $n\hat{q} = 244$ are both greater than 10, so the sample is large enough.
A one-proportion z-interval can be created for the proportion of freshmen that graduate on time from this college.

5. Conclusions.

a) Not correct. This statement implies certainty. There is no level of confidence in the statement.

b) Not correct. Different samples will give different results. Many fewer than 95% of samples are expected to have *exactly* 88% on-time orders.

c) Not correct. A confidence interval should say something about the unknown population proportion, not the sample proportion in different samples.

d) Not correct. We *know* that 88% of the orders arrived on time. There is no need to make an interval for the sample proportion.

e) Not correct. The interval should be about the proportion of on-time orders, not the days.

7. Confidence intervals.

a) False. For a given sample size, higher confidence means a *larger* margin of error.

b) True. Larger samples lead to smaller standard errors, which lead to smaller margins of error.

c) True. Larger samples are less variable, which makes us more confident that a given confidence interval succeeds in catching the population proportion.

d) False. The margin of error decreases as the square root of the sample size increases. Halving the margin of error requires a sample four times as large as the original.

9. Cars.

We are 90% confident that between 29.9% and 47.0% of cars are made in Japan.

11. Contaminated chicken.

a) $\hat{p} \pm z^* \sqrt{\frac{\hat{p}\hat{q}}{n}} = (0.83) \pm 1.960\sqrt{\frac{(0.83)(0.17)}{525}} = (0.798, 0.862)$

b) We are 95% confident that between 80% and 86% of all broiler chicken sold in U.S. food stores is infected with *campylobacter*.

c) The size of the population is irrelevant. If *Consumer Reports* had a random sample, 95% of intervals generated by studies like this are expected to capture the true contamination level.

13. Baseball fans.

a) $ME = z^* \times SE(\hat{p}) = z^* \times \sqrt{\frac{\hat{p}\hat{q}}{n}} = 1.645 \times \sqrt{\frac{(0.37)(0.63)}{1006}} \approx 0.025$

b) We're 90% confident that this poll's estimate is within 2.5% of the true proportion of people who are baseball fans.

c) The margin of error for 99% confidence would be larger. To be more certain, we must be less precise.

d) $ME = z^* \times SE(\hat{p}) = z^* \times \sqrt{\frac{\hat{p}\hat{q}}{n}} = 2.576 \times \sqrt{\frac{(0.37)(0.63)}{1006}} \approx 0.039$

e) Smaller margins of error involve less confidence. The narrower the confidence interval, the less likely we are to believe that we have succeeded in capturing the true proportion of people who are baseball fans.

f) These data provide no evidence of a change. With a margin of error of 2.5% for the 95% confidence interval, 37% is certainly still a plausible value for the proportion of people who are baseball fans.

15. Contributions please.

a) **Randomization condition:** Letters were sent to a random sample of 100,000 potential donors.
10% condition: We assume that the potential donor list has more than 1,000,000 names.
Success/Failure condition: $n\hat{p} = 4{,}781$ and $n\hat{q} = 95{,}219$ are both much greater than 10, so the sample is large enough.

$$\hat{p} \pm z^* \sqrt{\frac{\hat{p}\hat{q}}{n}} = \left(\frac{4{,}781}{100{,}000}\right) \pm 1.960\sqrt{\frac{\left(\frac{4{,}781}{100{,}000}\right)\left(\frac{95{,}219}{100{,}000}\right)}{100{,}000}} = (0.0465, 0.0491)$$

We are 95% confident that the between 4.65% and 4.91% of potential donors would donate.

b) The confidence interval gives the set of plausible values with 95% confidence. Since 5% is outside the interval, it seems to be a bit optimistic.

17. Teenage drivers.

a) **Independence assumption:** There is no reason to believe that accidents selected at random would be related to one another.
Randomization condition: The insurance company randomly selected 582 accidents.
10% condition: 582 accidents represent less than 10% of all accidents.
Success/Failure condition: $n\hat{p} = 91$ and $n\hat{q} = 491$ are both greater than 10, so the sample is large enough.

Since the conditions are met, we can use a one-proportion *z*-interval to estimate the percentage of accidents involving teenagers.

$$\hat{p} \pm z^* \sqrt{\frac{\hat{p}\hat{q}}{n}} = \left(\frac{91}{582}\right) \pm 1.960\sqrt{\frac{\left(\frac{91}{582}\right)\left(\frac{491}{582}\right)}{582}} = (12.7\%, 18.6\%)$$

b) We are 95% confident that between 12.7% and 18.6% of all accidents involve teenagers.

c) About 95% of random samples of size 582 will produce intervals that contain the true proportion of accidents involving teenagers.

d) Our confidence interval contradicts the assertion of the politician. The figure quoted by the politician, 1 out of every 5, or 20%, is outside the interval.

19. Safe food.

The grocer can conclude nothing about the opinions of all his customers from this survey. Those customers who bothered to fill out the survey represent a voluntary response sample, consisting of people who felt strongly one way or another about irradiated food. The random condition was not met.

21. Death penalty, again.

a) There may be response bias based on the wording of the question.

b) $\hat{p} \pm z^* \sqrt{\frac{\hat{p}\hat{q}}{n}} = (0.57) \pm 1.960\sqrt{\frac{(0.57)(0.43)}{1020}} = (54\%, 60\%)$

c) The margin of error based on the pooled sample is smaller, since the sample size is larger.

23. Rickets.

a) **Independence assumption:** It is reasonable to think that the randomly selected children are mutually independent in regards to vitamin D deficiency.
Randomization condition: The 2,700 children were chosen at random.
10% condition: 2,700 children are less than 10% of all English children.
Success/Failure condition: $n\hat{p} = (2{,}700)(0.20) = 540$ and $n\hat{q} = (2{,}700)(0.80) = 2160$ are both greater than 10, so the sample is large enough.

Since the conditions are met, we can use a one-proportion z-interval to estimate the proportion of the English children with vitamin D deficiency.

$$\hat{p} \pm z^* \sqrt{\frac{\hat{p}\hat{q}}{n}} = (0.20) \pm 2.326\sqrt{\frac{(0.20)(0.80)}{2700}} = (18.2\%, 21.8\%)$$

b) We are 98% confident that between 18.2% and 21.8% of English children are deficient in vitamin D.

c) About 98% of random samples of size 2,700 will produce confidence intervals that contain the true proportion of English children that are deficient in vitamin D.

25. Payments.

a) The confidence interval will be wider. The sample size is probably about one-fourth of the sample size of for all adults, so we would expect the confidence interval to be about twice as wide.

b) The second poll's margin of error should be smaller. The second poll used a slightly larger sample.

27. Deer ticks.

a) **Independence assumption:** Deer ticks are parasites. A deer carrying the parasite may spread it to others. Ticks may not be distributed evenly throughout the population.
Randomization condition: The sample is not random and may not represent all deer.
10% condition: 153 deer are less than 10% of all deer.
Success/Failure condition: $n\hat{p} = 32$ and $n\hat{q} = 121$ are both greater than 10, so the sample is large enough.

The conditions are not satisfied, so we should use caution when a one-proportion z-interval is used to estimate the proportion of deer carrying ticks.

$$\hat{p} \pm z^* \sqrt{\frac{\hat{p}\hat{q}}{n}} = \left(\frac{32}{153}\right) \pm 1.645\sqrt{\frac{\left(\frac{32}{153}\right)\left(\frac{121}{153}\right)}{153}} = (15.5\%, 26.3\%)$$

We are 90% confident that between 15.5% and 26.3% of deer have ticks.

b) In order to cut the margin of error in half, they must sample 4 times as many deer.
4(153) = 612 deer.

c) The incidence of deer ticks is not plausibly independent, and the sample may not be representative of all deer, since females and young deer are usually not hunted.

29. Graduation.

a)

$$ME = z^*\sqrt{\frac{\hat{p}\hat{q}}{n}}$$

$$0.06 = 1.645\sqrt{\frac{(0.25)(0.75)}{n}}$$

$$n = \frac{(1.645)^2(0.25)(0.75)}{(0.06)^2}$$

$$n \approx 141 \text{ people}$$

In order to estimate the proportion of non-graduates in the 25-to 30-year-old age group to within 6% with 90% confidence, we would need a sample of at least 141 people. All decimals in the final answer must be rounded up, to the next person.
(For a more cautious answer, let $\hat{p} = \hat{q} = 0.5$. This method results in a required sample of 188 people.)

b)

$$ME = z^*\sqrt{\frac{\hat{p}\hat{q}}{n}}$$

$$0.04 = 1.645\sqrt{\frac{(0.25)(0.75)}{n}}$$

$$n = \frac{(1.645)^2(0.25)(0.75)}{(0.04)^2}$$

$$n \approx 318 \text{ people}$$

In order to estimate the proportion of non-graduates in the 25-to 30-year-old age group to within 4% with 90% confidence, we would need a sample of at least 318 people. All decimals in the final answer must be rounded up, to the next person.
(For a more cautious answer, let $\hat{p} = \hat{q} = 0.5$. This method results in a required sample of 423 people.)
Alternatively, the margin of error is now 2/3 of the original, so the sample size must be increased by a factor of 9/4. 141(9/4) ≈ 318 people.

c)

$$ME = z^*\sqrt{\frac{\hat{p}\hat{q}}{n}}$$

$$0.03 = 1.645\sqrt{\frac{(0.25)(0.75)}{n}}$$

$$n = \frac{(1.645)^2(0.25)(0.75)}{(0.03)^2}$$

$$n \approx 564 \text{ people}$$

In order estimate the proportion of non-graduates in the 25-to 30-year-old age group to within 3% with 90% confidence, we would need a sample of at least 564 people. All decimals in the final answer must be rounded up, to the next person.
(For a more cautious answer, let $\hat{p} = \hat{q} = 0.5$. This method results in a required sample of 752 people.)

Alternatively, the margin of error is now half that of the original, so the sample size must be increased by a factor of 4. 141(4) ≈ 564 people.

31. Graduation, again.

$$ME = z^*\sqrt{\frac{\hat{p}\hat{q}}{n}}$$

$$0.02 = 1.960\sqrt{\frac{(0.25)(0.75)}{n}}$$

$$n = \frac{(1.960)^2(0.25)(0.75)}{(0.02)^2}$$

$$n \approx 1{,}801 \text{ people}$$

In order to estimate the proportion of non-graduates in the 25-to 30-year-old age group to within 2% with 95% confidence, we would need a sample of at least 1,801 people. All decimals in the final answer must be rounded up, to the next person.
(For a more cautious answer, let $\hat{p} = \hat{q} = 0.5$. This method results in a required sample of 2,401 people.)

33. Pilot study.

$$ME = z^*\sqrt{\frac{\hat{p}\hat{q}}{n}}$$

$$0.03 = 1.645\sqrt{\frac{(0.15)(0.85)}{n}}$$

$$n = \frac{(1.645)^2(0.15)(0.85)}{(0.03)^2}$$

$$n \approx 384 \text{ cars}$$

Use $\hat{p} = \frac{9}{60} = 0.15$ from the pilot study as an estimate.

In order to estimate the percentage of cars with faulty emissions systems to within 3% with 90% confidence, the state's environmental agency will need a sample of at least 384 cars. All decimals in the final answer must be rounded up, to the next car.

35. Approval rating.

$$ME = z^*\sqrt{\frac{\hat{p}\hat{q}}{n}}$$

$$0.025 = z^*\sqrt{\frac{(0.65)(0.35)}{972}}$$

$$z^* = \frac{0.025}{\sqrt{\frac{(0.65)(0.35)}{972}}}$$

$$z^* \approx 1.634$$

Since $z^* \approx 1.634$, which is close to 1.645, the pollsters were probably using 90% confidence. The slight difference in the z^* values is due to rounding of the governor's approval rating.

Chapter 20 – Testing Hypotheses about Proportions

1. Hypotheses.

a) H_0 : The governor's "negatives" are 30%. ($p = 0.30$)
H_A : The governor's "negatives" are less than 30%. ($p < 0.30$)

b) H_0 : The proportion of heads is 50%. ($p = 0.50$)
H_A : The proportion of heads is not 50%. ($p \neq 0.50$)

c) H_0 : The proportion of people who quit smoking is 20%. ($p = 0.20$)
H_A : The proportion of people who quit smoking is greater than 20%. ($p > 0.20$)

3. Negatives.

Statement d is the correct interpretation of a *P*-value.

5. Relief.

It is *not* reasonable to conclude that the new formula and the old one are equally effective. Furthermore, our inability to make that conclusion has nothing to do with the *P*-value. We can not prove the null hypothesis (that the new formula and the old formula are equally effective), but can only fail to find evidence that would cause us to reject it. All we can say about this *P*-value is that there is a 27% chance of seeing the observed effectiveness from natural sampling variation if the new formula and the old one are equally effective.

7. He cheats!

a) Two losses in a row aren't convincing. There is a 25% chance of losing twice in a row, and that is not unusual.

b) If the process is fair, three losses in a row can be expected to happen about 12.5% of the time. $(0.5)(0.5)(0.5) = 0.125$.

c) Three losses in a row is still not a convincing occurrence. We'd expect that to happen about once every eight times we tossed a coin three times.

d) Answers may vary. Maybe 5 times would be convincing. The chances of 5 losses in a row are only 1 in 32, which seems unusual.

9. Cell phones.

1) Null and alternative hypotheses should involve p, not $\hat{p}$.

2) The question is about *failing* to meet the goal. H_A should be $p < 0.96$.

3) The student failed to check $nq = (200)(0.04) = 8$. Since $nq < 10$, the Success/Failure condition is violated. Similarly, the 10% Condition is not verified.

4) $SD(\hat{p}) = \sqrt{\frac{pq}{n}} = \sqrt{\frac{(0.96)(0.04)}{200}} \approx 0.014$. The student used $\hat{p}$ and $\hat{q}$.

5) Value of z is incorrect. The correct value is $z = \frac{0.94 - 0.96}{0.014} \approx -1.43$.

6) P-value is incorrect. $P = P(z < -1.43) = 0.076$

7) For the P-value given, an incorrect conclusion is drawn. A P-value of 0.12 provides no evidence that the new system has failed to meet the goal. The correct conclusion for the corrected P-value is: Since the P-value of 0.076 is fairly low, there is weak evidence that the new system has failed to meet the goal.

11. Dowsing.

a) H_0 : The percentage of successful wells drilled by the dowser is 30%. ($p = 0.30$)
H_A : The percentage of successful wells drilled is greater than 30%. ($p > 0.30$)

b) **Independence assumption:** There is no reason to think that finding water in one well will affect the probability that water is found in another, unless the wells are close enough to be fed by the same underground water source.
Randomization condition: This sample is not random, so hopefully the customers you check with are representative of all of the dowser's customers.
10% condition: The 80 customers sampled may be considered less than 10% of all possible customers.
Success/Failure condition: $np = (80)(0.30) = 24$ and $nq = (80)(0.70) = 56$ are both greater than 10, so the sample is large enough.

c) The sample of customers may not be representative of all customers, so we will proceed cautiously. A Normal model can be used to model the sampling distribution of the proportion, with $\mu_{\hat{p}} = p = 0.30$ and

$SD(\hat{p}) = \sqrt{\frac{pq}{n}} = \sqrt{\frac{(0.30)(0.70)}{80}} \approx 0.0512$.

We can perform a one-proportion z-test. The observed proportion of successful wells is $\hat{p} = \frac{27}{80} = 0.3375$.

$$z = \frac{\hat{p} - p_0}{\sqrt{\frac{pq}{n}}}$$

$$z = \frac{0.3375 - 0.30}{\sqrt{\frac{(0.30)(0.70)}{80}}}$$

$$z \approx 0.73$$

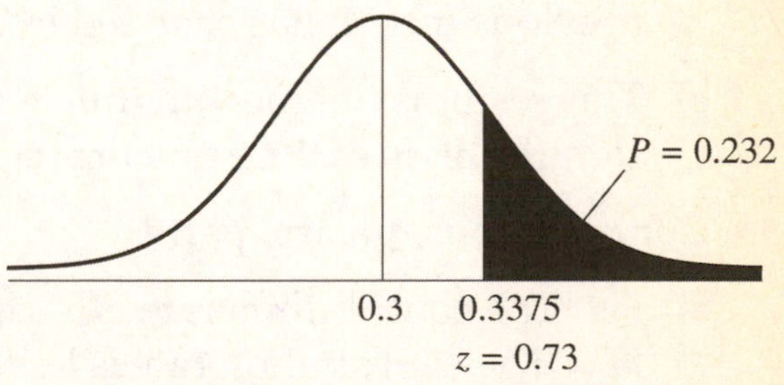

P = 0.232
0.3 0.3375
z = 0.73

d) If his dowsing has the same success rate as standard drilling methods, there is more than a 23% chance of seeing results as good as those of the dowser, or better, by natural sampling variation.

e) With a *P*-value of 0.232, we fail to reject the null hypothesis. There is no evidence to suggest that the dowser has a success rate any higher than 30%.

13. Absentees.

a) H_0 : The percentage of students in 2000 with perfect attendance the previous month is 34% ($p = 0.34$)
H_A : The percentage of students in 2000 with perfect attendance the previous month is different from 34% ($p \neq 0.34$)

b) **Independence assumption:** It is reasonable to think that the students' attendance records are independent of one another.
Randomization condition: Although not specifically stated, we can assume that the National Center for Educational Statistics used random sampling.
10% condition: The 8302 students are less than 10% of all students.
Success/Failure condition: $np = (8302)(0.34) = 2822.68$ and $nq = (8302)(0.66) = 5479.32$ are both greater than 10, so the sample is large enough.

c) Since the conditions for inference are met, a Normal model can be used to model the sampling distribution of the proportion, with $\mu_{\hat{p}} = p = 0.34$ and

$$SD(\hat{p}) = \sqrt{\frac{pq}{n}} = \sqrt{\frac{(0.34)(0.66)}{8302}} \approx 0.0052$$

We can perform a two-tailed one-proportion *z*-test. The observed proportion of perfect attendees is $\hat{p} = 0.33$.

d) With a *P*-value of 0.0544, we reject the null hypothesis. There is some evidence to suggest that the percentage of students with perfect attendance in the previous month has changed in 2000.

$$z = \frac{\hat{p} - p_0}{\sqrt{\frac{pq}{n}}}$$

$$z = \frac{0.33 - 0.34}{\sqrt{\frac{(0.34)(0.66)}{8302}}}$$

$$z \approx -1.923$$

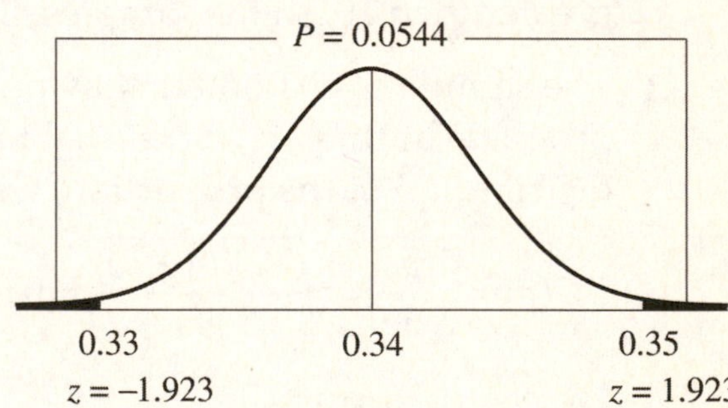

e) This result is not meaningful. A difference this small, although statistically significant, is of little practical significance.

15. Contributions, please, part II.

a) H_0 : The contribution rate is 5% ($p = 0.05$)
H_A : The contribution rate is less than 5% ($p < 0.05$)

b) **Independence assumption:** There is no reason to believe that one randomly selected potential donor's decision will affect another's decision.
Randomization condition: Potential donors were randomly selected.
10% condition: We will assume the entire mailing list has over 1,000,000 names.

Success/Failure condition: $np = 5000$ and $nq = 95{,}000$ are both greater than 10, so the sample is large enough.

The conditions have been satisfied, so a Normal model can be used to model the sampling distribution of the proportion, with $\mu_{\hat{p}} = p = 0.05$ and

$$SD(\hat{p}) = \sqrt{\frac{pq}{n}} = \sqrt{\frac{(0.05)(0.95)}{100{,}000}} \approx 0.0007.$$

We can perform a one-proportion z-test. The observed contribution rate is

$$\hat{p} = \frac{4{,}781}{100{,}000} = 0.04781.$$

c) Since the P-value = 0.0006 is low, we reject the null hypothesis. There is strong evidence that contribution rate for all potential donors is lower than 5%.

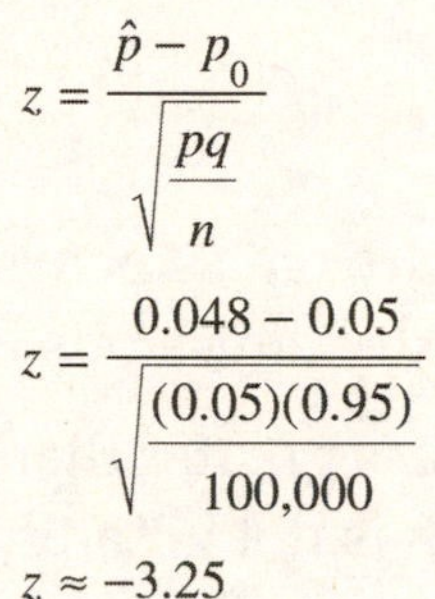

$$z = \frac{\hat{p} - p_0}{\sqrt{\frac{pq}{n}}}$$

$$z = \frac{0.048 - 0.05}{\sqrt{\frac{(0.05)(0.95)}{100{,}000}}}$$

$$z \approx -3.25$$

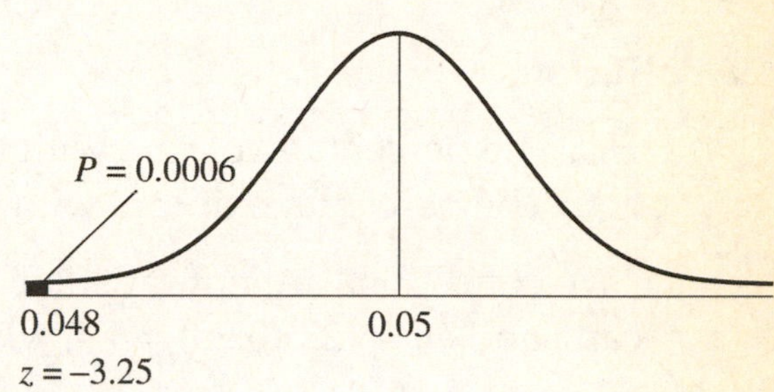

17. Law School.

a) H_0 : The law school acceptance rate for *LSATisfaction* is 63% ($p = 0.63$)
H_A : The law school acceptance rate is greater than 63% ($p > 0.63$)

b) **Randomization condition:** These 240 students may be considered representative of the population of law school applicants.
10% condition: There are certainly more than 2,400 law school applicants.
Success/Failure condition: $np = 151.2$ and $nq = 88.8$ are both greater than 10, so the sample is large enough.

The conditions have been satisfied, so a Normal model can be used to model the sampling distribution of the proportion, with $\mu_{\hat{p}} = p = 0.63$ and

$$SD(\hat{p}) = \sqrt{\frac{pq}{n}} = \sqrt{\frac{(0.63)(0.37)}{240}} \approx 0.0312.$$

We can perform a one-proportion z-test. The observed success rate is

$$\hat{p} = \frac{163}{240} = 0.6792.$$

$$z = \frac{\hat{p} - p_0}{\sqrt{\frac{pq}{n}}}$$

$$z = \frac{0.679 - 0.63}{\sqrt{\frac{(0.63)(0.37)}{240}}}$$

$$z \approx 1.58$$

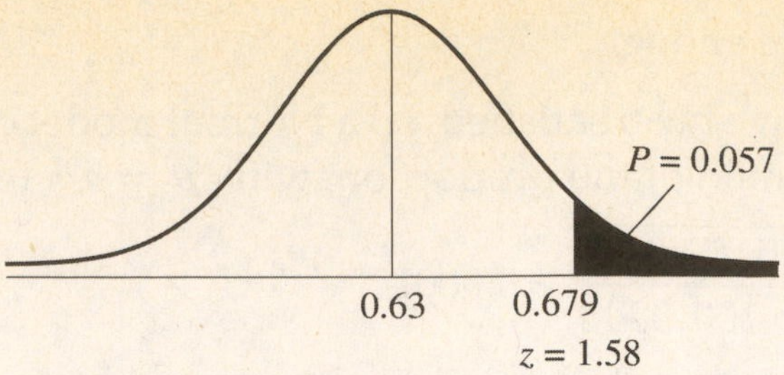

$P = 0.057$

0.63 0.679

$z = 1.58$

c) Since the *P*-value = 0.057 is fairly low, we reject the null hypothesis. There is weak evidence that the law school acceptance rate is higher for *LSATisfaction* applicants. Candidates should decide whether they can afford the time and expense.

19. Pollution.

H_0 : The percentage of cars with faulty emissions is 20%. ($p = 0.20$)
H_A : The percentage of cars with faulty emissions is greater than 20%. ($p > 0.20$)

Two conditions are not satisfied. 22 is greater than 10% of the population of 150 cars, and $np = (22)(0.20) = 4.4$, which is not greater than 10. It's not advisable to proceed with a test.

21. Twins.

H_0 : The percentage of twin births to teenage girls is 3%. ($p = 0.03$)
H_A : The percentage of twin births to teenage girls differs from 3%. ($p \neq 0.03$)

Independence assumption: One mother having twins will not affect another. Observations are plausibly independent.
Randomization condition: This sample may not be random, but it is reasonable to think that this hospital has a representative sample of teen mothers, with regards to twin births.
10% condition: The sample of 469 teenage mothers is less than 10% of all such mothers.
Success/Failure condition: $np = (469)(0.03) = 14.07$ and $nq = (469)(0.97) = 454.93$ are both greater than 10, so the sample is large enough.

The conditions have been satisfied, so a Normal model can be used to model the sampling distribution of the proportion, with $\mu_{\hat{p}} = p = 0.03$ and

$$SD(\hat{p}) = \sqrt{\frac{pq}{n}} = \sqrt{\frac{(0.03)(0.97)}{469}} \approx 0.0079.$$

We can perform a one-proportion *z*-test. The observed proportion of twin births to teenage mothers is $\hat{p} = \frac{7}{469} \approx 0.015$.

Since the *P*-value = 0.0556 is fairly low, we reject the null hypothesis. There is some evidence that the proportion of twin births for teenage mothers at this large city hospital is lower than the proportion of twin births for all mothers.

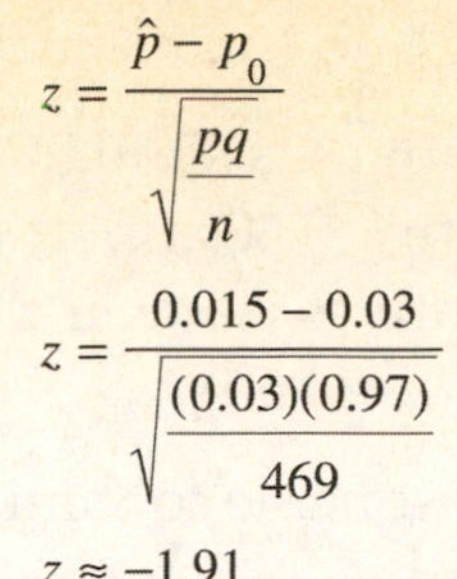

$$z = \frac{\hat{p} - p_0}{\sqrt{\frac{pq}{n}}}$$

$$z = \frac{0.015 - 0.03}{\sqrt{\frac{(0.03)(0.97)}{469}}}$$

$$z \approx -1.91$$

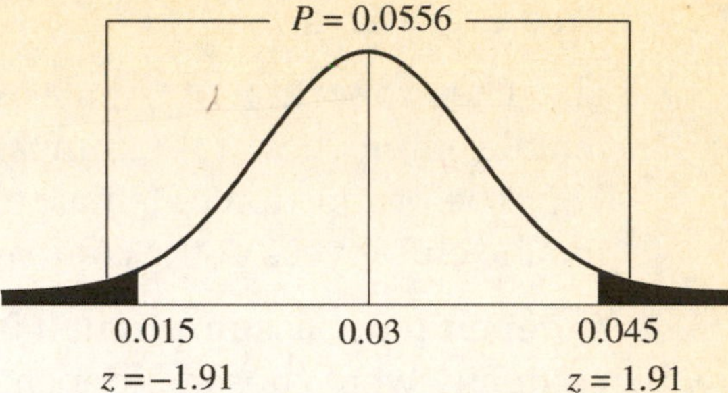

23. Webzine.

H_0 : The percentage of readers interested in an online edition is 25%. ($p = 0.25$)
H_A : The percentage of readers interested is greater than 25%. ($p > 0.25$)

Independence assumption: Interest of one reader should not affect interest of other readers.
Randomization condition: The magazine conducted an SRS of 500 current readers.
10% condition: 500 readers are less than 10% of all potential subscribers.
Success/Failure condition: $np = (500)(0.25) = 125$ and $nq = (500)(0.75) = 375$ are both greater than 10, so the sample is large enough.

The conditions have been satisfied, so a Normal model can be used to model the sampling distribution of the proportion, with $\mu_{\hat{p}} = p = 0.25$ and

$$SD(\hat{p}) = \sqrt{\frac{pq}{n}} = \sqrt{\frac{(0.25)(0.75)}{500}} \approx 0.0194.$$

We can perform a one-proportion *z*-test. The observed proportion of interested readers is $\hat{p} = \frac{137}{500} = 0.274$.

Since the *P*-value = 0.1076 is high, we fail to reject the null hypothesis. There is little evidence to suggest that the proportion of interested readers is greater than 25%. The magazine should not publish the online edition.

$$z = \frac{\hat{p} - p_0}{\sqrt{\frac{pq}{n}}}$$

$$z = \frac{0.274 - 0.25}{\sqrt{\frac{(0.25)(0.75)}{500}}}$$

$$z \approx 1.24$$

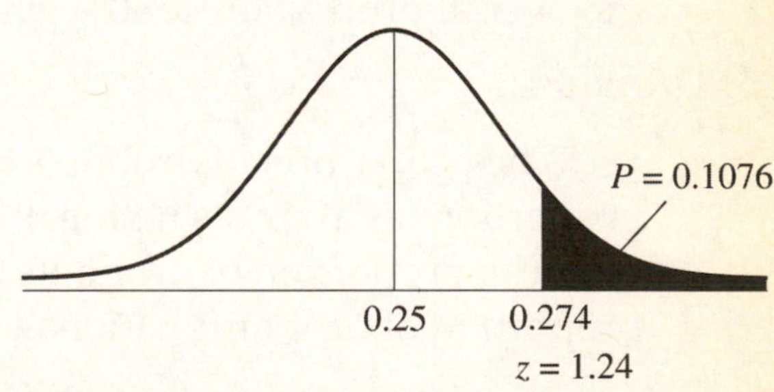

25. Women executives.

H_0 : The proportion of female executives is similar to the overall proportion of female employees at the company. ($p = 0.40$)
H_A : The proportion of female executives is lower than the overall proportion of female employees at the company. ($p < 0.40$)

Independence assumption: It is reasonable to think that executives at this company were chosen independently.
Randomization condition: The executives were not chosen randomly, but it is reasonable to think of these executives as representative of all potential executives over many years.
10% condition: 43 executives are less than 10% of all executives at the company.
Success/Failure condition: $np = (43)(0.40) = 17.2$ and $nq = (43)(0.60) = 25.8$ are both greater than 10, so the sample is large enough.

The conditions have been satisfied, so a Normal model can be used to model the sampling distribution of the proportion, with $\mu_{\hat{p}} = p = 0.40$ and

$$SD(\hat{p}) = \sqrt{\frac{pq}{n}} = \sqrt{\frac{(0.40)(0.60)}{43}} \approx 0.0747.$$

Perform a one-proportion z-test. The observed proportion is $\hat{p} = \frac{13}{43} \approx 0.302$.

$$z = \frac{\hat{p} - p_0}{\sqrt{\frac{pq}{n}}}$$

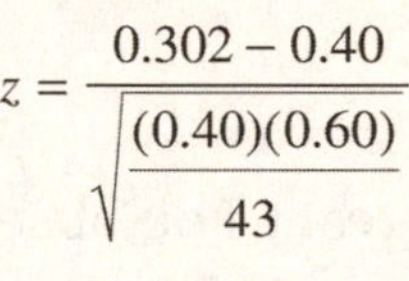

$$z = \frac{0.302 - 0.40}{\sqrt{\frac{(0.40)(0.60)}{43}}}$$

$$z \approx -1.31$$

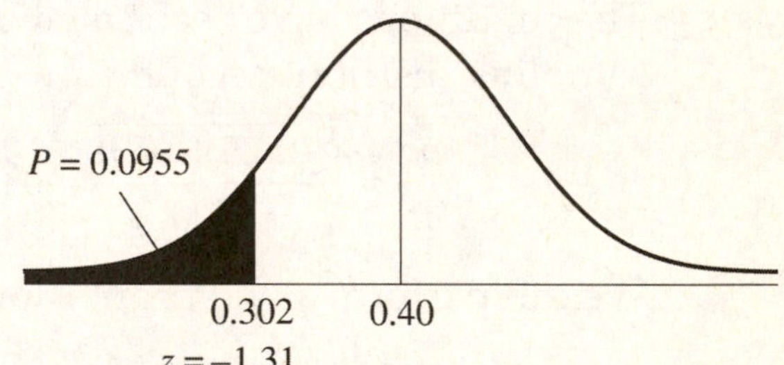

Since the P-value = 0.0955 is high, we fail to reject the null hypothesis. There is little evidence to suggest proportion of female executives is any different from the overall proportion of 40% female employees at the company.

27. Dropouts.

H_0 : The proportion of dropouts at this high school is similar to 10.3%, the proportion of dropouts nationally. ($p = 0.103$)
H_A : The proportion of dropouts at this high school is greater than 10.3%, the proportion of dropouts nationally. ($p > 0.103$)

Independence assumption /Randomization condition: Assume that the students at this high school are representative of all students nationally. This is really what we are testing. The dropout rate at this high school has traditionally been close to the national rate. If we reject the null hypothesis, we will have evidence that the dropout rate at this high school is no longer close to the national rate.

10% condition: 1782 students are less than 10% of all students nationally.
Success/Failure condition: $np = (1782)(0.103) = 183.546$ and $nq = (1782)(0.897) = 1598.454$ are both greater than 10, so the sample is large enough.

The conditions have been satisfied, so a Normal model can be used to model the sampling distribution of the proportion, $\mu_{\hat{p}} = p = 0.103$ and

$$SD(\hat{p}) = \sqrt{\frac{pq}{n}} = \sqrt{\frac{(0.103)(0.897)}{1782}} \approx 0.0072.$$

We can perform a one-proportion z-test. The observed proportion of dropouts is $\hat{p} = \frac{210}{1782} \approx 0.117845$.

Since the P-value = 0.02 is low, we reject the null hypothesis. There is evidence that the dropout rate at this high school is higher than 10.3%.

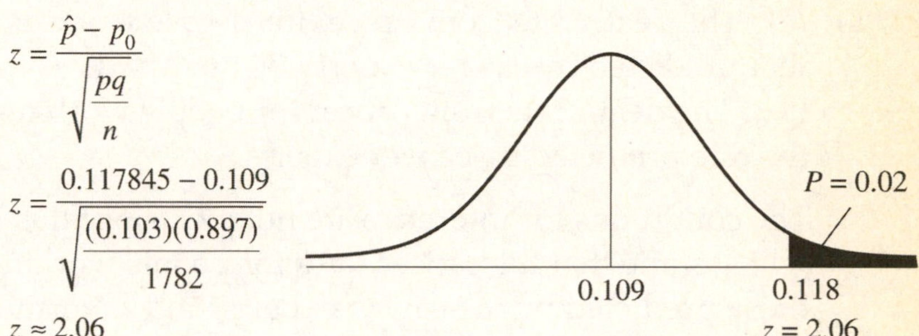

29. Lost luggage.

H_0: The proportion of lost luggage returned the next day is 90%. ($p = 0.90$)
H_A: The proportion of lost luggage returned is lower than 90%. ($p < 0.90$)

Independence assumption: It is reasonable to think that the people surveyed were independent with regards to their luggage woes.
Randomization condition: Although not stated, we will hope that the survey was conducted randomly, or at least that these air travelers are representative of all air travelers for that airline.
10% condition: 122 air travelers are less than 10% of all air travelers on the airline.
Success/Failure condition: $np = (122)(0.90) = 109.8$ and $nq = (122)(0.10) = 12.2$ are both greater than 10, so the sample is large enough.

The conditions have been satisfied, so a Normal model can be used to model the sampling distribution of the proportion, with $\mu_{\hat{p}} = p = 0.90$ and

$$SD(\hat{p}) = \sqrt{\frac{pq}{n}} = \sqrt{\frac{(0.90)(0.10)}{122}} \approx 0.0272.$$

We can perform a one-proportion z-test. The observed proportion of dropouts is $\hat{p} = \frac{103}{122} \approx 0.844$.

Since the P-value = 0.0201 is low, we reject the null hypothesis. There is evidence that the proportion of lost luggage returned the next day is lower than the 90% claimed by the airline.

$$z = \frac{\hat{p} - p_0}{\sqrt{\frac{pq}{n}}}$$

$$z = \frac{0.844 - 0.90}{\sqrt{\frac{(0.90)(0.10)}{122}}}$$

$$z \approx -2.05$$

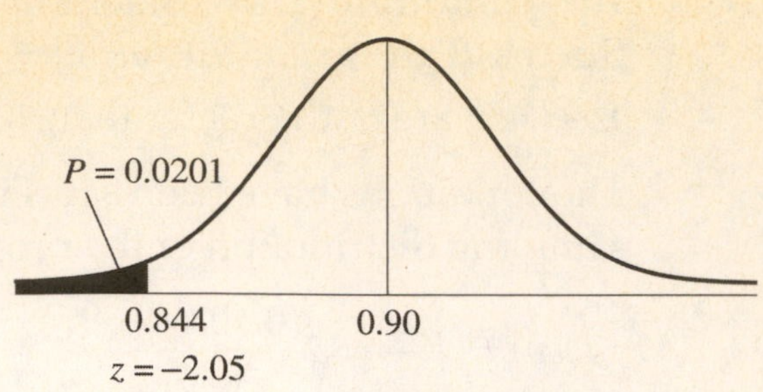

31. John Wayne.

a) H_0 : The death rate from cancer for people working on the film was similar to that predicted by cancer experts, 30 out of 220.
H_A : The death rate from cancer for people working on the film was higher than the rate predicted by cancer experts.

The conditions for inference are not met, since this is not a random sample. We will assume that the cancer rates for people working on the film are similar to those predicted by the cancer experts, and a Normal model can be used to model the sampling distribution of the rate, with $\mu_{\hat{p}} = p = 30/220$ and

$$SD(\hat{p}) = \sqrt{\frac{pq}{n}} = \sqrt{\frac{\left(\frac{30}{220}\right)\left(\frac{190}{220}\right)}{220}} \approx 0.0231.$$

We can perform a one-proportion z-test. The observed cancer rate is

$$\hat{p} = \frac{46}{220} \approx 0.209.$$

$$z = \frac{\hat{p} - p_0}{\sigma(\hat{p})}$$

$$z = \frac{\frac{46}{220} - \frac{30}{220}}{\sqrt{\frac{\left(\frac{30}{220}\right)\left(\frac{190}{220}\right)}{220}}}$$

$$z = 3.14$$

Since the P-value = 0.0008 is very low, we reject the null hypothesis. There is strong evidence that the cancer rate is higher than expected among the workers on the film.

b) This does not prove that exposure to radiation may increase the risk of cancer. This group of people may be atypical for reasons that have nothing to do with the radiation.

Chapter 21 – More About Tests and Intervals

1. One sided or two?

a) Two sided. Let p be the percentage of students who prefer Diet Coke.

H_0 : 50% of students prefer Diet Coke. ($p = 0.50$)
H_A : The percentage of students who prefer Diet Coke is not 50%. ($p \neq 0.50$)

b) One sided. Let p be the percentage of teenagers who prefer the new formulation.

H_0 : 50% of students prefer the new formulation. ($p = 0.50$)
H_A : More than 50%of students prefer the new formulation. ($p > 0.50$)

c) One sided. Let p be the percentage of people who plan to vote for the override.

H_0 : 2/3 of the residents intend to vote for the override. ($p = 2/3$)
H_A : More than 2/3 of the residents intend to vote for the override. ($p > 2/3$)

d) Two sided. Let p be the percentage of days the market goes up.

H_0 : The market goes up on 50% of days. ($p = 0.50$)
H_A : The percentage of days the market goes up is not 50%. ($p \neq 0.50$)

3. P-value.

If the effectiveness of the new poison ivy treatment is the same as the effectiveness of the old treatment, the chance of observing an effectiveness this large or larger in a sample of the same size is 4.7% by natural sampling variation alone.

5. Alpha.

Since the null hypothesis was rejected at $\alpha = 0.05$, the *P*-value for the researcher's test must have been less than 0.05. He would have made the same decision at $\alpha = 0.10$, since the *P*-value must also be less than 0.10. We can't be certain whether or not he would have made the same decision at $\alpha = 0.01$, since we only know that the *P*-value was less than 0.05. It may have been less than 0.01, but we can't be sure.

7. Significant?

a) If 90% of children have really been vaccinated, there is only a 1.1% chance of observing 89.4% of children (in a sample of 13,000) vaccinated by natural sampling variation alone.

b) We conclude that the proportion of children who have been vaccinated is below 90%, but a 95% confidence interval would show that the true proportion is between 88.9% and 89.9%. Most likely a decrease from 90% to 89.9% would not be considered important. The 90% figure was probably an approximate figure anyway.

9. Success.

a) Independence assumption: One man's response is not likely to have any effect on another man's response.
Randomization condition: The men were contacted through a random telephone poll.
10% condition: 1302 men represent less than 10% of all men.
Success/Failure condition: $n\hat{p} = 39$ and $n\hat{q} = 1263$ are both greater than 10, so the sample is large enough.

Since the conditions are met, we can use a one-proportion *z*-interval to estimate the percentage of men for whom work is their most important measure of success.

$$\hat{p} \pm z^* \sqrt{\frac{\hat{p}\hat{q}}{n}} = \left(\frac{39}{1302}\right) \pm 2.326 \sqrt{\frac{\left(\frac{39}{1302}\right)\left(\frac{1263}{1302}\right)}{1302}} = (1.9\%, 4.1\%)$$

We are 98% confident that between 1.9% and 4.1% of all men believe that work is their most important measure of success.

b) Since 5% is not in the interval, there is strong evidence that fewer than 5% of all men use work as their primary measure of success.

c) The significance level of this test is $\alpha = 0.01$. It's a lower tail test based on a 98% confidence interval.

11. Approval 2007.

a) Independence assumption: One response is not likely to effect on another response.
Randomization condition: The adults were randomly selected.
10% condition: 1125 adults represent less than 10% of all adults.
Success/Failure condition: $n\hat{p} = (1125)(0.30) = 338$ and $n\hat{q} = (1125)(0.70) = 787$ are both greater than 10, so the sample is large enough.

Since the conditions are met, we can use a one-proportion *z*-interval to estimate George W. Bush's approval rating.

$$\hat{p} \pm z^* \sqrt{\frac{\hat{p}\hat{q}}{n}} = (0.30) \pm 1.960 \sqrt{\frac{(0.30)(0.70)}{1125}} = (0.274, 0.327)$$

We are 95% confident that George W. Bush's approval rating is between 27.4% and 32.7%.

b) Since 27% is not within the interval, this is not a plausible value for George W. Bush's approval rating. There is evidence against the null hypothesis.

13. Dogs.

a) We cannot construct a confidence interval for the rate of occurrence of early hip dysplasia among 6-month old puppies because only 5 of 42 puppies were found with early hip dysplasia. The Success/Failure condition is not satisfied.

b) **Independence assumption:** If hip dysplasia is hereditary, and puppies brought to the vaccination clinic are from the same litter, independence might be an issue.
Randomization condition: The veterinarian considers the 42 puppies to be a random sample of all puppies.
10% condition: 42 puppies represent less than 10% of all puppies.
Success/Failure condition: As previously mentioned, this condition is not met, since there aren't at least 10 puppies with hip dysplasia in the sample.

Since the other conditions are met, we can construct a one-proportion plus-four *z*-interval to estimate the percentage of puppies with early hip dysplasia.

$$\tilde{p} = \frac{y+2}{n+4} = \frac{7}{46} = 0.152$$

$$\tilde{p} \pm z^{*}\sqrt{\frac{\tilde{p}\tilde{q}}{\tilde{n}}} = (0.152) \pm 1.96\sqrt{\frac{(0.152)(0.848)}{46}} = (4.8\%, 25.6\%)$$

We are 95% confident that between 4.8% and 25.6% of puppies have early hip dysplasia.

15. Loans.

a) The bank has made a Type II error. The person was not a good credit risk, and the bank failed to notice this.

b) The bank has made a Type I error. The person was a good credit risk, and the bank was convinced that he/she was not.

c) By making it easier to get a loan, the bank has reduced the alpha level. It takes less evidence to grant the person the loan.

d) The risk of Type I error is decreased and the risk of Type II error has increased.

17. Second loan.

a) Power is the probability that the bank denies a loan that could not have been repaid.

b) To increase power, the bank could raise the cutoff score.

c) If the bank raised the cutoff score, a larger number of trustworthy people would be denied credit, and the bank would lose the opportunity to collect the interest on these loans.

19. Homeowners 2005.

a) The null hypothesis is that the level of home ownership does not rise. The alternative hypothesis is that it rises.

b) In this context, a Type I error is when the city concludes that home ownership is on the rise, but in fact, the tax breaks don't help.

c) In this context, a Type II error is when the city abandons the tax breaks, thinking they don't help, when in fact they were helping.

d) A Type I error causes the city to forego tax revenue, while a Type II error withdraws help from those who might have otherwise been able to buy a house.

e) The power of the test is the city's ability to detect an actual increase in home ownership.

21. Testing cars.

H_0 : The shop is meeting the emissions standards.
H_A : The shop is not meeting the emissions standards.

a) Type I error is when the regulators decide that the shop is not meeting standards when they actually are meeting the standards.

b) Type II error is when the regulators certify the shop when they are not meeting the standards.

c) Type I would be more serious to the shop owners. They would lose their certification, even though they are meeting the standards.

d) Type II would be more serious to environmentalists. Shops are allowed to operate, even though they are allowing polluting cars to operate.

23. Cars again.

a) The power of the test is the probability of detecting that the shop is not meeting standards when they are not.

b) The power of the test will be greater when 40 cars are tested. A larger sample size increases the power of the test.

c) The power of the test will be greater when the level of significance is 10%. There is a greater chance that the null hypothesis will be rejected.

d) The power of the test will be greater when the shop is out of compliance "a lot". Larger problems are easier to detect.

25. Equal opportunity?

H_0 : The company is not discriminating against minorities.
H_A : The company is discriminating against minorities.

a) This is a one-tailed test. They wouldn't sue if "too many" minorities were hired.

b) Type I error would be deciding that the company is discriminating against minorities when they are not discriminating.

c) Type II error would be deciding that the company is not discriminating against minorities when they actually are discriminating.

d) The power of the test is the probability that discrimination is detected when it is actually occurring.

e) The power of the test will increase when the level of significance is increased from 0.01 to 0.05.

f) The power of the test is lower when the lawsuit is based on 37 employees instead of 87. Lower sample size leads to less power.

27. Dropouts.

a) The test is one-tailed. We are testing to see if a decrease in the dropout rate is associated with the software.

b) H_0 : The dropout rate does not change following the use of software. ($p = 0.13$)
H_A : The dropout rate decreases following the use of the software. ($p < 0.13$)

c) The professor makes a Type I error if he buys the software when the dropout rate has not actually decreased.

d) The professor makes a Type II error if he doesn't buy the software when the dropout rate has actually decreased.

e) The power of the test is the probability of buying the software when the dropout rate has actually decreased.

29. Dropouts, part II.

a) H_0 : The dropout rate does not change following the use of software. ($p = 0.13$)
H_A : The dropout rate decreases following the use of the software. ($p < 0.13$)

Independence assumption: One student's decision about dropping out should not influence another's decision.
Randomization condition: This year's class of 203 students is probably representative of all stats students.
10% condition: A sample of 203 students is less than 10% of all students.
Success/Failure condition: $np = (203)(0.13) = 26.39$ and $nq = (203)(0.87) = 176.61$ are both greater than 10, so the sample is large enough.

The conditions have been satisfied, so a Normal model can be used to model the sampling distribution of the proportion, with $\mu_{\hat{p}} = p = 0.13$ and

$$SD(\hat{p}) = \sqrt{\frac{pq}{n}} = \sqrt{\frac{(0.13)(0.87)}{203}} \approx 0.0236.$$

We can perform a one-proportion z-test. The observed proportion of dropouts is $\hat{p} = \dfrac{11}{203} \approx 0.054$.

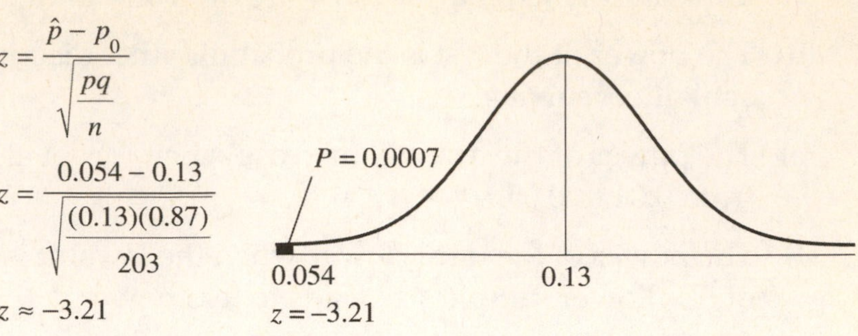

Since the P-value = 0.0007 is very low, we reject the null hypothesis. There is strong evidence that the dropout rate has dropped since use of the software program was implemented. As long as the professor feels confident that this class of stats students is representative of all potential students, then he should buy the program.

If you used a 95% confidence interval to assess the effectiveness of the program:

$$\hat{p} \pm z^* \sqrt{\frac{\hat{p}\hat{q}}{n}} = \left(\frac{11}{203}\right) \pm 1.960 \sqrt{\frac{\left(\frac{11}{203}\right)\left(\frac{192}{203}\right)}{203}} = (2.3\%, 8.5\%)$$

We are 95% confident that the dropout rate is between 2.3% and 8.5%. Since 15% is not contained in the interval, this provides evidence that the dropout rate has changed following the implementation of the software program.

b) The chance of observing 11 or fewer dropouts in a class of 203 is only 0.07% if the dropout rate in the population is really 13%.

31. Two coins.

a) The alternative hypothesis is that your coin produces 30% heads.

b) Reject the null hypothesis if the coin comes up tails. Otherwise, fail to reject.

c) There is a 10% chance that the coin comes up tails if the null hypothesis is true, so alpha is 10%.

d) Power is our ability to detect the 30% coin. That coin will come up tails 70% of the time. That's the power of our test.

e) To increase the power and lower the probability of Type I error at the same time, simply flip the coin more times.

33. Hoops.

H_0 : The player's foul-shot percentage is only 60%. ($p = 0.60$)
H_A : The player's foul-shot percentage is better than 60%. ($p > 0.60$)

a) The player's shots can be considered Bernoulli trials. There are only two possible outcomes, make the shot and miss the shot. The probability of making any shot is constant at $p = 0.60$. Assume that the shots are independent of each other. Use *Binom*(10, 0.60).

Let X = the number of shots made out of $n = 10$.

$$\begin{aligned} P(\text{makes at least 9 out of 10}) &= P(X \geq 9) \\ &= P(X = 9) + P(X = 10) \\ &= {}_{10}C_9(0.60)^9(0.40)^1 + {}_{10}C_{10}(0.60)^{10}(0.40)^0 \\ &\approx 0.0464 \end{aligned}$$

b) The coach made a Type I error.

c) The power of the test can be calculated for specific values of the new probability of success. Each true value of p has a power calculation associated with it. In this case, we are finding the power of the test to detect an 80% foul-shooter. Use *Binom*(10, 0.80).

Let X = the number of shots made out of $n = 10$.

$$\begin{aligned} P(\text{makes at least 9 out of 10}) &= P(X \geq 9) \\ &= P(X = 9) + P(X = 10) \\ &= {}_{10}C_9(0.80)^9(0.20)^1 + {}_{10}C_{10}(0.80)^{10}(0.20)^0 \approx 0.376 \end{aligned}$$

The power of the test to detect an increase in foul-shot percentage from 60% to 80% is about 37.6%.

d) The power of the test to detect improvement in foul-shooting can be increased by increasing the number of shots, or by keeping the number of shots at 10 but increasing the level of significance by declaring that 8, 9, or 10 shots made will convince the coach that the player has improved. In other words, the coach can increase the power of the test by lowering the standard of proof.

Chapter 22 – Comparing Two Proportions

1. **Online social networking.**

   It is very unlikely that samples would show an observed difference this large if, in fact, there was no real difference in the proportions of boys and girls who have used online social networks.

3. **Name recognition.**

   The ads may be effective. If there had been no real change in the name recognition, there would only be about a 3% chance the percentage of voters who had heard of this candidate would be at least this much higher in a different sample.

5. **Revealing information.**

   This test is not appropriate for these data, since the responses are not from independent groups, but are from the same individuals. The independent samples condition has been violated.

7. **Gender gap.**

   a) This is a stratified random sample, stratified by gender.

   b) We would expect the difference in proportions in the sample to be the same as the difference in proportions in the population, with the percentage of respondents with a favorable impression of the candidate 6% higher among males.

   c) The standard deviation of the difference in proportions is:

$$\sigma(\hat{p}_M - \hat{p}_F) = \sqrt{\frac{\hat{p}_M\hat{q}_M}{n_M} + \frac{\hat{p}_F\hat{q}_F}{n_F}} = \sqrt{\frac{(0.59)(0.41)}{300} + \frac{(0.53)(0.47)}{300}} \approx 4\%$$

   d)

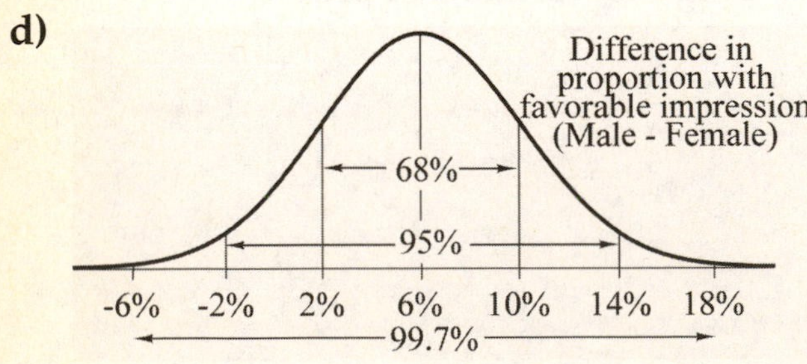

   e) The campaign could certainly be misled by the poll. According to the model, a poll showing little difference could occur relatively frequently. That result is only 1.5 standard deviations below the expected difference in proportions.

9. Arthritis.

a) Randomization condition: Americans age 65 and older were selected randomly.
10% condition: 1012 men and 1062 women are less than 10% of all men and women.
Independent samples condition: The sample of men and the sample of women were drawn independently of each other.
Success/Failure condition: $n\hat{p}$ (men) = 411, $n\hat{q}$ (men) = 601, $n\hat{p}$ (women) = 535, and $n\hat{q}$ (women) = 527 are all greater than 10, so the samples are both large enough.

Since the conditions have been satisfied, we will find a two-proportion z-interval.

b)

$$(\hat{p}_F - \hat{p}_M) \pm z^* \sqrt{\frac{\hat{p}_F\hat{q}_F}{n_F} + \frac{\hat{p}_M\hat{q}_M}{n_M}} = \left(\tfrac{535}{1062} - \tfrac{411}{1012}\right) \pm 1.960\sqrt{\frac{\left(\frac{535}{1062}\right)\left(\frac{527}{1062}\right)}{1062} + \frac{\left(\frac{411}{1012}\right)\left(\frac{601}{1012}\right)}{1012}}$$

$$= (0.055,\ 0.140)$$

c) We are 95% confident that the proportion of American women age 65 and older who suffer from arthritis is between 5.5% and 14.0% higher than the proportion of American men the same age who suffer from arthritis.

d) Since the interval for the difference in proportions of arthritis sufferers does not contain 0, there is strong evidence that arthritis is more likely to afflict women than men.

11. Pets.

a) $$SE(\hat{p}_{Herb} - \hat{p}_{None}) = \sqrt{\frac{\hat{p}_{Herb}\hat{q}_{Herb}}{n_{Herb}} + \frac{\hat{p}_{None}\hat{q}_{None}}{n_{None}}} = \sqrt{\frac{\left(\frac{473}{827}\right)\left(\frac{354}{827}\right)}{827} + \frac{\left(\frac{19}{130}\right)\left(\frac{111}{130}\right)}{130}} = 0.035$$

b) Randomization condition: Assume that the dogs studied were representative of all dogs.
10% condition: 827 dogs from homes with herbicide used regularly and 130 dogs from homes with no herbicide used are less than 10% of all dogs.
Independent samples condition: The samples were drawn independently of each other.
Success/Failure condition: $n\hat{p}$ (herb) = 473, $n\hat{q}$ (herb) = 354, $n\hat{p}$ (none) = 19, and $n\hat{q}$ (none) = 111 are all greater than 10, so the samples are both large enough.

Since the conditions have been satisfied, we will find a two-proportion z-interval.

$$(\hat{p}_{Herb} - \hat{p}_{None}) \pm z^* \sqrt{\frac{\hat{p}_{Herb}\hat{q}_{Herb}}{n_{Herb}} + \frac{\hat{p}_{None}\hat{q}_{None}}{n_{None}}}$$

$$= \left(\tfrac{473}{827} - \tfrac{19}{130}\right) \pm 1.960\sqrt{\frac{\left(\frac{473}{827}\right)\left(\frac{354}{827}\right)}{827} + \frac{\left(\frac{19}{130}\right)\left(\frac{111}{130}\right)}{130}} = (0.356,\ 0.495)$$

c) We are 95% confident that the proportion of pets with a malignant lymphoma in homes where herbicides are used is between 35.6% and 49.5% higher than the proportion of pets with lymphoma in homes where no pesticides are used.

13. Ear infections.

a) **Randomization condition:** The babies were randomly assigned to the two treatment groups.
Independent samples condition: The groups were assigned randomly, so the groups are not related.
Success/Failure condition: $n\hat{p}$(vaccine) = 333, $n\hat{q}$(vaccine) = 2122, $n\hat{p}$(none) = 499, and $n\hat{q}$(none) = 1953 are all greater than 10, so the samples are both large enough.

Since the conditions have been satisfied, we will find a two-proportion *z*-interval.

b) $$(\hat{p}_{None} - \hat{p}_{Vacc}) \pm z^* \sqrt{\frac{\hat{p}_{None}\hat{q}_{None}}{n_{None}} + \frac{\hat{p}_{Vacc}\hat{q}_{Vacc}}{n_{Vacc}}}$$

$$= \left(\tfrac{499}{2452} - \tfrac{333}{2455}\right) \pm 1.960\sqrt{\frac{\left(\frac{499}{2452}\right)\left(\frac{1953}{2452}\right)}{2452} + \frac{\left(\frac{333}{2455}\right)\left(\frac{2122}{2455}\right)}{2455}} = (0.047, 0.089)$$

c) We are 95% confident that the proportion of unvaccinated babies who develop ear infections is between 4.7% and 8.9% higher than the proportion of vaccinated babies who develop ear infections. The vaccinations appear to be effective, especially considering the 20% infection rate among the unvaccinated. A reduction of 5% to 9% is meaningful.

15. Another ear infection.

a) H_0: The proportion of vaccinated babies who get ear infections is the same as the proportion of unvaccinated babies who get ear infections.
$(p_{Vacc} = p_{None} \text{ or } p_{Vacc} - p_{None} = 0)$

H_A: The proportion of vaccinated babies who get ear infections is the lower than the proportion of unvaccinated babies who get ear infections.
$(p_{Vacc} < p_{None} \text{ or } p_{Vacc} - p_{None} < 0)$

b) Since 0 is not in the confidence interval, reject the null hypothesis. There is evidence that the vaccine reduces the rate of ear infections.

c) Since a 95% confidence interval was used originally, the alpha level is half of 5%, or 2.5%

d) If we think that the vaccine really reduces the rate of ear infections and it really does not reduce the rate of ear infections, then we have committed a Type I error.

e) Babies would be given ineffective vaccines.

17. Teen smoking, part I.

a) This is a prospective observational study.

b) H_0: The proportion of teen smokers among the group whose parents disapprove of smoking is the same as the proportion of teen smokers among the group whose parents are lenient about smoking.

$$\left(p_{Dis} = p_{Len} \text{ or } p_{Dis} - p_{Len} = 0\right)$$

H_A: The proportion of teen smokers among the group whose parents disapprove of smoking is lower than the proportion of teen smokers among the group whose parents are lenient about smoking.

$$\left(p_{Dis} < p_{Len} \text{ or } p_{Dis} - p_{Len} < 0\right)$$

c) **Randomization condition:** Assume that the teens surveyed are representative of all teens.
10% condition: 284 and 41 are both less than 10% of all teens.
Independent samples condition: The groups were surveyed independently.
Success/Failure condition: $n\hat{p}$(disapprove) = 54, $n\hat{q}$(disapprove) = 230, $n\hat{p}$(lenient) = 11, and $n\hat{q}$(lenient) = 30 are all greater than 10, so the samples are both large enough.

Since the conditions have been satisfied, we will model the sampling distribution of the difference in proportion with a Normal model with mean 0 and standard deviation estimated by

$$SE_{\text{pooled}}\left(\hat{p}_{Dis} - \hat{p}_{Len}\right) = \sqrt{\frac{\hat{p}_{\text{pooled}}\hat{q}_{\text{pooled}}}{n_{Dis}} + \frac{\hat{p}_{\text{pooled}}\hat{q}_{\text{pooled}}}{n_{Len}}} = \sqrt{\frac{\left(\frac{65}{325}\right)\left(\frac{260}{325}\right)}{284} + \frac{\left(\frac{65}{325}\right)\left(\frac{260}{325}\right)}{41}} = 0.0668.$$

d) The observed difference between the proportions is 0.190 – 0.268 = – 0.078.

Since the P-value = 0.1211 is high, we fail to reject the null hypothesis. There is little evidence to suggest that parental attitudes influence teens' decisions to smoke.

$$z = \frac{-0.078 - 0}{0.0668}$$

$$z \approx -1.17$$

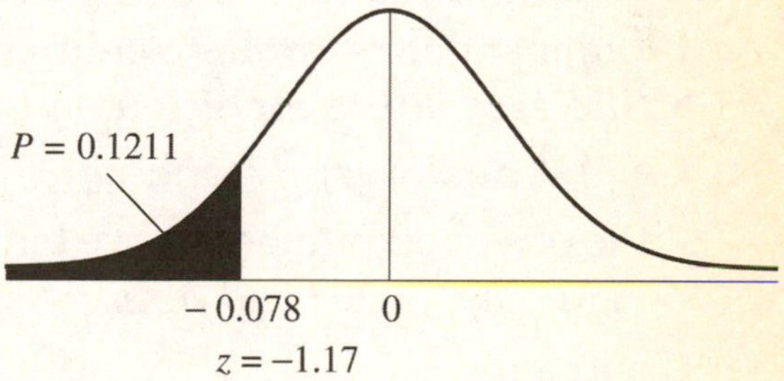

P = 0.1211
– 0.078
0
z = –1.17

e) If there is no difference in the proportions, there is about a 12% chance of seeing the observed difference or larger by natural sampling variation.

f) If teens' decisions about smoking *are* influenced, we have committed a Type II error.

19. Teen smoking, part II.

a) Since the conditions have already been satisfied in a previous exercise, we will find a two-proportion *z*-interval.

$$(\hat{p}_{Dis} - \hat{p}_{Len}) \pm z^* \sqrt{\frac{\hat{p}_{Dis}\hat{q}_{Dis}}{n_{Dis}} + \frac{\hat{p}_{Len}\hat{q}_{Len}}{n_{Len}}}$$

$$= \left(\tfrac{54}{284} - \tfrac{11}{41}\right) \pm 1.960\sqrt{\frac{\left(\frac{54}{284}\right)\left(\frac{230}{284}\right)}{284} + \frac{\left(\frac{11}{41}\right)\left(\frac{30}{41}\right)}{41}} = (-0.065, 0.221)$$

b) We are 95% confident that the proportion of teens whose parents disapprove of smoking who will eventually smoke is between 6.5% less and 22.1% more than for teens with parents who are lenient about smoking.

c) We expect 95% of random samples of this size to produce intervals that contain the true difference between the proportions.

21. Pregnancy.

a) This is not an experiment, since subjects were not assigned to treatments. This is an observational study.

b) H_0: The proportion of live births is the same for women under the age of 38 as it is for women 38 or older. $(p_{<38} = p_{\geq 38}$ or $p_{<38} - p_{\geq 38} = 0)$

H_A: The proportion of live births is different for women under the age of 38 than for women 38 or older. $(p_{<38} \neq p_{\geq 38}$ or $p_{<38} - p_{\geq 38} \neq 0)$

Randomization condition: Assume that the women studied are representative of all women.
10% condition: 157 and 89 are both less than 10% of all women.
Independent samples condition: The groups are not associated.
Success/Failure condition: $n\hat{p}$ (under 38) = 42, $n\hat{q}$ (under 38) = 115, $n\hat{p}$ (38 and over) = 7, and $n\hat{q}$ (38 and over) = 82 are not all greater than 10, since the observed number of live births is only 7. However, if we check the pooled value, $n\hat{p}_{\text{pooled}}$ (38 and over) = (89)(0.1992) = 18. All of the samples are large enough.

Since the conditions have been satisfied, we will model the sampling distribution of the difference in proportion with a Normal model with mean 0 and standard deviation estimated by

$$SE_{\text{pooled}}(\hat{p}_{<38} - \hat{p}_{\geq 38}) = \sqrt{\frac{\hat{p}_{\text{pooled}}\hat{q}_{\text{pooled}}}{n_{<38}} + \frac{\hat{p}_{\text{pooled}}\hat{q}_{\text{pooled}}}{n_{\geq 38}}} = \sqrt{\frac{\left(\frac{49}{246}\right)\left(\frac{197}{246}\right)}{157} + \frac{\left(\frac{49}{246}\right)\left(\frac{197}{246}\right)}{89}} \approx 0.0530.$$

The observed difference between the proportions is: 0.2675 – 0.0787 = 0.1888.

Since the P-value = 0.0004 is low, we reject the null hypothesis. There is strong evidence to suggest a difference in the proportion of live births for women under 38 and women 38 and over at this clinic. In fact, the evidence suggests that women under 38 have a higher proportion of live births.

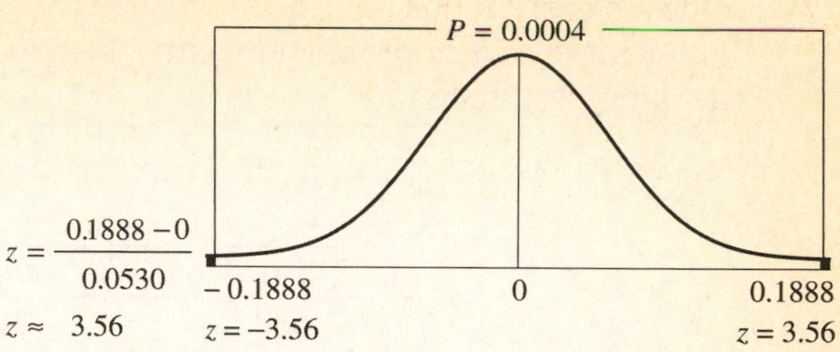

c) $(\hat{p}_{<38} - \hat{p}_{\geq 38}) \pm z^* \sqrt{\dfrac{\hat{p}_{<38}\hat{q}_{<38}}{n_{<38}} + \dfrac{\hat{p}_{\geq 38}\hat{q}_{\geq 38}}{n_{\geq 38}}}$

$$= \left(\tfrac{42}{157} - \tfrac{7}{89}\right) \pm 1.960 \sqrt{\frac{\left(\frac{42}{157}\right)\left(\frac{115}{157}\right)}{157} + \frac{\left(\frac{7}{89}\right)\left(\frac{82}{89}\right)}{89}} = (0.100,\ 0.278)$$

We are 95% confident that the proportion of live births for patients at this clinic is between 10.0% and 27.8% higher for women under 38 than for women 38 and over. However, the Success/Failure condition is not met for the older women, so we should be cautious when using this interval. (The expected number of successes from the pooled proportion cannot be used for a condition for a confidence interval. It's based upon an assumption that the proportions are the same. We don't make that assumption in a confidence interval. In fact, we are implicitly assuming a *difference*, by finding an interval for the difference in proportion.)

23. Politics and sex.

a) H_0: The proportion of voters in support of the candidate is the same before and after news of his extramarital affair got out. $(p_B = p_A \text{ or } p_B - p_A = 0)$

H_A: The proportion of voters in support of the candidate has decreased after news of his extramarital affair got out. $(p_B > p_A \text{ or } p_B - p_A > 0)$

Randomization condition: Voters were randomly selected.
10% condition: 630 and 1010 are both less than 10% of all voters.
Independent samples condition: Since the samples were random, the groups are independent.
Success/Failure condition: $n\hat{p}$ (before) = (630)(0.54) = 340, $n\hat{q}$ (before) = (630)(0.46) = 290, $n\hat{p}$ (after) = (1010)(0.51) = 515, and $n\hat{q}$ (after) = (1010)(0.49) = 505 are all greater than 10, so both samples are large enough.

Since the conditions have been satisfied, we will model the sampling distribution of the difference in proportion with a Normal model with mean 0 and standard deviation estimated by:

$$SE_{\text{pooled}}\left(\hat{p}_B - \hat{p}_A\right) = \sqrt{\frac{\hat{p}_{\text{pooled}}\hat{q}_{\text{pooled}}}{n_B} + \frac{\hat{p}_{\text{pooled}}\hat{q}_{\text{pooled}}}{n_A}}$$

$$= \sqrt{\frac{(0.5215)(0.4785)}{630} + \frac{(0.5215)(0.4785)}{1010}} \approx 0.02536$$

The observed difference between the proportions is: 0.54 – 0.51 = 0.03.

Since the *P*-value = 0.118 is fairly high, we fail to reject the null hypothesis. There is little evidence of a decrease in the proportion of voters in support of the candidate after the news of his extramarital affair got out.

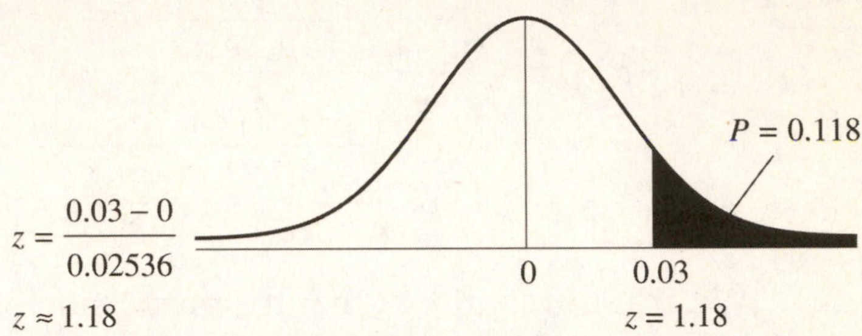

b) No evidence of a decrease in the proportion of voters in support of the candidate was found. If there is actually a decrease, and we failed to notice, that's a Type II error.

25. Twins.

a) H_0: The proportion of multiple births is the same for white women and black women. $\left(p_W = p_B \text{ or } p_W - p_B = 0\right)$

H_A: The proportion of multiple births is different for white women and black women. $\left(p_W \neq p_B \text{ or } p_W - p_B \neq 0\right)$

Randomization condition: Assume that these women are representative of all women.
10% condition: 3132 and 606 are both less than 10% of all people.
Independent samples condition: The groups are independent.
Success/Failure condition: $n\hat{p}$ (white) = 94, $n\hat{q}$ (white) = 3038, $n\hat{p}$ (black) = 20, and $n\hat{q}$ (black) = 586 are all greater than 10, so both samples are large enough.

Since the conditions have been satisfied, we will model the sampling distribution of the difference in proportion with a Normal model with mean 0 and standard deviation estimated by:

$$SE_{\text{pooled}}\left(\hat{p}_W - \hat{p}_B\right) = \sqrt{\frac{\hat{p}_{\text{pooled}}\hat{q}_{\text{pooled}}}{n_W} + \frac{\hat{p}_{\text{pooled}}\hat{q}_{\text{pooled}}}{n_B}} = \sqrt{\frac{\left(\frac{114}{3738}\right)\left(\frac{3624}{3738}\right)}{3132} + \frac{\left(\frac{114}{3738}\right)\left(\frac{3624}{3738}\right)}{606}} \approx 0.007631$$

The observed difference between the proportions is: 0.030 – 0.033 = –0.003.

Since the P-value = 0.6951 is high, we fail to reject the null hypothesis. There is no evidence of a difference between the proportions of multiple births for white women and black women.

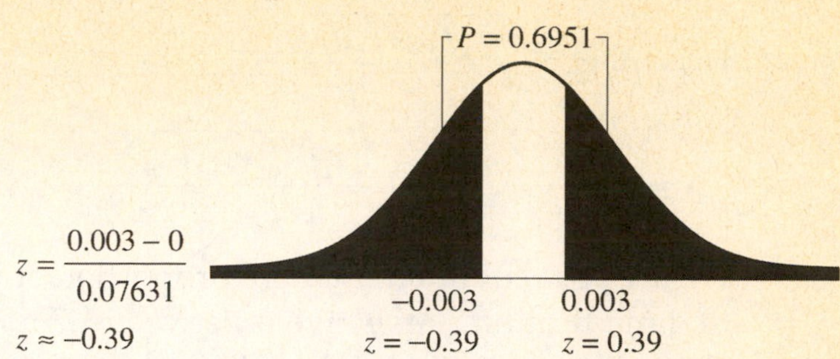

b) If there is actually a difference between the proportions of multiple births for white women and black women, then we have committed a Type II error.

27. Pain.

a) **Randomization condition:** The patients were randomly selected AND randomly assigned to treatment groups.
10% condition: 112 and 108 are both less than 10% of all people with joint pain.
Success/Failure condition: $n\hat{p}$ (A) = 84, $n\hat{q}$ (A) = 28, $n\hat{p}$ (B) = 66, and $n\hat{q}$ (B) = 42 are all greater than 10, so both samples are large enough.

Since the conditions are met, we can use a one-proportion z-interval to estimate the percentage of patients who may get relief from medication A.

$$\hat{p} \pm z^* \sqrt{\frac{\hat{p}\hat{q}}{n}} = \left(\frac{84}{112}\right) \pm 1.960 \sqrt{\frac{\left(\frac{84}{112}\right)\left(\frac{28}{112}\right)}{112}} = (67.0\%, 83.0\%)$$

We are 95% confident that between 67.0% and 83.0% of patients with joint pain will find medication A to be effective.

b) Since the conditions were met in part a, we can use a one-proportion z-interval to estimate the percentage of patients who may get relief from medication B.

$$\hat{p} \pm z^* \sqrt{\frac{\hat{p}\hat{q}}{n}} = \left(\frac{66}{108}\right) \pm 1.960 \sqrt{\frac{\left(\frac{66}{108}\right)\left(\frac{42}{108}\right)}{108}} = (51.9\%, 70.3\%)$$

We are 95% confident that between 51.9% and 70.3% of patients with joint pain will find medication B to be effective.

c) The 95% confidence intervals overlap, which might lead one to believe that there is no evidence of a difference in the proportions of people who find each medication effective. However, if one was lead to believe that, one should proceed to part…

d) Most of the conditions were checked in part a. We only have one more to check:
Independent samples condition: The groups were assigned randomly, so there is no reason to believe there is a relationship between them.

Since the conditions have been satisfied, we will find a two-proportion z-interval.

$$(\hat{p}_A - \hat{p}_B) \pm z^* \sqrt{\frac{\hat{p}_A \hat{q}_A}{n_A} + \frac{\hat{p}_B \hat{q}_B}{n_B}}$$

$$= \left(\tfrac{84}{112} - \tfrac{66}{108}\right) \pm 1.960 \sqrt{\frac{\left(\frac{84}{112}\right)\left(\frac{28}{112}\right)}{112} + \frac{\left(\frac{66}{108}\right)\left(\frac{42}{108}\right)}{112}} = (0.017, 0.261)$$

We are 95% confident that the proportion of patients with joint pain who will find medication A effective is between 1.70% and 26.1% higher than the proportion of patients who will find medication B effective.

e) The interval does not contain zero. There is evidence that medication A is more effective than medication B.

f) The two-proportion method is the proper method. By attempting to use two, separate, confidence intervals, you are adding standard deviations when looking for a difference in proportions. We know from our previous studies that *variances* add when finding the standard deviation of a difference. The two-proportion method does this.

29. Sensitive men.

H_0: The proportion of 18-24-year-old men who are comfortable talking about their problems is the same as the proportion of 25-34-year old men. $\left(p_{Young} = p_{Old} \text{ or } p_{Young} - p_{Old} = 0\right)$

H_A: The proportion of 18-24-year-old men who are comfortable talking about their problems is higher than the proportion of 25-34-year old men. $\left(p_{Young} > p_{Old} \text{ or } p_{Young} - p_{Old} > 0\right)$

Randomization condition: We assume the respondents were chosen randomly.
10% condition: 129 and 184 are both less than 10% of all people.
Independent samples condition: The groups were chosen independently.
Success/Failure condition: $n\hat{p}$ (young) = 80, $n\hat{q}$ (young) = 49, $n\hat{p}$ (old) = 98, and $n\hat{q}$ (old) = 86 are all greater than 10, so both samples are large enough.

Since the conditions have been satisfied, we will model the sampling distribution of the difference in proportion with a Normal model with mean 0 and standard deviation estimated
by:.

$$SE_{pooled}\left(\hat{p}_{Young} - \hat{p}_{Old}\right) = \sqrt{\frac{\hat{p}_{pooled}\hat{q}_{pooled}}{n_Y} + \frac{\hat{p}_{pooled}\hat{q}_{pooled}}{n_O}} = \sqrt{\frac{\left(\frac{178}{313}\right)\left(\frac{135}{313}\right)}{129} + \frac{\left(\frac{178}{313}\right)\left(\frac{135}{313}\right)}{184}} \approx 0.05687$$

The observed difference between the proportions is: 0.620 – 0.533 = 0.087.

Since the P-value = 0.0619 is high, we fail to reject the null hypothesis. There is little evidence that the proportion of 18-24-year-old men who are comfortable talking about their problems is higher than the proportion of 25-34-year-old men who are comfortable. *Time* magazine's interpretation is questionable.

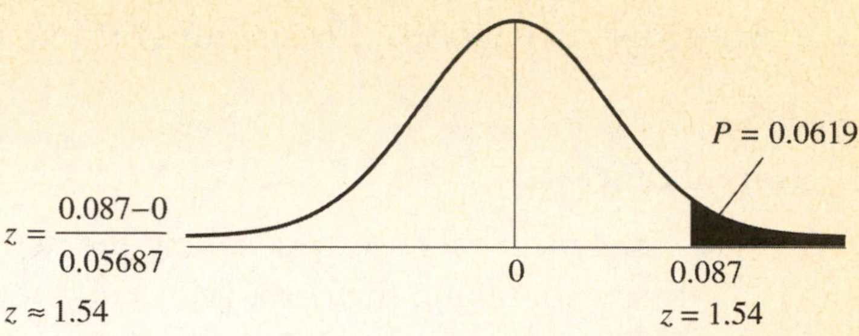

31. Online activity checks.

H_0: The proportion of teens who say their parents check to see what websites they visited is the same in 2006 as it was in 2004.
$(p_{2006} = p_{2004} \text{ or } p_{2006} - p_{2004} = 0)$

H_A: The proportion of teens who say their parents check to see what websites they visited is higher in 2006 than it was in 2004. $(p_{2006} > p_{2004} \text{ or } p_{2006} - p_{2004} > 0)$

Randomization condition: The samples were random.
10% condition: 811 and 868 are both less than 10% of all teens.
Independent samples condition: The samples were taken independently.
Success/Failure condition: $n\hat{p}$ (2006) = 333, $n\hat{q}$ (2006) = 478, $n\hat{p}$ (2004) = 286, and $n\hat{q}$ (2004) = 582 are all greater than 10, so both samples are large enough.

Since the conditions have been satisfied, we will model the sampling distribution of the difference in proportion with a Normal model with mean 0 and standard deviation estimated by:

$$SE_{\text{pooled}}(\hat{p}_{2006} - \hat{p}_{2004}) = \sqrt{\frac{\hat{p}_{\text{pool}}\hat{q}_{\text{pool}}}{n_{2006}} + \frac{\hat{p}_{\text{pool}}\hat{q}_{\text{pool}}}{n_{2004}}}$$

$$= \sqrt{\frac{(0.369)(0.631)}{811} + \frac{(0.369)(0.631)}{868}} \approx 0.02355$$

The observed difference between the proportions is: 0.41 – 0.33 = 0.08. We will perform a 2-proportion z-test.

The value of $z = 3.44$ and the P-value = 0.0003. Since the P-value is low, we reject the null hypothesis. There is strong evidence that a greater proportion of teens in 2006 say their parents checked in to see what web sites they visited than said this in 2004.

Review of Part V – From the Data at Hand to the World at Large

1. Herbal cancer.

H0: The cancer rate for those taking the herb is the same as the cancer rate for those not taking the herb. $(p_{Herb} = p_{Not}$ or $p_{Herb} - p_{Not} = 0)$

HA: The cancer rate for those taking the herb is higher than the cancer rate for those not taking the herb. $(p_{Herb} > p_{Not}$ or $p_{Herb} - p_{Not} > 0)$

3. Birth days.

a) If births are distributed uniformly across all days, we expect the number of births on each day to be $np = (72)(1/7) \approx 10.29$.

b) **Randomization condition:** The 72 births are likely to be representative of all births at the hospital with regards to day of birth.
10% condition: 72 births are less than 10% of the births.
Success/Failure condition: The expected number of births on a particular day of the week is $np = (72)(1/7) \approx 10.29$ and the expected number of births not on that particular day is $nq = (72)(6/7) \approx 61.71$. These are both greater than 10, so the sample is large enough.

Since the conditions have been satisfied, a Normal model can be used to model the sampling distribution of the proportion of 72 births that occur on a given day of the week.

$$\mu_{\hat{p}} = p = 1/7 \approx 0.1429$$

$$\sigma(\hat{p}) = \sqrt{\frac{pq}{n}} = \sqrt{\frac{(1/7)(6/7)}{72}} \approx 0.04124$$

There were 7 births on Mondays, so $\hat{p} = \frac{7}{72} \approx 0.09722$. This is only about a 1.11 standard deviations below the expected proportion, so there's no evidence that this is unusual.

c) The 17 births on Tuesdays represent an unusual occurrence. For Tuesdays, $\hat{p} = \frac{17}{72} \approx 0.2361$, which is about 2.26 standard deviations above the expected proportion of births. There is evidence to suggest that the proportion of births on Tuesdays is higher than expected, if births are distributed uniformly across days.

d) Some births are scheduled for the convenience of the doctor and/or the mother.

5. Leaky gas tanks.

a) H_0: The proportion of leaky gas tanks is 40%. ($p = 0.40$)
H_A: The proportion of leaky gas tanks is less than 40%. ($p < 0.40$)

b) **Randomization condition:** A random sample of 27 service stations in California was taken.
10% condition: 27 stations are less than 10% of all service stations in California.
Success/Failure condition: $np = (27)(0.40) = 10.8$ and $nq = (27)(0.60) = 16.2$ are both greater than 10, so the sample is large enough.

c) Since the conditions have been satisfied, a Normal model can be used to model the sampling distribution of the proportion, with $\mu_{\hat{p}} = p = 0.40$ and

$\sigma(\hat{p}) = \sqrt{\frac{pq}{n}} = \sqrt{\frac{(0.40)(0.60)}{27}} \approx 0.09428$. We can perform a one-proportion z-test.

The observed proportion of leaky gas tanks is $\hat{p} = \frac{7}{27} \approx 0.2593$.

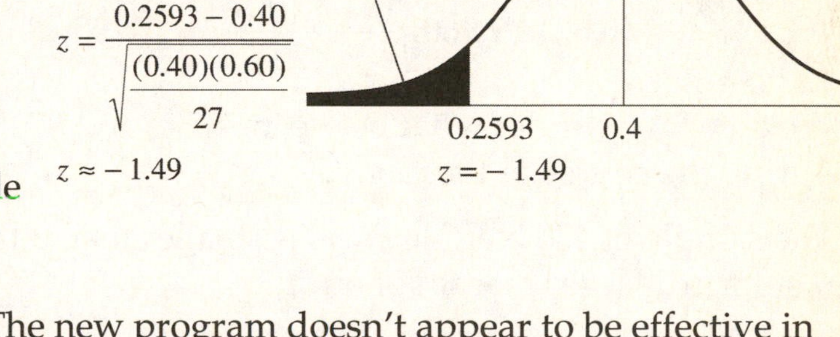

d) Since the P-value = 0.0677 is relatively high, we fail to reject the null hypothesis. There is little evidence that the proportion of leaky gas tanks is less than 40%. The new program doesn't appear to be effective in decreasing the proportion of leaky gas tanks.

e) If the program actually works, we haven't done anything *wrong*. Our methods are correct. Statistically speaking, we have committed a Type II error.

f) In order to decrease the probability of making this type of error, we could lower our standards of proof, by raising the level of significance. This will increase the power of the test to detect a decrease in the proportion of leaky gas tanks. Another way to decrease the probability that we make a Type II error is to sample more service stations. This will decrease the variation in the sample proportion, making our results more reliable.

g) Increasing the level of significance is advantageous, since it decreases the probability of making a Type II error, and increases the power of the test. However, it also increases the probability that a Type I error is made, in this case, thinking that the program is effective when it really is not effective. Increasing the sample size decreases the probability of making a Type II error and increases power, but can be costly and time-consuming.

7. Scrabble.

a) The researcher believes that the true proportion of As is within 10% of the estimated 54%, namely, between 44% and 64%.

b) A large margin of error is usually associated with a small sample, but the sample consisted of "many" hands. The margin of error is large because the standard error of the sample is large. This occurs because the true proportion of As in a hand is close to 50%, the most difficult proportion to predict.

c) This provides no evidence that the simulation is faulty. The true proportion of As is contained in the confidence interval. The researcher's results are consistent with 63% As.

9. Net-Newsers.

a) The Pew Research Foundation believes that the true proportion of people who obtain news from the Internet is between 11% and 15%.

b) The smaller sample size in the cell sample would result in a larger standard error. This would make the margin of error larger, as well.

c) $\hat{p} \pm z^* \sqrt{\dfrac{\hat{p}\hat{q}}{n}} = (0.82) \pm 1.960 \sqrt{\dfrac{(0.82)(0.18)}{470}} = (78.5\%, 85.5\%)$

We are 95% confident that between 78.5% and 85.5% of Net-Newsers get news during the course of the day.

d) The sample of 470 Net-Newsers is smaller than either of the earlier samples. This results in a larger margin of error.

11. Bimodal.

a) The *sample's* distribution (NOT the *sampling* distribution), is expected to look more and more like the distribution of the population, in this case, bimodal.

b) The expected value of the sample's mean is expected to be μ, the population mean, regardless of sample size.

c) The variability of the sample mean, $\sigma(\bar{y})$, is $\dfrac{\sigma}{\sqrt{n}}$, the population standard deviation divided by the square root of the sample size, regardless of the sample size.

d) As the sample size increases, the sampling distribution model becomes closer and closer to a Normal model.

13. Archery.

a) $\mu_{\hat{p}} = p = 0.80$ $\qquad$ $\sigma(\hat{p}) = \sqrt{\dfrac{pq}{n}} = \sqrt{\dfrac{(0.80)(0.20)}{200}} \approx 0.028$

b) $np = (200)(0.80) = 160$ and $nq = (200)(0.20) = 40$ are both greater than 10, so the Normal model is appropriate.

c) The Normal model of the sampling distribution of the proportion of bull's-eyes she makes out of 200 is at the right.

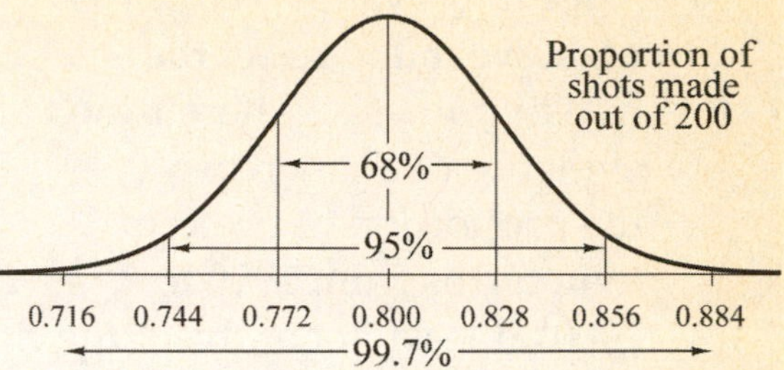

Approximately 68% of the time, we expect her to hit the bull's-eye on between 77.2% and 82.8% of her shots. Approximately 95% of the time, we expect her to hit the bull's-eye on between 74.4% and 85.6% of her shots. Approximately 99.7% of the time, we expect her to hit the bull's-eye on between 71.6% and 88.4% of her shots.

d) According to the Normal model, the probability that she hits the bull's-eye in at least 85% of her 200 shots is approximately 0.039.

$$z = \frac{0.85 - 0.80}{0.028}$$

$$z \approx 1.79$$

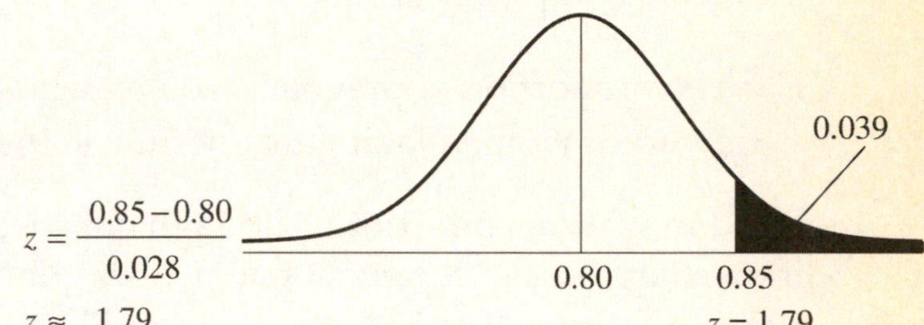

15. Twins.

H_0: The proportion of preterm twin births in 1990 is the same as the proportion of preterm twin births in 2000. $(p_{1990} = p_{2000} \text{ or } p_{1990} - p_{2000} = 0)$

H_A: The proportion of preterm twin births in 1990 is the less than the proportion of preterm twin births in 2000. $(p_{1990} < p_{2000} \text{ or } p_{1990} - p_{2000} < 0)$

Randomization condition: Assume that these births are representative of all twin births.
10% condition: 43 and 48 are both less than 10% of all twin births.
Independent samples condition: The samples are from different years, so they are unlikely to be related.
Success/Failure condition: $n\hat{p}(1990) = 20$, $n\hat{q}(1990) = 23$, $n\hat{p}(2000) = 26$, and $n\hat{q}(2000) = 22$ are all greater than 10, so both samples are large enough.

Since the conditions have been satisfied, we will perform a two-proportion *z*-test. We will model the sampling distribution of the difference in proportion with a Normal model with mean 0 and standard deviation estimated by:

$$SE_{\text{pooled}}(\hat{p}_{1990} - \hat{p}_{2000}) = \sqrt{\frac{\hat{p}_{\text{pooled}}\hat{q}_{\text{pooled}}}{n_{1900}} + \frac{\hat{p}_{\text{pooled}}\hat{q}_{\text{pooled}}}{n_{2000}}} = \sqrt{\frac{\left(\frac{46}{91}\right)\left(\frac{45}{91}\right)}{43} + \frac{\left(\frac{46}{91}\right)\left(\frac{45}{91}\right)}{48}} \approx 0.1050$$

The observed difference between the proportions is:
0.4651 – 0.5417 = – 0.0766

Since the P-value = 0.2329 is high, we fail to reject the null hypothesis. There is no evidence of an increase in the proportion of preterm twin births from 1990 to 2000, at least not at this large city hospital.

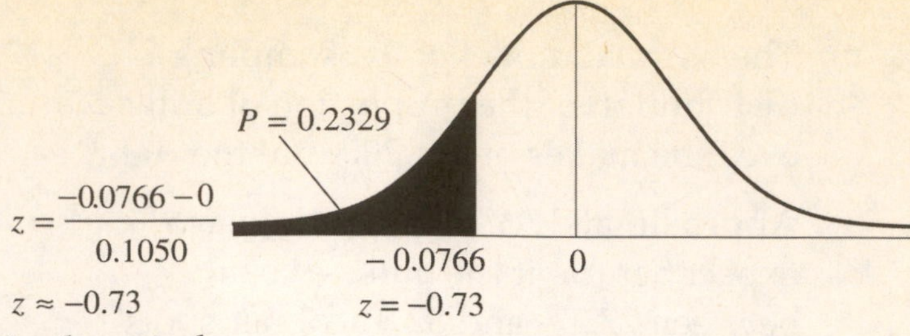

17. Eclampsia.

a) H_0: The proportion of pregnant women who die after developing eclampsia is the same for women taking magnesium sulfide as it is for women not taking magnesium sulfide. $(p_{MS} = p_N \text{ or } p_{MS} - p_N = 0)$

H_A: The proportion of pregnant women who die after developing eclampsia is lower for women taking magnesium sulfide. $(p_{MS} < p_N \text{ or } p_{MS} - p_N < 0)$

b) Randomization condition: Although not specifically stated, these results are from a large-scale experiment, which was undoubtedly properly randomized.
10% condition: 40 and 96 are less than 10% of all pregnant women.
Independent samples condition: Subjects were randomly assigned to the treatments.
Success/Failure condition: $n\hat{p}$ (mag. sulf.) = 11, $n\hat{q}$ (mag. sulf.) = 29, $n\hat{p}$ (placebo) = 20, and $n\hat{q}$ (placebo) = 76 are all greater than 10, so both samples are large enough.

Since the conditions have been satisfied, we will perform a two-proportion z-test. We will model the sampling distribution of the difference in proportion with a Normal model with mean 0 and standard deviation estimated by:

$$SE_{\text{pooled}}(\hat{p}_{MS} - \hat{p}_N) = \sqrt{\frac{\hat{p}_{\text{pooled}}\hat{q}_{\text{pooled}}}{n_{MS}} + \frac{\hat{p}_{\text{pooled}}\hat{q}_{\text{pooled}}}{n_N}} = \sqrt{\frac{\left(\frac{31}{136}\right)\left(\frac{105}{136}\right)}{40} + \frac{\left(\frac{31}{136}\right)\left(\frac{105}{136}\right)}{96}} \approx 0.07895.$$

c) The observed difference between the proportions is: 0.275 – 0.2083 = 0.0667

Since the P-value = 0.8008 is high, we fail to reject the null hypothesis. There is no evidence that the proportion of women who may die after developing eclampsia is lower for women taking magnesium sulfide than for women who are not taking the drug.

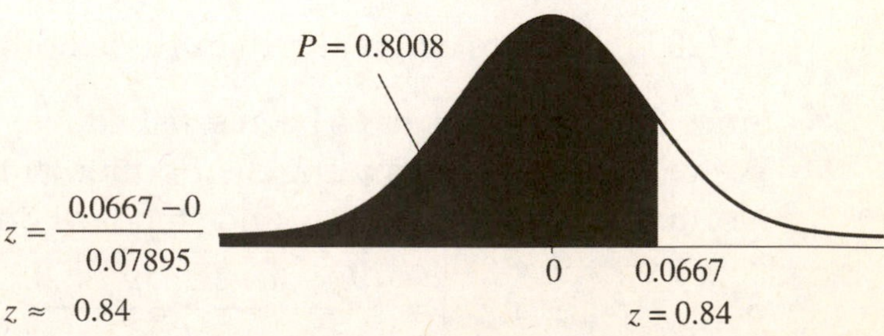

d) There is not sufficient evidence to conclude that magnesium sulfide is effective in preventing death when eclampsia develops.

e) If magnesium sulfide is effective in preventing death when eclampsia develops, then we have made a Type II error.

f) To increase the power of the test to detect a decrease in death rate due to magnesium sulfide, we could increase the sample size or increase the level of significance.

g) Increasing the sample size lowers variation in the sampling distribution, but may be costly. The sample size is already quite large. Increasing the level of significance increases power by increasing the likelihood of rejecting the null hypothesis, but increases the chance of making a Type I error, namely thinking that magnesium sulfide is effective when it is not.

19. Polling disclaimer.

a) It is not clear what specific question the pollster asked. Otherwise, they did a great job of identifying the W's.

b) A sample that was stratified by age, sex, region, and education was used.

c) The margin of error was 4%.

d) Since "no more than 1 time in 20 should chance variations in the sample cause the results to vary by more than 4 percentage points", the confidence level is $19/20 = 95\%$.

e) The subgroups had smaller sample sizes than the larger group. The standard errors in these subgroups were larger as a result, and this caused the margins of error to be larger.

f) They cautioned readers about response bias due to wording and order of the questions.

21. Teen deaths.

a) H_0 : The percentage of fatal accidents involving teenage girls is 14.3%, the same as the overall percentage of fatal accidents involving teens . ($p = 0.143$)
H_A : The percentage of fatal accidents involving teenage girls is lower than 14.3%, the overall percentage of fatal accidents involving teens . ($p < 0.143$)

Independence assumption: It is reasonable to think that accidents occur independently.
Randomization condition: Assume that the 388 accidents observed are representative of all accidents.
10% condition: The sample of 388 accidents is less than 10% of all accidents.
Success/Failure condition: $np = (388)(0.143) = 55.484$ and $nq = (388)(0.857) = 332.516$ are both greater than 10, so the sample is large enough.

The conditions have been satisfied, so a Normal model can be used to model the sampling distribution of the proportion, with $\mu_{\hat{p}} = p = 0.143$ and

$$\sigma(\hat{p}) = \sqrt{\frac{pq}{n}} = \sqrt{\frac{(0.143)(0.857)}{388}} \approx 0.01777.$$

We can perform a one-proportion z-test. The observed proportion of fatal accidents involving teen girls is $\hat{p} = \frac{44}{388} \approx 0.1134$.

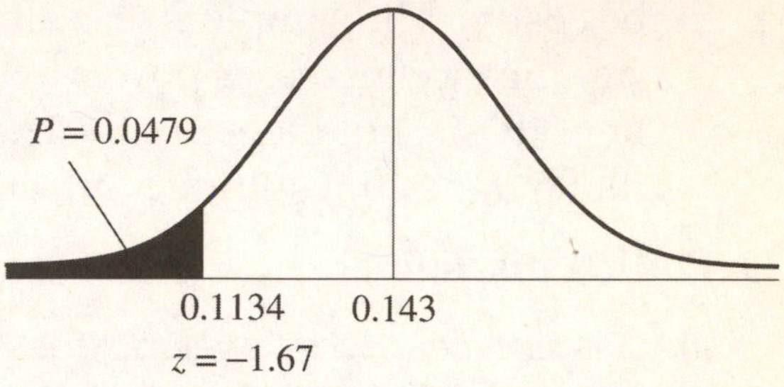

Since the P-value = 0.0479 is low, we reject the null hypothesis. There is some evidence that the proportion of fatal accidents involving teen girls is less than the overall proportion of fatal accidents involving teens.

$$z = \frac{\hat{p} - p_0}{\sqrt{\frac{pq}{n}}}$$

$$z = \frac{0.1134 - 0.143}{\sqrt{\frac{(0.143)(0.857)}{388}}}$$

$$z \approx -1.67$$

$P = 0.0479$

0.1134 0.143

$z = -1.67$

b) If the proportion of fatal accidents involving teenage girls is really 14.3%, we expect to see the observed proportion, 11.34%, in about 4.79% of samples of size 388 simply due to sampling variation.

23. Largemouth bass.

a) One would expect many small fish, with a few large fish.

b) We cannot determine the probability that a largemouth bass caught from the lake weighs over 3 pounds because we don't know the exact shape of the distribution. We know that it is NOT Normal.

c) It would be quite risky to attempt to determine whether or not the mean weight of 5 fish was over 3 pounds. With a skewed distribution, a sample of size 5 is not large enough for the Central Limit Theorem to guarantee that a Normal model is appropriate to describe the distribution of the mean.

d) A sample of 60 randomly selected fish is large enough for the Central Limit Theorem to guarantee that a Normal model is appropriate to describe the sampling distribution of the mean, as long as 60 fish is less than 10% of the population of all the fish in the lake.

The mean weight is $\mu = 3.5$ pounds, with standard deviation $\sigma = 2.2$ pounds. Since the sample size is sufficiently large, we can model the sampling distribution of the mean weight of 60 fish with a Normal model, with $\mu_{\bar{y}} = 3.5$ pounds and standard deviation $\sigma(\bar{y}) = \frac{2.2}{\sqrt{60}} \approx 0.284$ pounds.

According to the Normal model, the probability that 60 randomly selected fish average more than 3 pounds is approximately 0.961.

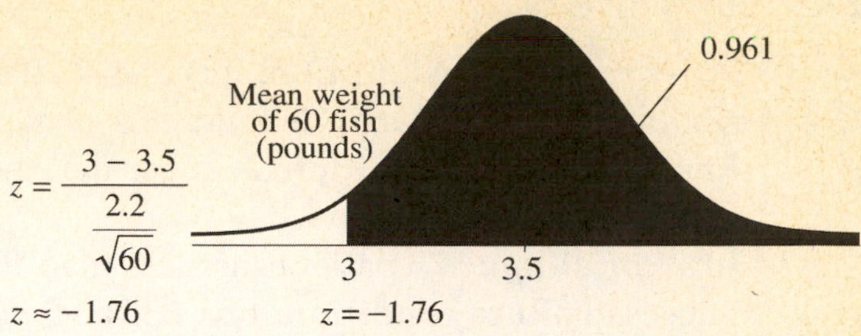

25. Language.

a) **Randomization condition:** 60 people were selected at random.
10% condition: The 60 people represent less than 10% of all people.
Success/Failure condition: $np = (60)(0.80) = 48$ and $nq = (60)(0.20) = 12$ are both greater than 10.

Therefore, the sampling distribution model for the proportion of 60 randomly selected people who have left-brain language control is Normal, with

$\mu_{\hat{p}} = p = 0.80$ and standard deviation $\sigma(\hat{p}) = \sqrt{\dfrac{pq}{n}} = \sqrt{\dfrac{(0.80)(0.20)}{60}} \approx 0.0516$.

b) According to the Normal model, the probability that over 75% of these 60 people have left-brain language control is approximately 0.834.

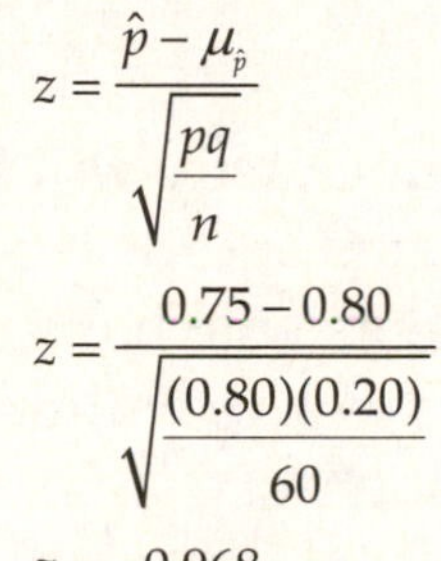

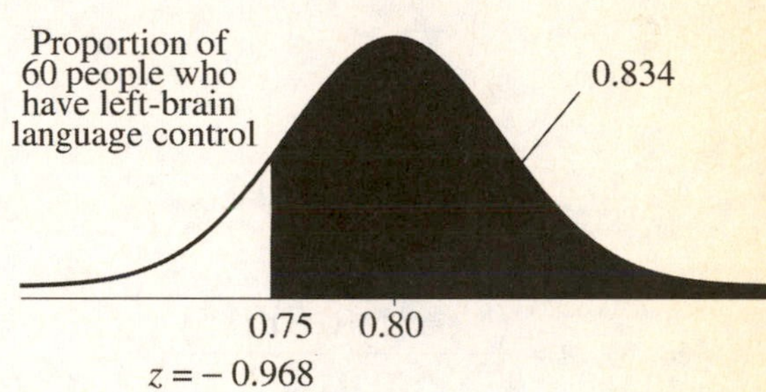

c) If the sample had consisted of 100 people, the probability would have been higher. A larger sample results in a smaller standard deviation for the sample proportion.

d) Answers may vary. Let's consider three standard deviations below the expected proportion to be "almost certain". It would take a sample of (exactly!) 576 people to make sure that 75% would be 3 standard deviations below the expected percentage of with left-brain language control.

$$z = \frac{\hat{p} - \mu_{\hat{p}}}{\sqrt{\frac{pq}{n}}}$$

$$-3 = \frac{0.75 - 0.80}{\sqrt{\frac{(0.80)(0.20)}{n}}}$$

$$n = \frac{(-3)^2(0.80)(0.20)}{(0.75 - 0.80)^2} = 576$$

Using round numbers for n instead of z, about 500 people in the sample would make the probability of choosing a sample with at least 75% of the people having left-brain language control is a whopping 0.997. It all depends on what "almost certain" means to you.

27. Crohn's disease.

a) Independence assumption: It is reasonable to think that the patients would respond to infliximab independently of each other.
Randomization condition: Assume that the 573 patients are representative of all Crohn's disease sufferers.
10% condition: 573 patients are less than 10% of all sufferers of Crohn's disease.
Success/Failure condition: $n\hat{p} = 335$ and $n\hat{q} = 238$ are both greater than 10.

Since the conditions are met, we can use a one-proportion z-interval to estimate the percentage of Crohn's disease sufferers who respond positively to infliximab.

$$\hat{p} \pm z^* \sqrt{\frac{\hat{p}\hat{q}}{n}} = \left(\frac{335}{573}\right) \pm 1.960 \sqrt{\frac{\left(\frac{335}{573}\right)\left(\frac{238}{573}\right)}{573}} = (54.4\%, 62.5\%)$$

b) We are 95% confident that between 54.4% and 62.5% of Crohn's disease sufferers would respond positively to infliximab.

c) 95% of random samples of size 573 will produce intervals that contain the true proportion of Crohn's disease sufferers who respond positively to infliximab.

29. Alcohol abuse.

$$ME = z^* \sqrt{\frac{\hat{p}\hat{q}}{n}}$$

$$0.04 = 1.645 \sqrt{\frac{(0.5)(0.5)}{n}}$$

$$n = \frac{(1.645)^2 (0.5)(0.5)}{(0.04)^2}$$

$$n \approx 423$$

The university will have to sample at least 423 students in order to estimate the proportion of students who have been drunk with in the past week to within ± 4%, with 90% confidence.

31. Preemies.

a) Randomization condition: Assume that these kids are representative of all kids.
10% condition: 242 and 233 are less than 10% of all kids.
Independent samples condition: The groups are independent.
Success/Failure condition: $n\hat{p}$ (preemies) = (242)(0.74) = 179, $n\hat{q}$ (preemies) = (242)(0.26) = 63, $n\hat{p}$ (normal weight) = (233)(0.83) = 193, and $n\hat{q}$ (normal weight) = 40 are all greater than 10, so the samples are both large enough.

Since the conditions have been satisfied, we will find a two-proportion z-interval.

$$(\hat{p}_N - \hat{p}_P) \pm z^* \sqrt{\frac{\hat{p}_N\hat{q}_N}{n_N} + \frac{\hat{p}_P\hat{q}_P}{n_P}}$$

$$= (0.83 - 0.74) \pm 1.960 \sqrt{\frac{(0.83)(0.17)}{233} + \frac{(0.74)(0.26)}{242}} = (0.017, 0.163)$$

We are 95% confident that between 1.7% and 16.3% more normal birth-weight children graduated from high school than children who were born premature.

b) Since the interval for the difference in percentage of high school graduates is above 0, there is evidence normal birth-weight children graduate from high school at a greater rate than premature children.

c) If preemies do not have a lower high school graduation rate than normal birth-weight children, then we made a Type I error. We rejected the null hypothesis of "no difference" when we shouldn't have.

33. Fried PCs.

a) H_0: The computer is undamaged.
H_A: The computer is damaged.

b) The biggest advantage is that all of the damaged computers will be detected, since, historically, damaged computers never pass all the tests. The disadvantage is that only 80% of undamaged computers pass all the tests. The engineers will be classifying 20% of the undamaged computers as damaged.

c) In this example, a Type I error is rejecting an undamaged computer. To allow this to happen only 5% of the time, the engineers would reject any computer that failed 3 or more tests, since 95% of the undamaged computers fail two or fewer tests.

d) The power of the test in part c is 20%, since only 20% of the damaged machines fail 3 or more tests.

e) By declaring computers "damaged" if the fail 2 or more tests, the engineers will be rejecting only 7% of undamaged computers. From 5% to 7% is an increase of 2% in α. Since 90% of the damaged computers fail 2 or more tests, the power of the test is now 90%, a substantial increase.

35. Approval 2007.

H_0 : George W. Bush's May 2007 disapproval rating was 66%. ($p = 0.66$)
H_A : George W. Bush's disapproval rating was lower than 66%. ($p < 0.66$)

Independence assumption: One adult's response will not affect another's.
Randomization condition: The adults were chosen randomly.
10% condition: 1000 adults are less than 10% of all adults.
Success/Failure condition: $np = (1000)(0.66) = 660$ and $nq = (1000)(0.44) = 440$ are both greater than 10, so the sample is large enough.

The conditions have been satisfied, so a Normal model can be used to model the sampling distribution of the proportion, with $\mu_{\hat{p}} = p = 0.66$ and

$$\sigma(\hat{p}) = \sqrt{\frac{pq}{n}} = \sqrt{\frac{(0.66)(0.44)}{1000}} \approx 0.017.$$

We can perform a one-proportion z-test. The observed approval rating is $\hat{p} = 0.63$.

The value of z is -2.00. Since the P-value = 0.023 is low, we reject the null hypothesis. There is strong evidence that President George W. Bush's May 2007 disapproval rating was lower than the 66% disapproval rating of President Richard Nixon.

37. Name Recognition.

a) The company wants evidence that the athlete's name is recognized more often than 25%.

b) Type I error means that fewer than 25% of people will recognize the athlete's name, yet the company offers the athlete an endorsement contract anyway. In this case, the company is employing an athlete that doesn't fulfill their advertising needs.

Type II error means that more than 25% of people will recognize the athlete's name, but the company doesn't offer the contract to the athlete. In this case, the company is letting go of an athlete that meets their advertising needs.

c) If the company uses a 10% level of significance, the company will hire more athletes that don't have high enough name recognition for their needs. The risk of committing a Type I error is higher.

At the same level of significance, the company is less likely to lose out on athletes with high name recognition. They will commit fewer Type II errors.

39. NIMBY.

Randomization condition: Not only was the sample random, but Gallup randomly divided the respondents into groups.
10% condition: 502 and 501 are less than 10% of all adults.
Independent samples condition: The groups are independent.
Success/Failure condition: The number of respondents in favor and opposed in both groups are all greater than 10, so the samples are both large enough.

Since the conditions have been satisfied, we will find a two-proportion z-interval.

$$(\hat{p}_1 - \hat{p}_2) \pm z^* \sqrt{\frac{\hat{p}_1\hat{q}_1}{n_1} + \frac{\hat{p}_2\hat{q}_2}{n_2}}$$

$$= (0.53 - 0.40) \pm 1.960\sqrt{\frac{(0.53)(0.47)}{502} + \frac{(0.40)(0.60)}{501}} = (0.07, 0.19)$$

We are 95% confident that the proportion of U.S. adults who favor nuclear energy is between 7 and 19 percentage points higher than the proportion that would accept a nuclear plant near their area.

Chapter 23 – Inferences About Means

1. *t*-models, part I.

a) 1.74 **b)** 2.37 **c)** 0.0524 **d)** 0.0889

3. *t*-models, part III.

As the number of degrees of freedom increases, the shape and center of *t*-models do not change. The spread of *t*-models decreases as the number of degrees of freedom increases, and the shape of the distribution becomes closer to Normal.

5. Cattle.

a) Not correct. A confidence interval is for the mean weight gain of the population of all cows. It says nothing about individual cows. This interpretation also appears to imply that there is something special about the interval that was generated, when this interval is actually one of many that could have been generated, depending on the cows that were chosen for the sample.

b) Not correct. A confidence interval is for the mean weight gain of the population of all cows, not individual cows.

c) Not correct. We don't need a confidence interval about the average weight gain for cows in this study. We are certain that the mean weight gain of the cows in this study is 56 pounds. Confidence intervals are for the mean weight gain of the population of all cows.

d) Not correct. This statement implies that the average weight gain varies. It doesn't. We just don't know what it is, and we are trying to find it. The average weight gain is either between 45 and 67 pounds, or it isn't.

e) Not correct. This statement implies that there is something special about our interval, when this interval is actually one of many that could have been generated, depending on the cows that were chosen for the sample. The correct interpretation is that 95% of samples of this size will produce an interval that will contain the mean weight gain of the population of all cows.

7. Meal plan.

a) Not correct. The confidence interval is not about the individual students in the population.

b) Not correct. The confidence interval is not about individual students in the sample. In fact, we know exactly what these students spent, so there is no need to estimate.

c) Not correct. We know that the mean cost for students in this sample was $1196.

d) Not correct. A confidence interval is not about other sample means.

e) This is the correct interpretation of a confidence interval. It estimates a population parameter.

9. Pulse rates.

a) We are 95% confident the interval 70.9 to 74.5 beats per minute contains the true mean heart rate.

b) The width of the interval is about 74.5 – 70.9 = 3.6 beats per minute. The margin of error is half of that, about 1.8 beats per minute.

c) The margin of error would have been larger. More confidence requires a larger critical value of *t*, which increases the margin of error.

11. CEO compensation.

We should be hesitant to trust this confidence interval, since the conditions for inference are not met. The distribution is highly skewed and there is an outlier.

13. Normal temperature.

a) **Randomization condition:** The adults were randomly selected.
10% condition: 52 adults are less than 10% of all adults.
Nearly Normal condition: The sample of 52 adults is large, and the histogram shows no serious skewness, outliers, or multiple modes.

The people in the sample had a mean temperature of 98.2846° and a standard deviation in temperature of 0.682379°. Since the conditions are satisfied, the sampling distribution of the mean can be modeled by a Student's *t* model, with 52 – 1 = 51 degrees of freedom. We will use a one-sample *t*-interval with 98% confidence for the mean body temperature.
(By hand, use $t_{50}^{*} \approx 2.403$ from the table.)

b) $\bar{y} \pm t_{n-1}^{*}\left(\frac{s}{\sqrt{n}}\right) = 98.2846 \pm t_{51}^{*}\left(\frac{0.682379}{\sqrt{52}}\right) \approx (98.06, 98.51)$

c) We are 98% confident that the interval 98.06°F to 98.51°F contains the true mean body temperature for adults. (If you calculated the interval by hand, using $t_{50}^{*} \approx 2.403$ from the table, your interval may be slightly different than intervals calculated using technology. With the rounding used here, they are identical. Even if they aren't, it's not a big deal.)

d) 98% of all random samples of size 52 will produce intervals that contain the true mean body temperature of adults.

e) Since the interval is completely below the body temperature of 98.6°F, there is strong evidence that the true mean body temperature of adults is lower than 98.6°F.

15. Normal temperatures, part II.

a) The 90% confidence interval would be narrower than the 98% confidence interval. We can be more precise with our interval when we are less confident.

b) The 98% confidence interval has a greater chance of containing the true mean body temperature of adults than the 90% confidence interval, but the 98% confidence interval is less precise (wider) than the 90% confidence interval.

c) The 98% confidence interval would be narrower if the sample size were increased from 52 people to 500 people. The smaller standard error would result in a smaller margin of error.

d) Our sample of 52 people gave us a 98% confidence interval with a margin of error of (98.51 – 98.05)/2 = 0.225°F. In order to get a margin of error of 0.1, less than half of that, we need a sample over 4 times as large. It should be safe to use $t^*_{100} \approx 2.364$ from the table, since the sample will need to be larger than 101. Or we could use $z^* \approx 2.326$, since we expect the sample to be large. We need a sample of about 252 people in order to estimate the mean body temperature of adults to within 0.1°F.

$$ME = t^*_{n-1}\left(\frac{s}{\sqrt{n}}\right)$$

$$0.1 = 2.326\left(\frac{0.682379}{\sqrt{n}}\right)$$

$$n = \frac{(2.326)^2(0.682379)^2}{(0.1)^2}$$

$$n \approx 252$$

17. Speed of Light.

a) $\bar{y} \pm t^*_{n-1}\left(\frac{s}{\sqrt{n}}\right) = 756.22 \pm t^*_{22}\left(\frac{107.12}{\sqrt{23}}\right) \approx (709.9, 802.5)$

b) We are 95% confident that the interval 299,709.9 to 299,802.5 km/sec contains the speed of light.

c) We have assumed that the measurements are independent of each other and that the distribution of the population of all possible measurements is Normal. The assumption of independence seems reasonable, but it might be a good idea to look at a display of the measurements made by Michelson to verify that the Nearly Normal Condition is satisfied.

19. Departures 2009.

a) **Randomization condition:** Since there is no time trend, the monthly on-time departure rates should be independent. This is not a random sample, but should be representative.
10% condition: These months represent fewer than 10% of all months.
Nearly Normal condition: The histogram looks unimodal, and slightly skewed to the left. Since the sample size is 172, this should not be of concern.

b) The on-time departure rates in the sample had a mean of 80.759%, and a standard deviation in of 4.706%. Since the conditions have been satisfied, construct a one-sample t-interval, with 172 – 1 = 171 degrees of freedom, at 90% confidence.

$$\bar{y} \pm t^*_{n-1}\left(\frac{s}{\sqrt{n}}\right) = 80.759 \pm t^*_{171}\left(\frac{4.706}{\sqrt{172}}\right) \approx (80.17, 81.35)$$

c) We are 90% confident that the interval from 80.17% to 81.35% contains the true mean monthly percentage of on-time flight departures.

21. For example, second look.

The 95% confidence interval lies entirely above the 0.08 ppm limit. This is evidence that mirex contamination is too high and consistent with rejecting the null hypothesis. We used an upper-tail test, so the *P*-value should be smaller than $\frac{1}{2}(1-0.95) = 0.025$, and it was.

23. Pizza.

If in fact the mean cholesterol of pizza eaters does not indicate a health risk, then only 7 out of every 100 samples would be expected to have mean cholesterol as high or higher than the mean cholesterol observed in the sample.

25. TV Safety.

a) The inspectors are performing an upper-tail test. They need to prove that the stands will support 500 pounds (or more) easily.

b) The inspectors commit a Type I error if they certify the stands as safe, when they are not.

c) The inspectors commit a Type II error if they decide the stands are not safe, when they are.

27. TV safety revisited.

a) The value of α should be decreased. This means a smaller chance of declaring the stands safe under the null hypothesis that they are not safe.

b) The power of the test is the probability of correctly detecting that the stands can safely hold over 500 pounds.

c) **1)** The company could redesign the stands so that their strength is more consistent, as measured by the standard deviation. Redesigning the manufacturing process is likely to be quite costly.
2) The company could increase the number of stands tested. This costs them both time to perform the test and money to pay the quality control personnel.
3) The company could increase α, effectively lowering their standards for what is required to certify the stands "safe". This is a big risk, since there is a greater chance of Type I error, namely allowing unsafe stands to be sold.

4) The company could make the stands stronger, increasing the mean amount of weight that the stands can safely hold. This type of redesign is expensive.

29. Marriage.

a) H_0: The mean age at which American men first marry is 23.3 years. $(\mu = 23.3)$
H_A: The mean age is greater than 23.3 years. $(\mu > 23.3)$

b) **Randomization condition:** The 40 men were selected randomly.
10% condition: 40 men are less than 10% of all recently married men.
Nearly Normal condition: The population of ages of men at first marriage is likely to be skewed to the right. It is much more likely that there are men who marry for the first time at an older age than at an age that is very young. We should examine the distribution of the sample to check for serious skewness and outliers, but with a large sample of 40 men, it should be safe to proceed.

c) Since the conditions for inference are satisfied, we can model the sampling distribution of the mean age of men at first marriage with $N\left(23.3, \frac{\sigma}{\sqrt{n}}\right)$. Since we do not know σ, the standard deviation of the population, $\sigma(\bar{y})$ will be estimated by $SE(\bar{y}) = \frac{s}{\sqrt{n}}$, and we will use a Student's *t* model, with $40 - 1 = 39$ degrees of freedom, $t_{39}\left(23.3, \frac{s}{\sqrt{40}}\right)$.

d) The mean age at first marriage in the sample was 24.2 years, with a standard deviation in age of 5.3 years. Use a one-sample *t*-test, modeling the sampling distribution of $\bar{y}$ with $t_{39}\left(23.3, \frac{5.3}{\sqrt{40}}\right)$. The *P*-value is 0.1447.

$$t = \frac{\bar{y} - \mu_0}{SE(\bar{y})}$$

$$t = \frac{24.2 - 23.3}{\frac{5.3}{\sqrt{40}}}$$

$$t \approx 1.07$$

P = 0.1447
23.3
24.2
t = 1.07
t_{39}

e) If the mean age at first marriage is still 23.3 years, there is a 14.5% chance of getting a sample mean of 24.2 years or older simply from natural sampling variation.

f) Since the *P*-value = 0.1447 is high, we fail to reject the null hypothesis. There is no evidence to suggest that the mean age of men at first marriage has changed from 23.3 years, the mean in 1960.

31. Ruffles.

a) **Randomization condition:** The 6 bags were not selected at random, but it is reasonable to think that these bags are representative of all bags of chips.
10% condition: 6 bags are less than 10% of all bags of chips.
Nearly Normal condition: The histogram of the weights of chips in the sample is nearly normal.

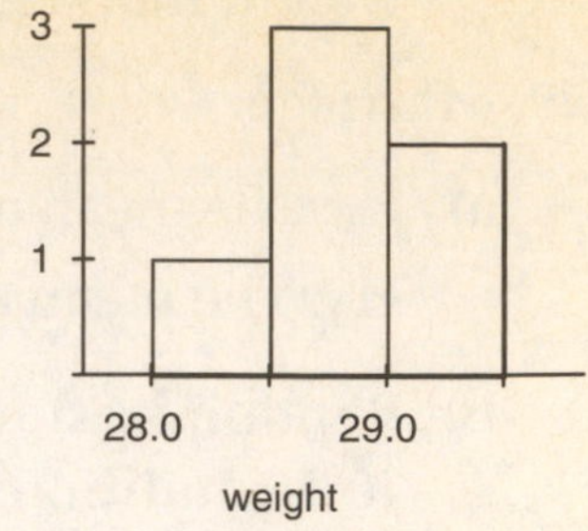

b) $\bar{y} \approx 28.78$ grams, $s \approx 0.40$ grams

c) Since the conditions for inference have been satisfied, use a one-sample t-interval, with $6 - 1 = 5$ degrees of freedom, at 95% confidence.

$$\bar{y} \pm t_{n-1}^{*}\left(\frac{s}{\sqrt{n}}\right) = 28.78 \pm t_{5}^{*}\left(\frac{0.40}{\sqrt{6}}\right) \approx (28.36,\ 29.21)$$

d) We are 95% confident that the mean weight of the contents of Ruffles bags is between 28.36 and 29.21 grams.

e) Since the interval is above the stated weight of 28.3 grams, there is evidence that the company is filling the bags to more than the stated weight, on average.

33. Popcorn.

a) Hopps made a Type I error. He mistakenly rejected the null hypothesis that the proportion of unpopped kernels was 10% (or higher).

b) H_0: The mean proportion of unpopped kernels is 10%. ($\mu = 10$)

H_A: The mean proportion of unpopped kernels is lower than 10%. ($\mu < 10$)

Randomization condition: The 8 bags were randomly selected.
10% condition: 8 bags are less than 10% of all bags.
Nearly Normal condition: The histogram of the percentage of unpopped kernels is unimodal and roughly symmetric.

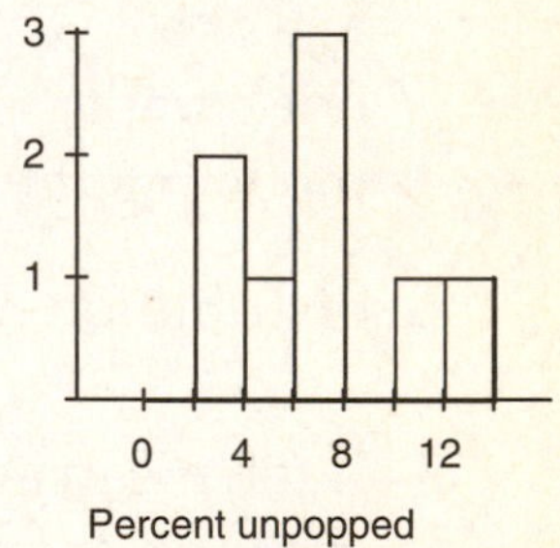

The bags in the sample had a mean percentage of unpopped kernels of 6.775 percent and a standard deviation in percentage of unpopped kernels of 3.637 percent. Since the conditions for inference are satisfied, we can model the sampling distribution of the mean percentage of unpopped kernels with a Student's t model, with $8 - 1 = 7$ degrees of freedom, $t_7\left(6.775, \frac{3.637}{\sqrt{8}}\right)$.

We will perform a one-sample t-test.

Since the P-value = 0.0203 is low, we reject the null hypothesis. There is evidence to suggest the mean percentage of unpopped kernels is less than 10% at this setting.

$$t = \frac{\bar{y} - \mu_0}{SE(\bar{y})}$$

$$t = \frac{6.775 - 10}{\frac{3.637}{\sqrt{8}}}$$

$$t \approx -2.51$$

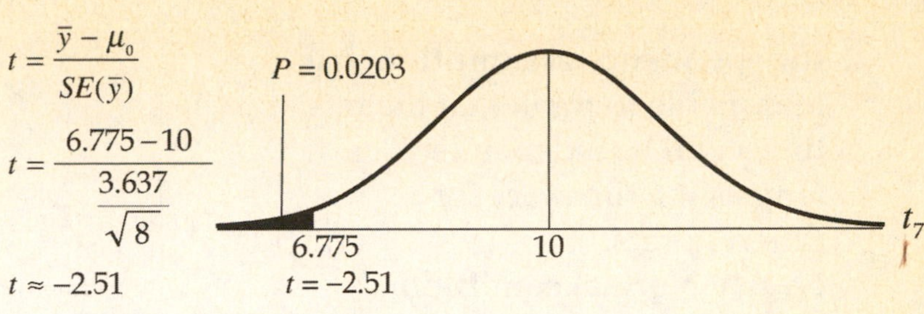

35. Chips ahoy.

a) **Randomization condition: The bags of cookies were randomly selected.**
10% condition: 16 bags are less than 10% of all bags
Nearly Normal condition: The Normal probability plot is reasonably straight, and the histogram of the number of chips per bag is unimodal and symmetric.

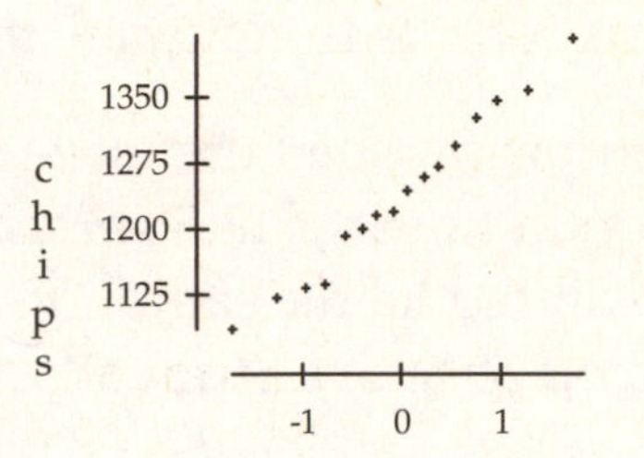

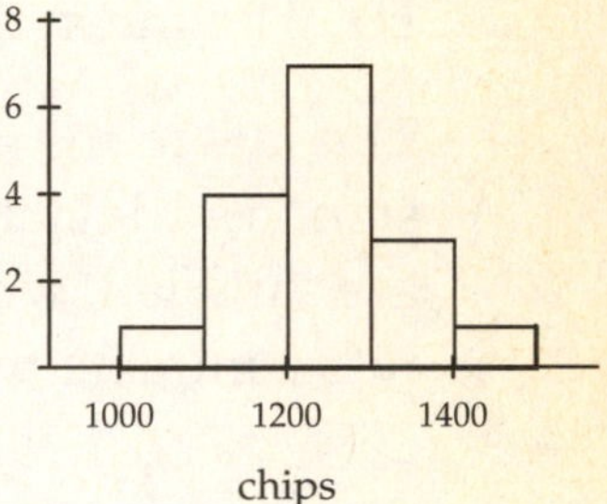

b) The bags in the sample had with a mean number of chips or 1238.19, and a standard deviation of 94.282 chips. Since the conditions for inference have been satisfied, use a one-sample t-interval, with 16 – 1 = 15 degrees of freedom, at 95% confidence.

$$\bar{y} \pm t_{n-1}^{*}\left(\frac{s}{\sqrt{n}}\right) = 1238.19 \pm t_{15}^{*}\left(\frac{94.282}{\sqrt{16}}\right) \approx (1187.9, 1288.4)$$

We are 95% confident that the mean number of chips in an 18-ounce bag of Chips Ahoy cookies is between 1187.9 and 1288.4.

c) H_0: The mean number of chips per bag is 1000. ($\mu = 1000$)

H_A: The mean number of chips per bag is greater than 1000. ($\mu > 1000$)

Since the confidence interval is well above 1000, there is strong evidence that the mean number of chips per bag is well above 1000.

However, since the "1000 Chip Challenge" is about individual bags, not means, the claim made by Nabisco may not be true. If the mean was around 1188 chips, the low end of our confidence interval, and the standard deviation of the population was about 94 chips, our best estimate obtained from our sample, a bag containing 1000 chips would be about 2 standard deviations below the mean. This is not likely to happen, but not an outrageous occurrence. These data do not provide evidence that the "1000 Chip Challenge" is true.

37. Maze.

a) **Independence assumption:** It is reasonable to think that the rats' times will be independent, as long as the times are for different rats.
Nearly Normal condition: There is an outlier in both the Normal probability plot and the histogram that should probably be eliminated before continuing the test. One rat took a long time to complete the maze.

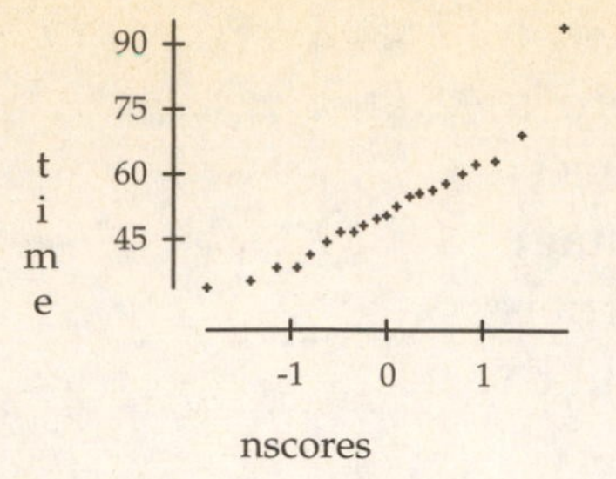

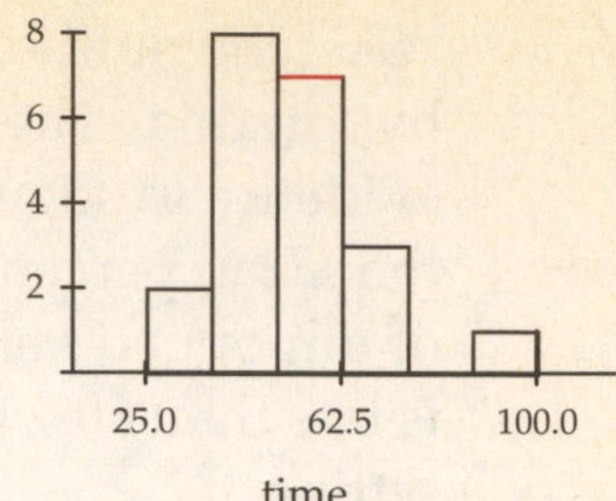

b) H_0: The mean time for rats to complete this maze is 60 seconds. $(\mu = 60)$

H_A: The mean time for rats to complete this maze is not 60 seconds. $(\mu \neq 60)$

The rats in the sample finished the maze with a mean time of 52.21 seconds and a standard deviation in times of 13.5646 seconds. Since the conditions for inference are satisfied, we can model the sampling distribution of the mean time in which rats complete the maze with a Student's t model, with $21 - 1 = 20$ degrees of freedom, $t_{20}\left(60, \frac{13.5646}{\sqrt{21}}\right)$. We will perform a one-sample t-test.

Since the P-value = 0.0160 is low, we reject the null hypothesis. There is evidence that the mean time required for rats to finish the maze is not 60 seconds. Our evidence suggests the mean time is less than 60 seconds.

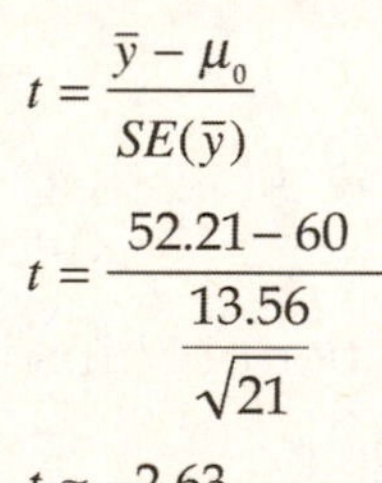

$$t = \frac{\bar{y} - \mu_0}{SE(\bar{y})}$$

$$t = \frac{52.21 - 60}{\frac{13.56}{\sqrt{21}}}$$

$$t \approx -2.63$$

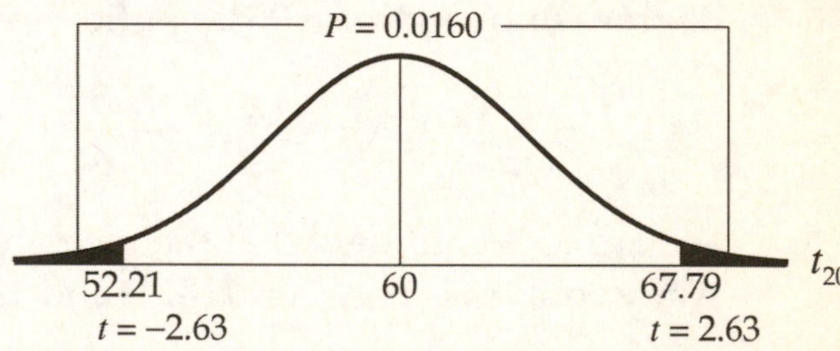

c) Without the outlier, the rats in the sample finished the maze with a mean time of 50.13 seconds and standard deviation in times of 9.90 seconds. Since the conditions for inference are satisfied, we can model the sampling distribution of the mean time in which rats complete the maze with a Student's t model, with $20 - 1 = 19$ degrees of freedom, $t_{19}\left(60, \frac{9.90407}{\sqrt{20}}\right)$. We will use a one-sample t-test.

This test results in a value of $t = -4.46$, and a two-sided P-value = 0.0003. Since the P-value is low, we reject the null hypothesis. There is evidence that the mean time required for rats to finish the maze is not 60 seconds. Our evidence suggests that the mean time is actually less than 60 seconds.

d) According to both tests, there is evidence that the mean time required for rats to complete the maze is different than 60 seconds. The maze does not meet the "one-minute average" requirement. It should be noted that the test without the outlier is the appropriate test. The one slow rat made the mean time required seem much higher than it probably was.

39. Driving distance.

a) $\bar{y} \pm t^*_{n-1}\left(\frac{s}{\sqrt{n}}\right) = 286.36 \pm t^*_{191}\left(\frac{8.627}{\sqrt{192}}\right) \approx (285.1,\ 287.6)$

b) These data are not a random sample of golfers. The top professionals are not representative of all golfers and were not selected at random. We might consider the 2009 data to represent the population of all professional golfers, past, present, and future.

c) The data are means for each golfer, so they are less variable than if we looked at separate drives.

Chapter 24 – Comparing Means

1. Dogs and calories.

Yes, the 95% confidence interval would contain 0. The high P-value means that we lack evidence of a difference, so 0 is a possible value for $\mu_{Meat} - \mu_{Beef}$.

3. Dogs and fat.

a) Plausible values for $\mu_{Meat} - \mu_{Beef}$ are all negative, so the mean fat content is probably higher for beef hot dogs.

b) The fact that the confidence interval does not contain 0 indicates that the difference is significant.

c) The corresponding alpha level is 10%.

5. Dogs and fat, second helping.

a) False. The confidence interval is about means, not about individual hot dogs.

b) False. The confidence interval is about means, not about individual hot dogs.

c) True.

d) False. Confidence intervals based on other samples will also try to estimate the true difference in population means. There's not reason to expect other samples to conform to this result.

e) True.

7. Learning math.

a) The margin of error of this confidence interval is (11.427 – 5.573)/2 = 2.927 points.

b) The margin of error for a 98% confidence interval would have been larger. The critical value of t^* is larger for higher confidence levels. We need a wider interval to increase the likelihood that we catch the true mean difference in test scores within our interval. In other words, greater confidence comes at the expense of precision.

c) We are 95% confident that the mean score for the CPMP math students will be between 5.573 and 11.427 points higher on this assessment than the mean score of the traditional students.

d) Since the entire interval is above 0, there is strong evidence that students who learn with CPMP will have higher mean scores is algebra than those in traditional programs.

9. CPMP, again.

a) H_0: The mean score of CPMP students is the same as the mean score of traditional students. $(\mu_C = \mu_T \text{ or } \mu_C - \mu_T = 0)$

H_A: The mean score of CPMP students is different from the mean score of traditional students. $(\mu_C \neq \mu_T \text{ or } \mu_C - \mu_T \neq 0)$

b) **Independent groups assumption:** Scores of students from different classes should be independent.
Randomization condition: Although not specifically stated, classes in this experiment were probably randomly assigned to either CPMP or traditional curricula.
10% condition: 312 and 265 are less than 10% of all students.
Nearly Normal condition: We don't have the actual data, so we can't check the distribution of the sample. However, the samples are large. The Central Limit Theorem allows us to proceed.

Since the conditions are satisfied, we can use a two-sample *t*-test with 583 degrees of freedom (from the computer).

c) If the mean scores for the CPMP and traditional students are really equal, there is less than a 1 in 10,000 chance of seeing a difference as large or larger than the observed difference just from natural sampling variation.

d) Since the *P*-value < 0.0001, reject the null hypothesis. There is strong evidence that the CPMP students have a different mean score than the traditional students. The evidence suggests that the CPMP students have a higher mean score.

11. Commuting.

a) **Independent groups assumption:** Since the choice of route was determined at random, the commuting times for Route A are independent of the commuting times for Route B.
Randomization condition: The man randomly determined which route he would travel on each day.
Nearly Normal condition: The histograms of travel times for the routes are roughly unimodal and symmetric. (Given)

Since the conditions are satisfied, it is appropriate to model the sampling distribution of the difference in means with a Student's *t*-model, with 33.1 degrees of freedom (from the approximation formula). We will construct a two-sample *t*-interval, with 95% confidence.

$$(\bar{y}_B - \bar{y}_A) \pm t^*_{df}\sqrt{\frac{s_B^2}{n_B} + \frac{s_A^2}{n_A}} = (43 - 40) \pm t^*_{33.1}\sqrt{\frac{2^2}{20} + \frac{3^2}{20}} \approx (1.36,\ 4.64)$$

We are 95% confident that Route B has a mean commuting time between 1.36 and 4.64 minutes longer than the mean commuting time of Route A.

b) Since 5 minutes is beyond the high end of the interval, there is no evidence that the Route B is an average of 5 minutes longer than Route A. It appears that the old-timer may be exaggerating the average difference in commuting time.

13. Cereal.

Independent groups assumption: The percentage of sugar in the children's cereals is unrelated to the percentage of sugar in adult's cereals.
Randomization condition: It is reasonable to assume that the cereals are representative of all children's cereals and adult cereals in sugar content.
Nearly Normal condition: The histogram of adult cereal sugar content is skewed to the right, but the sample sizes are of reasonable size. The Central Limit Theorem allows us to proceed.

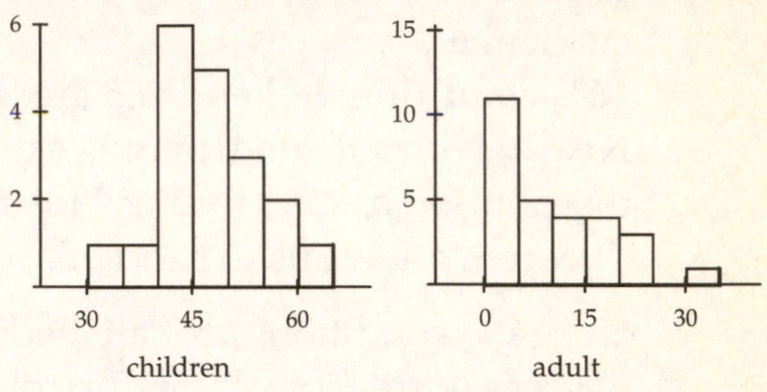

Since the conditions are satisfied, it is appropriate to model the sampling distribution of the difference in means with a Student's *t*-model, with 42 degrees of freedom (from the approximation formula). We will construct a two-sample *t*-interval, with 95% confidence.

$$(\bar{y}_C - \bar{y}_A) \pm t^*_{df}\sqrt{\frac{s_C^2}{n_C}+\frac{s_A^2}{n_A}} = (46.8 - 10.1536) \pm t^*_{42}\sqrt{\frac{6.41838^2}{19}+\frac{7.61239^2}{28}} \approx (32.49,\ 40.80)$$

We are 95% confident that children's cereals have a mean sugar content that is between 32.49% and 40.80% higher than the mean sugar content of adult cereals.

15. Reading.

H_0: The mean reading comprehension score of students who learn by the new method is the same as the mean score of students who learn by traditional methods. $(\mu_N = \mu_T \text{ or } \mu_N - \mu_T = 0)$

H_A: The mean reading comprehension score of students who learn by the new method is greater than the mean score of students who learn by traditional methods. $(\mu_N > \mu_T \text{ or } \mu_N - \mu_T > 0)$

Independent groups assumption: Student scores in one group should not have an impact on the scores of students in the other group.

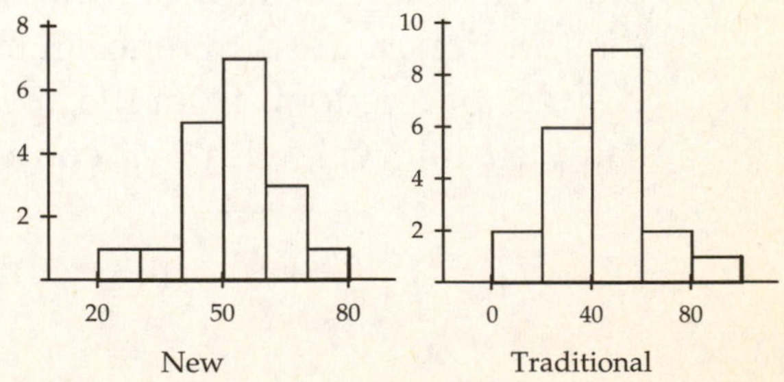

Randomization condition: Students were randomly assigned to classes.
Nearly Normal condition: The histograms of the scores are unimodal and symmetric.

Since the conditions are satisfied, it is appropriate to model the sampling distribution of the difference in means with a Student's *t*-model, with 33 degrees of freedom (from the approximation formula). We will perform a two-sample *t*-test. We know:

$$\bar{y}_N = 51.7222 \qquad \bar{y}_T = 41.8$$
$$s_N = 11.7062 \qquad s_T = 17.4495$$
$$n_N = 18 \qquad n_T = 20$$

The sampling distribution model has mean 0, with standard error:

$$SE(\bar{y}_N - \bar{y}_T) = \sqrt{\frac{11.7062^2}{18} + \frac{17.4495^2}{20}} \approx 4.779.$$

The observed difference between the mean scores is $51.7222 - 41.8 \approx 9.922$.

$$t = \frac{(\bar{y}_N - \bar{y}_T) - (0)}{SE(\bar{y}_N - \bar{y}_T)}$$
$$t \approx \frac{9.922}{4.779}$$
$$t \approx 2.076$$

P = 0.0228
0
9.92
t = 2.076
t_{33}

Since the *P*-value = 0.0228 is low, we reject the null hypothesis. There is evidence that the students taught using the new activities have a higher mean score on the reading comprehension test than the students taught using traditional methods.

Tukey's test is not appropriate because one group has both the highest and lowest scores.

To calculate W for the rank sum test, rank the data, then add the ranks for the students taught using new activities.

$$W = 4 + 7.5 + 16.5(3) + 19 + 21 + 22 + 23.5 + 25 + 28 + 29 + 30 + 31 + 33 + 34.5 + 36 + 37 = 430$$

$$\mu_W = \frac{n_1(N+1)}{2} = \frac{18(38+1)}{2} = 351$$

$$SD(W) = \sqrt{Var(W)} = \sqrt{\frac{n_1 n_2 (N+1)}{12}} = \sqrt{\frac{(18)(20)(38+1)}{12}} \approx 34.205$$

$$z = \frac{W - \mu_W}{SD(W)} = \frac{430 - 351}{\sqrt{\frac{(18)(20)(38+1)}{12}}} \approx 2.310$$

This z-score results in a 1-sided P-value of 0.010. Since the p-value is so low, we conclude that the mean reading comprehension score is higher for the students taught using the new activities. This is the same conclusion reached with the two-sample t-test.

17. Baseball 2006.

a) The boxplots of the average number of runs scored at the ballparks in the two leagues are at the right. Both distributions appear at least roughly symmetric, with roughly the same center, around 9.5 runs. The distribution of average runs appears a bit more spread out for the American League.

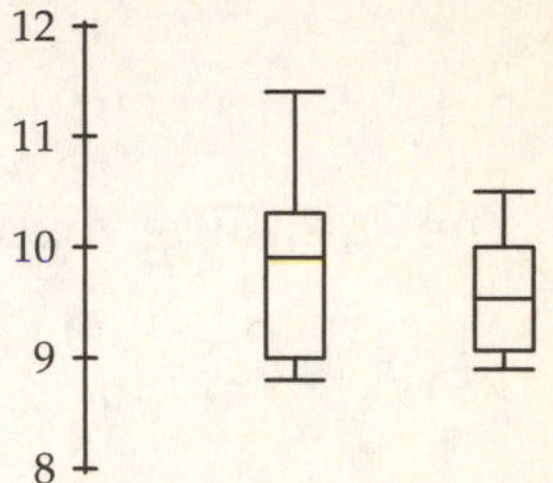

b) $\bar{y} \pm t^*_{n-1}\left(\frac{s}{\sqrt{n}}\right) = 9.79286 \pm t^*_{13}\left(\frac{0.757998}{\sqrt{14}}\right) \approx (9.36,\ 10.23)$

We are 95% confident that the mean number of runs scored per game in American League stadiums is between 9.36 and 10.23.

c) The average of 10.5 runs scored per game in Coors Field is not unusual. It is the highest average in the National League, but by no means an outlier.

d) If you attempt to use two confidence intervals to assess a difference in means, you are actually adding standard deviations. But it's the variances that add, not the standard deviations. The two-sample difference of means procedure takes this into account.

19. Double header 2006.

a)

$$(\bar{y}_A - \bar{y}_N) \pm t^*_{df}\sqrt{\frac{s_A^2}{n_A} + \frac{s_N^2}{n_N}}$$

$$= (9.79286 - 9.43750) \pm t^*_{23}\sqrt{\frac{0.757998^2}{14} + \frac{0.638618^2}{16}} \approx (-0.18,\ 0.89)$$

b) We are 95% confident that the mean number of runs scored in American League stadiums is between 0.18 runs lower and 0.89 runs higher than the mean number of runs scored in National League stadiums.

c) Since the interval contains 0, there is no evidence of a difference in the mean number of runs scored per game in the stadiums of the two leagues.

21. Job satisfaction.

A two-sample *t*-procedure is not appropriate for these data, because the two groups are not independent. They are before and after satisfaction scores for the same workers. Workers that have high levels of job satisfaction before the exercise program is implemented may tend to have higher levels of job satisfaction than other workers after the program as well.

23. Sex and violence.

a) Since the *P*-value = 0.136 is high, we fail to reject the null hypothesis. There is no evidence of a difference in the mean number of brands recalled by viewers of sexual content and viewers of violent content.

b) H_0: The mean number of brands recalled is the same for viewers of sexual content and viewers of neutral content. $(\mu_S = \mu_N$ or $\mu_S - \mu_N = 0)$

H_A: The mean number of brands recalled is different for viewers of sexual content and viewers of neutral content. $(\mu_S \neq \mu_N$ or $\mu_S - \mu_N \neq 0)$

Independent groups assumption: Recall of one group should not affect recall of another.
Randomization condition: Subjects were randomly assigned to groups.
Nearly Normal condition: The samples are large.

Since the conditions are satisfied, it is appropriate to model the sampling distribution of the difference in means with a Student's *t*-model, with 214 degrees of freedom (from the approximation formula). We will perform a two-sample *t*-test.

The sampling distribution model has mean 0, with standard error:

$$SE(\bar{y}_S - \bar{y}_N) = \sqrt{\frac{1.76^2}{108} + \frac{1.77^2}{108}} \approx 0.24.$$

The observed difference between the mean scores is 1.71 – 3.17 = – 1.46.

$$t = \frac{(\bar{y}_S - \bar{y}_N) - (0)}{SE(\bar{y}_S - \bar{y}_N)}$$

$$t \approx \frac{-1.46}{0.24}$$

$$t \approx -6.08$$

Since the *P*-value = 5.5×10^{-9} is low, we reject the null hypothesis. There is strong evidence that the mean number of brand names recalled is different for viewers of sexual content and viewers of neutral content. The evidence suggests that viewers of neutral ads remember more brand names on average than viewers of sexual content.

25. Sex and violence II.

a) H_0: The mean number of brands recalled is the same for viewers of violent content and viewers of neutral content. $(\mu_V = \mu_N \text{ or } \mu_V - \mu_N = 0)$

H_A: The mean number of brands recalled is different for viewers of violent content and viewers of neutral content. $(\mu_V \neq \mu_N \text{ or } \mu_V - \mu_N \neq 0)$

Independent groups assumption: Recall of one group should not affect recall of another.
Randomization condition: Subjects were randomly assigned to groups.
Nearly Normal condition: The samples are large.

Since the conditions are satisfied, it is appropriate to model the sampling distribution of the difference in means with a Student's t-model, with 201.96 degrees of freedom (from the approximation formula). We will perform a two-sample t-test.

The sampling distribution model has mean 0, with standard error:

$$SE(\bar{y}_V - \bar{y}_N) = \sqrt{\frac{1.61^2}{101} + \frac{1.62^2}{103}} \approx 0.226.$$

The observed difference between the mean scores is $3.02 - 4.65 = -1.63$.

$$t = \frac{(\bar{y}_V - \bar{y}_N) - (0)}{SE(\bar{y}_V - \bar{y}_N)}$$

$$t \approx \frac{-1.63}{0.226}$$

$$t \approx -7.21$$

Since the P-value $= 1.1 \times 10^{-11}$ is low, we reject the null hypothesis. There is strong evidence that the mean number of brand names recalled is different for viewers of violent content and viewers of neutral content. The evidence suggests that viewers of neutral ads remember more brand names on average than viewers of violent content.

b) $$(\bar{y}_N - \bar{y}_S) \pm t^*_{df}\sqrt{\frac{s_N^2}{n_N} + \frac{s_S^2}{n_S}} = (4.65 - 2.72) \pm t^*_{204.8}\sqrt{\frac{1.62^2}{103} + \frac{1.85^2}{106}} \approx (1.456,\ 2.404)$$

We are 95% confident that the mean number of brand names recalled 24 hours later is between 1.46 and 2.40 higher for viewers of shows with neutral content than for viewers of shows with sexual content.

27. Hungry?

H_0: The mean number of ounces of ice cream people scoop is the same for large and small bowls. $(\mu_{big} = \mu_{small} \text{ or } \mu_{big} - \mu_{small} = 0)$

H_A: The mean number of ounces of ice cream people scoop is the different for large and small bowls. $(\mu_{big} \neq \mu_{small} \text{ or } \mu_{big} - \mu_{small} \neq 0)$

Independent groups assumption: The amount of ice cream scooped by individuals should be independent.
Randomization condition: Subjects were randomly assigned to groups.
Nearly Normal condition: Assume that this condition is met.

Since the conditions are satisfied, it is appropriate to model the sampling distribution of the difference in means with a Student's t-model, with 34 degrees of freedom (from the approximation formula). We will perform a two-sample t-test.

The sampling distribution model has mean 0, with standard error:

$$SE(\bar{y}_{big} - \bar{y}_{small}) = \sqrt{\frac{2.91^2}{22} + \frac{1.84^2}{26}} \approx 0.7177 \text{ oz.}$$

The observed difference between the mean amounts is 6.58 – 5.07 = 1.51 oz.

$$t = \frac{(\bar{y}_{big} - \bar{y}_{small}) - (0)}{SE(\bar{y}_{big} - \bar{y}_{small})}$$

$$t \approx \frac{1.51}{0.7177}$$

$$t \approx 2.104$$

Since the P-value of 0.0428 is low, we reject the null hypothesis. There is strong evidence that the that the mean amount of ice cream people put into a bowl is related to the size of the bowl. People tend to put more ice cream into the large bowl, on average, than the small bowl.

29. Lower scores?

a) Assuming that the conditions for inference were met by the NAEP, a 95% confidence interval for the difference in mean score is:

$$(\bar{y}_{1996} - \bar{y}_{2000}) \pm t^*_{df}\sqrt{\frac{s^2_{1996}}{n_{1996}} + \frac{s^2_{2000}}{n_{2000}}} = (150 - 147) \pm 1.960(1.22) \approx (0.61, 5.39)$$

Since the samples sizes are very large, it should be safe to use $z^* = 1.960$ for the critical value of t. We are 95% confident that the mean score in 2000 was between 0.61 and 5.39 points lower than the mean score in 1996. Since 0 is not contained in the interval, this provides evidence that the mean score has decreased from 1996 to 2000.

b) Both sample sizes are very large, which will make the standard errors of these samples very small. They are both likely to be very accurate. The difference in sample size shouldn't make you any more certain or any less certain.

However, these results are completely dependent upon whether or not the conditions for inference were met. If, by sampling more students, the NAEP sampled from a different population, then the two years are incomparable.

31. Running heats.

H0: The mean time to finish is the same for heats 2 and 5. $(\mu_2 = \mu_5 \text{ or } \mu_2 - \mu_5 = 0)$

HA: The mean time is not the same for heats 2 and 5. $(\mu_2 \neq \mu_5 \text{ or } \mu_2 - \mu_5 \neq 0)$

Independent groups assumption: The two heats were independent.
Randomization condition: Runners were randomly assigned.
Nearly Normal condition: The boxplots show an outlier in the distribution of times in heat 2. We will perform the test twice, once with the outlier and once without.

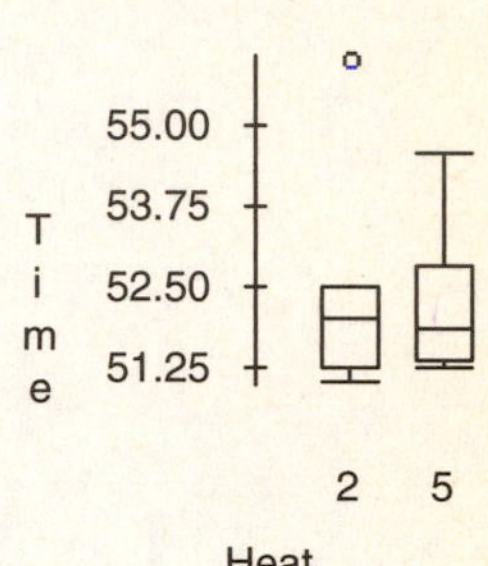

Since the conditions are satisfied, it is appropriate to model the sampling distribution of the difference in means with a Student's t-model, with 10.82 degrees of freedom (from the approximation formula). We will perform a two-sample t-test.

The sampling distribution model has mean 0, with standard error:

$$SE(\bar{y}_2 - \bar{y}_5) = \sqrt{\frac{1.69319^2}{7} + \frac{1.20055^2}{7}} \approx 0.7845.$$

The observed difference between mean times is 52.3557 – 52.3286 = 0.0271.

$$t = \frac{(\bar{y}_2 - \bar{y}_5) - (0)}{SE(\bar{y}_2 - \bar{y}_5)}$$

$$t \approx \frac{0.0271}{0.7845}$$

$$t \approx 0.035$$

Since the P-value = 0.97 is high, we fail to reject the null hypothesis. There is no evidence that the mean time to finish differs between the two heats.

Without the outlier, it is appropriate to model the sampling distribution of the difference in means with a Student's t-model, with 8.83 degrees of freedom (from the approximation formula). We will perform a two-sample t-test.

The sampling distribution model has mean 0, with standard error:

$$SE(\bar{y}_2 - \bar{y}_5) = \sqrt{\frac{0.56955^2}{6} + \frac{1.20055^2}{7}} \approx 0.5099.$$

The observed difference between mean times is 51.7467 – 52.3286 = –0.5819.

Since the P-value = 0.2837 is high, we fail to reject the null hypothesis. There is no evidence that the mean time to finish differs between the two heats.

$$t = \frac{(\bar{y}_2 - \bar{y}_5) - (0)}{SE(\bar{y}_2 - \bar{y}_5)}$$

$$t \approx \frac{-0.58}{0.5099}$$

$$t \approx -1.14$$

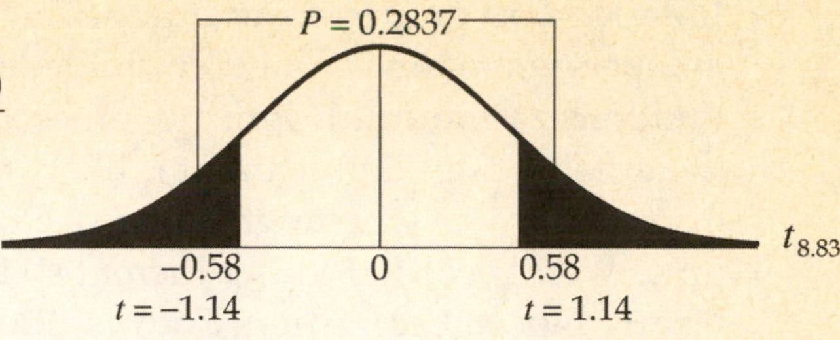

33. Tees.

H0: The mean ball velocity is the same for regular and Stinger tees.
$(\mu_S = \mu_R \text{ or } \mu_S - \mu_R = 0)$

HA: The mean ball velocity is higher for the Stinger tees. $(\mu_S > \mu_R \text{ or } \mu_S - \mu_R > 0)$

Assuming the conditions are satisfied, it is appropriate to model the sampling distribution of the difference in means with a Student's t-model, with 7.03 degrees of freedom (from the approximation formula). We will perform a two-sample t-test.

The sampling distribution model has mean 0, with standard error:

$$SE(\bar{y}_S - \bar{y}_R) = \sqrt{\frac{0.41^2}{6} + \frac{0.89^2}{6}} \approx 0.4000.$$

The observed difference between the mean velocities is 128.83 – 127 = 1.83.

$$t = \frac{(\bar{y}_S - \bar{y}_R) - (0)}{SE(\bar{y}_S - \bar{y}_R)}$$

$$t \approx \frac{1.83}{0.4000}$$

$$t \approx 4.57$$

Since the P-value = 0.0013, we reject the null hypothesis. There is strong evidence that the mean ball velocity for stinger tees is higher than the mean velocity for regular tees.

35. Crossing Ontario.

a)

$$(\bar{y}_M - \bar{y}_W) \pm t^*_{df}\sqrt{\frac{s_M^2}{n_M} + \frac{s_W^2}{n_W}}$$

$$= (1196.75 - 1271.59) \pm t^*_{37.67}\sqrt{\frac{304.369^2}{20} + \frac{261.111^2}{22}} \approx (-252.89, 103.21)$$

We are 95% confident that the interval –252.89 to 103.20 minutes (–74.84 ± 178.05 minutes) contains the true difference in mean crossing times between men and women. Because the interval includes zero, we cannot be confident that there is any difference at all.

b) **Independent groups assumption:** The times from the two groups are likely to be independent of one another, provided that these were all individual swims.
Randomization condition: The times are not a random sample from any identifiable population, but it is likely that the times are representative of times from swimmers who might attempt a challenge such as this. Hopefully, these times were recorded from different swimmers.
Nearly Normal condition: The distributions of times are both unimodal, with no outliers. The distribution of men's times is somewhat skewed to the left.

37. Rap.

a) H_0: The mean memory test score is the same for those who listen to rap as it is for those who listen to no music. $(\mu_R = \mu_N \text{ or } \mu_R - \mu_N = 0)$

H_A: The mean memory test score is lower for those who listen to rap than it is for those who listen to no music. $(\mu_R < \mu_N \text{ or } \mu_R - \mu_N < 0)$

Independent groups assumption: The groups are not related in regards to memory score.
Randomization condition: Subjects were randomly assigned to groups.
Nearly Normal condition: We don't have the actual data. We will assume that the distributions of the populations of memory test scores are Normal.

Since the conditions are satisfied, it is appropriate to model the sampling distribution of the difference in means with a Student's *t*-model, with 20.00 degrees of freedom (from the approximation formula). We will perform a two-sample *t*-test.

The sampling distribution model has mean 0, with standard error:

$$SE(\bar{y}_R - \bar{y}_N) = \sqrt{\frac{3.99^2}{29} + \frac{4.73^2}{13}} \approx 1.5066.$$

The observed difference between the mean number of objects remembered is $10.72 - 12.77 = -2.05$.

$$t = \frac{(\bar{y}_R - \bar{y}_N) - (0)}{SE(\bar{y}_R - \bar{y}_N)}$$

$$t \approx \frac{-2.05}{1.5066}$$

$$t \approx -1.36$$

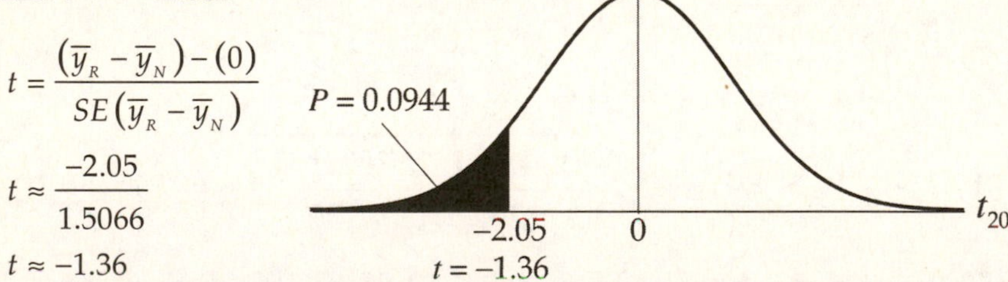

Since the *P*-value = 0.0944 is high, we fail to reject the null hypothesis. There is little evidence that the mean number of objects remembered by those who listen to rap is lower than the mean number of objects remembered by those who listen to no music.

b) We did not conclude that there was a difference in the number of items remembered.

Chapter 25 – Paired Samples and Blocks

1. More eggs?

a) Randomly assign 50 hens to each of the two kinds of feed. Compare the mean egg production of the two groups at the end of one month.

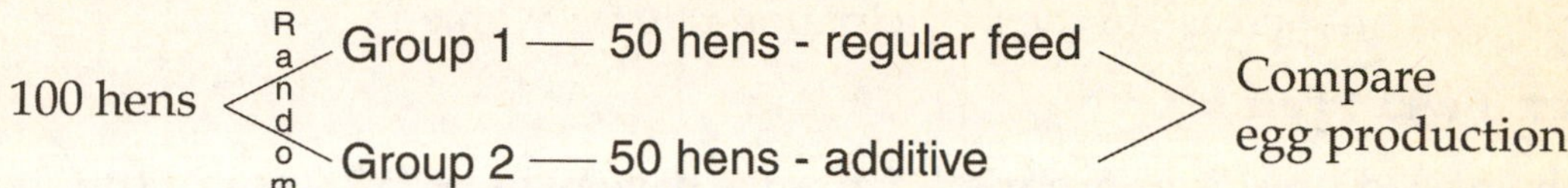

b) Randomly divide the 100 hens into two groups of 50 hens each. Feed the hens in the first group the regular feed for two weeks, then switch to the additive for 2 weeks. Feed the hens in the second group the additive for two weeks, and then switch to the regular feed for two weeks. Subtract each hen's "regular" egg production from her "additive" egg production, and analyze the mean difference in egg production.

100 hens — Random
Group 1 — additive first 2 weeks, regular second 2 weeks
Group 2 — regular first 2 weeks, additive second 2 weeks
Analyze differences in egg production

c) The matched pairs design in part b is the stronger design. Hens vary in their egg production regardless of feed. This design controls for that variability by matching the hens with themselves.

3. Sex sells.

a) Randomly assign half of the volunteers to watch ads with sexual images, and assign the other half to watch ads without the sexual images. Record the number of items remembered. Then have each group watch the other type of ad. Record the number of items recalled. Examine the difference in the number of items remembered for each person.

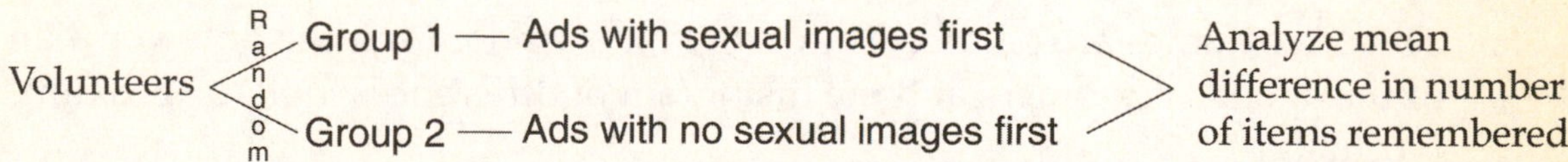

b) Randomly assign half of the volunteers to watch ads with sexual images, and assign the other half to watch ads without the sexual images. Record the number of items remembered. Compare the mean number of products remembered by each group.

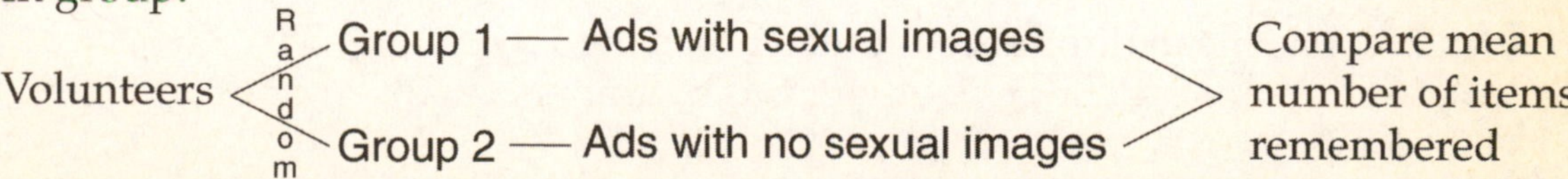

5. Women.

a) The paired t-test is appropriate. The labor force participation rate for two different years was paired by city.

b) Since the P-value = 0.0244, there is evidence of a difference in the average labor force participation rate for women between 1968 and 1972. The evidence suggests an increase in the participation rate for women.

7. Friday the 13th Part I:

a) The paired t-test is appropriate, since we have pairs of Fridays in 5 different months. Data from adjacent Fridays within a month may be more similar than randomly chosen Fridays.

b) Since the P-value = 0.0212, there is evidence that the mean number of cars on the M25 motorway on Friday the 13th is less than the mean number of cars on the previous Friday.

c) We don't know if these Friday pairs were selected at random. Obviously, if these are the Fridays with the largest differences, this will affect our conclusion. The Nearly Normal condition appears to be met by the differences, but the sample size of five pairs is small.

9. Online insurance I.

Adding variances requires that the variables be independent. These price quotes are for the same cars, so they are paired. Drivers quoted high insurance premiums by the local company will be likely to get a high rate from the online company, too.

11. Online insurance II.

a) The histogram would help you decide whether the online company offers cheaper insurance. We are concerned with the difference in price, not the distribution of each set of prices.

b) Insurance cost is based on risk, so drivers are likely to see similar quotes from each company, making the differences relatively smaller.

c) The price quotes are paired. They were for a random sample of fewer than 10% of the agent's customers and the histogram of differences looks approximately Normal.

13. Online insurance 3.

H_0: The mean difference between online and local insurance rates is zero. $(\mu_{Local-Online} = 0)$

H_A: The mean difference is greater than zero. $(\mu_{Local-Online} > 0)$

Since the conditions are satisfied (in a previous exercise), the sampling distribution of the difference can be modeled with a Student's t-model with 10 – 1 = 9 degrees of freedom, $t_9\left(0, \frac{175.663}{\sqrt{10}}\right)$.

We will use a paired t-test, with $\bar{d} = 45.9$.

Since the P-value = 0.215 is high, we fail to reject the null hypothesis. There is no evidence that online insurance premiums are lower on average.

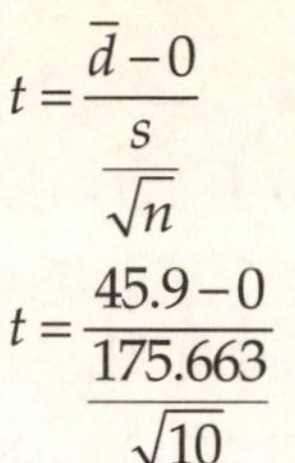

$$t = \frac{\bar{d} - 0}{\frac{s}{\sqrt{n}}}$$

$$t = \frac{45.9 - 0}{\frac{175.663}{\sqrt{10}}}$$

$$t \approx 0.83$$

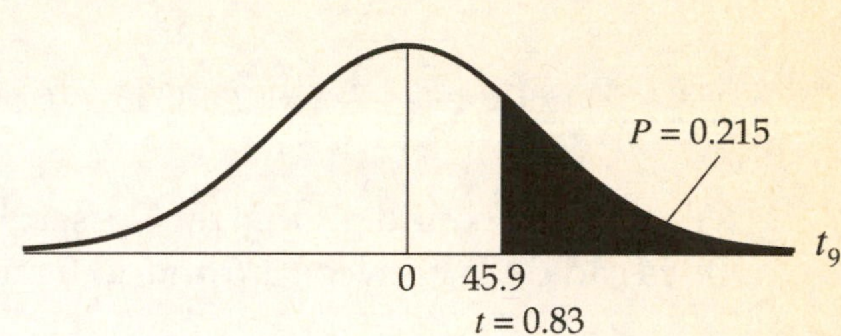

15. Temperatures.

Paired data assumption: The data are paired by city.
Randomization condition: These cities might not be representative of all European cities, so be cautious in generalizing the results.
10% condition: 12 cities are less than 10% of all European cities.
Normal population assumption: The histogram of differences between January and July mean temperature is roughly unimodal and symmetric.

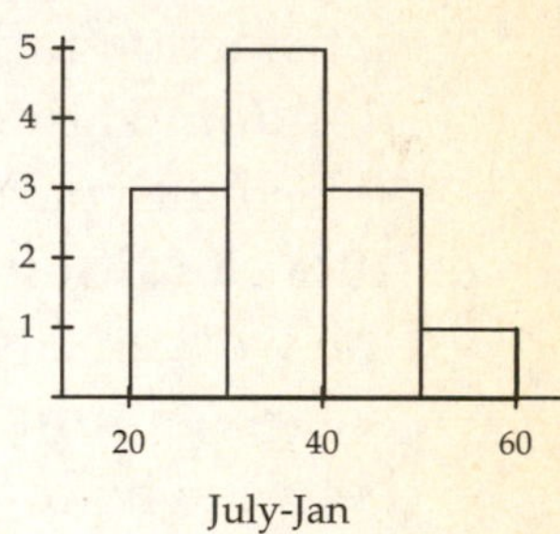

Since the conditions are satisfied, the sampling distribution of the difference can be modeled with a Student's t-model with 12 – 1 = 11 degrees of freedom. We will find a paired t-interval, with 90% confidence.

$$\bar{d} \pm t^*_{n-1}\left(\frac{s_d}{\sqrt{n}}\right) = 36.8333 \pm t^*_{11}\left(\frac{8.66375}{\sqrt{12}}\right) \approx (32.3, 41.3)$$

We are 90% confident that the average high temperature in European cities in July is an average of between 32.3° to 41.4° higher than in January.

17. Push-ups.

Independent groups assumption: The group of boys is independent of the group of girls.
Randomization condition: Assume that students are assigned to gym classes at random.
10% condition: 12 boys and 12 girls are less than 10% of all kids.

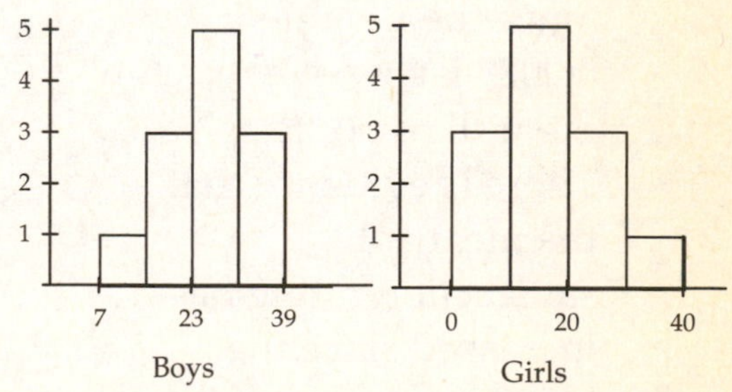

Nearly Normal condition: The histograms of the number of push-ups from each group are roughly unimodal and symmetric.

Since the conditions are satisfied, it is appropriate to model the sampling distribution of the difference in means with a Student's *t*-model, with 21 degrees of freedom (from the approximation formula). We will construct a two-sample *t*-interval, with 90% confidence.

$$(\bar{y}_B - \bar{y}_G) \pm t^*_{df}\sqrt{\frac{s_B^2}{n_B} + \frac{s_G^2}{n_G}} = (23.8333 - 16.5000) \pm t^*_{21}\sqrt{\frac{7.20900^2}{12} + \frac{8.93919^2}{12}} \approx (1.6, 13.0)$$

We are 90% confident that, at Gossett High, the mean number of push-ups that boys can do is between 1.6 and 13.0 more than the mean for the girls.

19. Job satisfaction.

a) Use a paired *t*-test.

Paired data assumption: The data are before and after job satisfaction rating for the same workers.
Randomization condition: The workers were randomly selected to participate.
10% condition: Assume that 10 workers are less than 10% of the workers at the company.
Nearly Normal conditon: The histogram of differences between before and after job satisfaction ratings is roughly unimodal and symmetric.

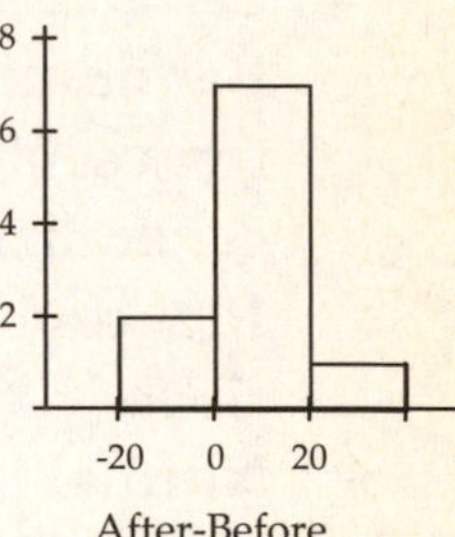

b) H$_0$: The mean difference in before and after job satisfaction scores is zero, and the exercise program is not effective at improving job satisfaction. $(\mu_d = 0)$

H$_A$: The mean difference is greater than zero. $(\mu_d > 0)$

Since the conditions are satisfied, the sampling distribution of the difference can be modeled with a Student's *t*-model with 10 – 1 = 9 degrees of freedom, $t_9\left(0, \frac{7.47217}{\sqrt{10}}\right)$. We will use a paired *t*-test, with $\bar{d} = 8.5$.

Since the *P*-value = 0.0029 is low, we reject the null hypothesis. There is evidence that the mean job satisfaction rating has increased since the implementation of the exercise program.

$$t = \frac{\bar{d} - 0}{\frac{s_d}{\sqrt{n}}}$$

$$t = \frac{8.5 - 0}{\frac{7.47217}{\sqrt{10}}}$$

$$t \approx 3.60$$

$P = 0.0029$

0   8.5 t_9

$t = 3.60$

c) We concluded that there was an increase job satisfaction rating. If we are wrong, and there actually was no increase, we have committed a Type I error.

d) 8 out of 10 people's satisfaction increased.

H_0: Half of the workers had increased job satisfaction after the exercise program was implemented. $(p = 0.5)$

H_A: More than half of the workers had increased job satisfaction after the exercise program was implemented. $(p > 0.5)$

Since the number of workers is less than 20, use the model $Binom(10, 0.5)$. The 1-sided *P*-value is 0.0547. Since the *P*-value is fairly low, there is some evidence that the mean job satisfaction index has increased after the implementation of the exercise program. This is the same conclusion reached using the matched pairs *t*-test.

21. Yogurt.

H_0: The mean difference in calories between servings of strawberry and vanilla yogurt is zero. $(\mu_d = 0)$

H_A: The mean difference in calories between servings of strawberry and vanilla yogurt is different from zero. $(\mu_d \neq 0)$

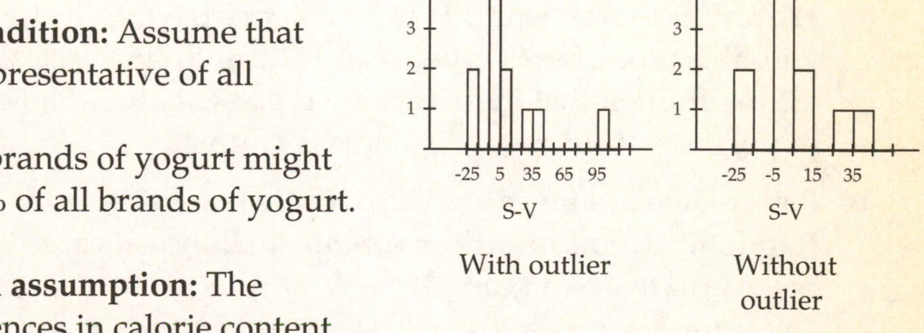

With outlier   Without outlier

Paired data assumption: The yogurt is paired by brand.
Randomization condition: Assume that these brands are representative of all brands.
10% condition: 12 brands of yogurt might not be less than 10% of all brands of yogurt. Proceed cautiously.
Normal population assumption: The histogram of differences in calorie content between strawberry and vanilla shows an outlier, Great Value. When the outlier is eliminated, the histogram of differences is roughly unimodal and symmetric.

When Great Value yogurt is removed, the conditions are satisfied. The sampling distribution of the difference can be modeled with a Student's *t*-model with

$11 - 1 = 10$ degrees of freedom, $t_{10}\left(0, \frac{18.0907}{\sqrt{11}}\right)$.

We will use a paired *t*-test, with $\bar{d} \approx 4.54545$.

Since the P-value = 0.4241 is high, we fail to reject the null hypothesis. There is no evidence of a mean difference in calorie content between strawberry yogurt and vanilla yogurt.

$$t = \frac{\bar{d} - 0}{\frac{s_d}{\sqrt{n}}}$$

$$t = \frac{4.54545 - 0}{\frac{18.0907}{\sqrt{11}}}$$

$$t \approx 0.833$$

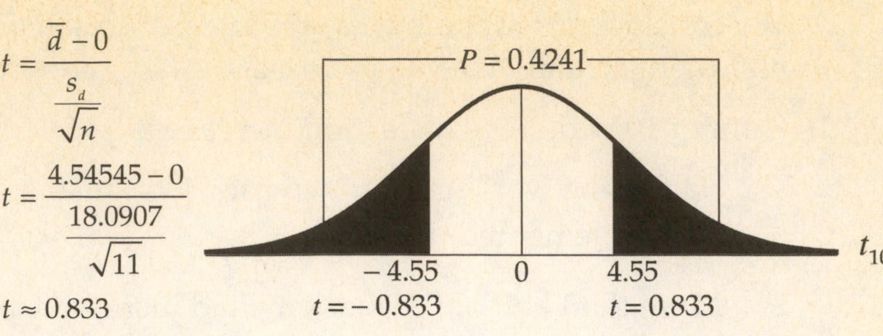

23. Braking.

a) **Randomization Condition:** These cars are not a random sample, but are probably representative of all cars in terms of stopping distance.
10% Condition: These 10 cars are less than 10% of all cars.
Nearly Normal Condition: A histogram of the stopping distances is skewed to the right, but this may just be sampling variation from a Normal population. The "skew" is only a couple of stopping distances. We will proceed cautiously.

The cars in the sample had a mean stopping distance of 138.7 feet and a standard deviation of 9.66149 feet. Since the conditions have been satisfied, construct a one-sample t-interval, with 10 – 1 = 9 degrees of freedom, at 95% confidence.

$$\bar{y} \pm t^*_{n-1}\left(\frac{s}{\sqrt{n}}\right) = 138.7 \pm t^*_9\left(\frac{9.66149}{\sqrt{10}}\right) \approx (131.8, 145.6)$$

We are 95% confident that the mean dry pavement stopping distance for cars with this type of tires is between 131.8 and 145.6 feet. This estimate is based on an assumption that these cars are representative of all cars and that the population of stopping distances is Normal.

b) **Paired data assumption:** The data are paired by car.
Randomization condition: Assume that the cars are representative of all cars.
10% condition: The 10 cars tested are less than 10% of all cars.
Normal population assumption: The difference in stopping distance for car #4 is an outlier, at only 12 feet. After excluding this difference, the histogram of differences is unimodal and symmetric.

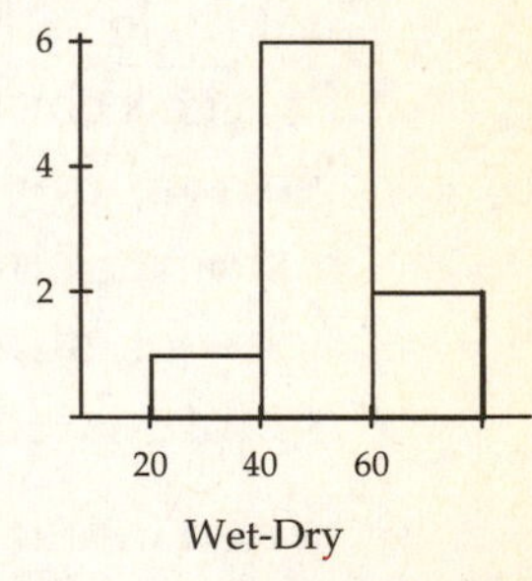

Since the conditions are satisfied, the sampling distribution of the difference can be modeled with a Student's t-model with 9 – 1 = 8 degrees of freedom. We will find a paired t-interval, with 95% confidence.

$$\bar{d} \pm t^*_{n-1}\left(\frac{s_d}{\sqrt{n}}\right) = 55 \pm t^*_8\left(\frac{10.2103}{\sqrt{9}}\right) \approx (47.2, 62.8)$$

With car #4 removed, we are 95% confident that the mean increase in stopping distance on wet pavement is between 47.2 and 62.8 feet. (If you leave the outlier in, the interval is 38.8 to 62.6 feet, but you should remove it! This procedure is sensitive to outliers!)

25. Tuition 2006.

a) **Paired data assumption:** The data are paired by college.
Randomization condition: The colleges were selected randomly.
Normal population assumption: The tuition difference for UC Irvine, at $9300, is an outlier. Once it has been set aside, the histogram of the differences is roughly unimodal and symmetric.

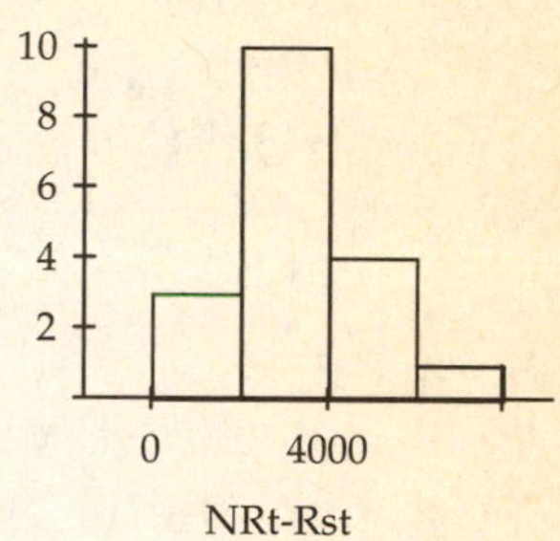

Since the conditions are satisfied, the sampling distribution of the difference can be modeled with a Student's t-model with 18 – 1 = 17 degrees of freedom. We will find a paired t-interval, with 90% confidence.

$$\bar{d} \pm t_{n-1}^{*}\left(\frac{s_d}{\sqrt{n}}\right) = 3266.67 \pm t_{17}^{*}\left(\frac{1588.56}{\sqrt{18}}\right) \approx (2615.31,\ 3918.02)$$

b) With UC Irvine removed, we are 90% confident that the mean increase in tuition for nonresidents versus residents is between about $2615 and $3918. (If you left UC Irvine in your data, the interval is about $2759 to $4409, but you should set it aside! This procedure is sensitive to the presence of outliers!)

c) There is no evidence to suggest that the magazine made a false claim. An increase of $3500 for nonresidents is contained within our 90% confidence interval.

27. Strikes.

a) Since 60% of 50 pitches is 30 pitches, the Little Leaguers would have to throw an average of more than 30 strikes in order to give support to the claim made by the advertisements.

H_0: The mean number of strikes thrown by Little Leaguers who have completed the training is 30. $(\mu_A = 30)$

H_A: The mean number of strikes thrown by Little Leaguers who have completed the training is greater than 30. $(\mu_A > 30)$

Randomization Condition: Assume that these players are representative of all Little League pitchers.
10% Condition: 20 pitchers are less than 10% of all Little League pitchers.

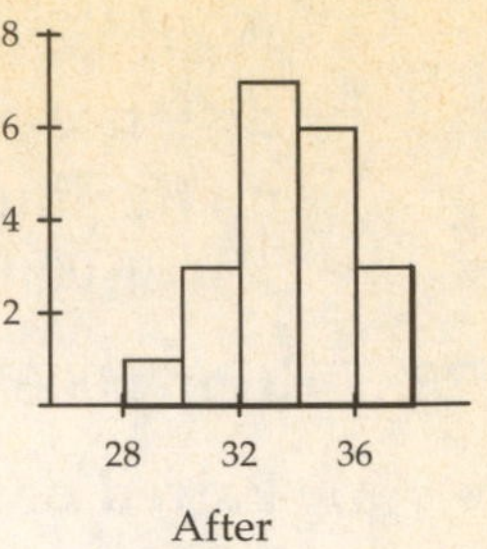

Nearly Normal Condition: The histogram of the number of strikes thrown is roughly unimodal and symmetric.

The pitchers in the sample threw a mean of 33.15 strikes, with a standard deviation of 2.32322 strikes. Since the conditions for inference are satisfied, we can model the sampling distribution of the mean number of strikes thrown with a Student's t model, with $20 - 1 = 19$ degrees of freedom, $t_{19}\left(30, \frac{2.32322}{\sqrt{20}}\right)$. We will perform a one-sample t-test.

$$t = \frac{\bar{y}_A - \mu_0}{\frac{s_A}{\sqrt{n_A}}}$$

$$t = \frac{33.15 - 30}{\frac{2.32322}{\sqrt{20}}}$$

$$t = 6.06$$

Since the P-value $= 3.92 \times 10^{-6}$ is very low, we reject the null hypothesis. There is strong evidence that the mean number of strikes that Little Leaguers can throw after the training is more than 30. (This test says nothing about the effectiveness of the training; just that Little Leaguers can throw more than 60% strikes on average after completing the training. This might not be an improvement.)

b) H_0: The mean difference in number of strikes thrown before and after the training is zero. $(\mu_d = 0)$

H_A: The mean difference in number of strikes is greater than zero. $(\mu_d > 0)$

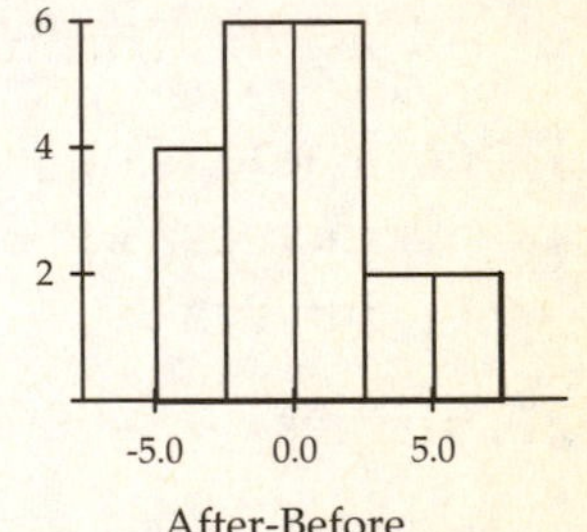

Paired data assumption: The data are paired by pitcher.
Randomization condition: Assume that these players are representative of all Little League pitchers.
10% condition: 20 players are less than 10% of all players.
Normal population assumption: The histogram of differences is roughly unimodal and symmetric.

Since the conditions are satisfied, the sampling distribution of the difference can be modeled with a Student's t-model with $20 - 1 = 19$ degrees of freedom, $t_{19}\left(0, \frac{3.32297}{\sqrt{19}}\right)$. We will use a paired t-test, with $\bar{d} = 0.1$.

Since the P-value = 0.4472 is high, we fail to reject the null hypothesis. There is no evidence of a mean difference in number of strikes thrown before and after the training. The training does not appear to be effective.

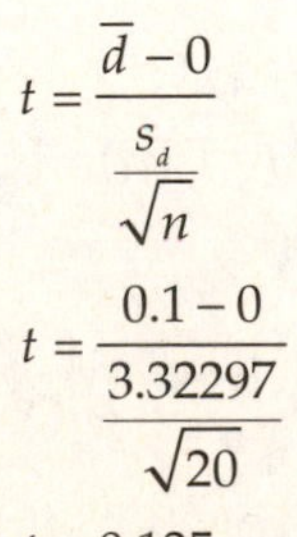

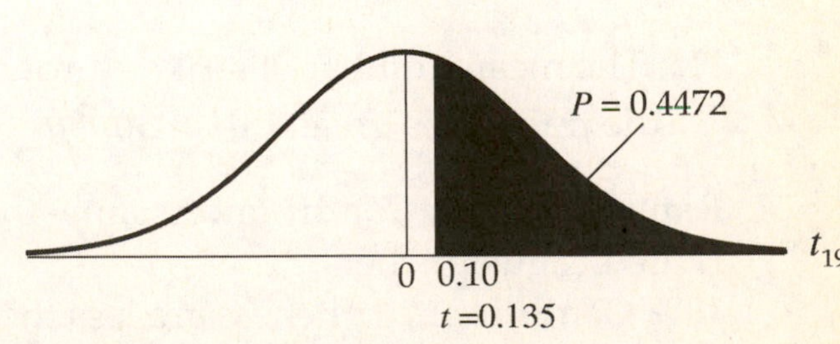

c) With the matched-pairs sign test, we can test whether the training is effective in improving a players ability to throw strikes, but not whether players can throw strikes more than 60% of the time.

Of the 20 players, 9 increased their number of strikes, and 9 decreased their number of strikes. Since we don't count "no change", the sample proportion is $\frac{9}{18}=0.5$.

H_0: Half of the players increased their number of strikes. $(p=0.5)$

H_A: More than half of the players increased their number of strikes. $(p>0.5)$

Since the number of players is less than 20, use the model $Binom(18, 0.5)$. The 1-sided P-value is 0.593. Since the P-value is high, there is no evidence of a mean increase in the number of strikes thrown after the training program. This is the same conclusion reached using the matched pairs t-test.

29. Wheelchair marathon 2009.

a) Even if the individual times show a trend of improving speed over time, the differences may well be independent of each other. They are subject to random year-to-year fluctuations, and we may believe that these data are representative of similar races. We don't have any information with which to check the Nearly Normal condition.

b) $\bar{d} \pm t^*_{n-1}\left(\frac{s_d}{\sqrt{n}}\right) = -3.00 \pm t^*_{32}\left(\frac{35.4674}{\sqrt{33}}\right) \approx (-15.58, 9.58)$

We are 95% confident that the interval –15.58 to 9.58 minutes contains the true mean time difference for women's wheelchair times and men's running times.

c) The interval contains zero, so we would not reject a null hypothesis of no mean difference at a significance level of 0.05. We are unable to discern a difference between the female wheelchair times and the male running times.

31. BST.

a) **Paired data assumption:** We are testing the same cows, before and after injections of BST.
Randomization condition: These cows are likely to be representative of all Ayrshires.
10% condition: 60 cows are less than 10% of all cows.
Normal population assumption: We don't have the list of individual differences, so we can't look at a histogram. The sample is large, so we may proceed.

Since the conditions are satisfied, the sampling distribution of the difference can be modeled with a Student's t-model with $60 - 1 = 59$ degrees of freedom. We will find a paired t-interval, with 95% confidence.

b) $\bar{d} \pm t^*_{n-1}\left(\frac{s_d}{\sqrt{n}}\right) = 14 \pm t^*_{59}\left(\frac{5.2}{\sqrt{60}}\right) \approx (12.66, 15.34)$

c) We are 95% confident that the mean increase in daily milk production for Ayshire cows after BST injection is between 12.66 and 15.34 pounds.

d) 25% of 47 pounds is 11.75 pounds. According to the interval generated in part b, the average increase in milk production is more than this, so the farmer can justify the extra expense for BST.

Chapter 26 – Comparing Counts

1. Which test?

a) Chi-square test of Independence. We have one sample and two variables. We want to see if the variable *account type* is independent of the variable *trade type.*

b) Some other statistics test. The variable *account size* is quantitative, not counts.

c) Chi-square test of Homogeneity. We have two samples (residential and non-residential students), and one variable, *courses.* We want to see if the distribution of *courses* is the same for the two groups.

3. Dice.

a) If the die were fair, you'd expect each face to show 10 times.

b) Use a chi-square test for goodness-of-fit. We are comparing the distribution of a single variable (outcome of a die roll) to an expected distribution.

c) H_0: The die is fair. (All faces have the same probability of coming up.)

H_A: The die is not fair. (Some faces are more or less likely to come up than others.)

d) **Counted data condition:** We are counting the number of times each face comes up.
Randomization condition: Die rolls are random and independent of each other.
Expected cell frequency condition: We expect each face to come up 10 times, and 10 is greater than 5.

e) Under these conditions, the sampling distribution of the test statistic is χ^2 on 6 – 1 = 5 degrees of freedom. We will use a chi-square goodness-of-fit test.

f)

Face	Observed	Expected	Residual = $(Obs - Exp)$	$(Obs - Exp)^2$	Component = $\frac{(Obs - Exp)^2}{Exp}$
1	11	10	1	1	0.1
2	7	10	– 3	9	0.9
3	9	10	– 1	1	0.1
4	15	10	5	25	2.5
5	12	10	2	4	0.4
6	6	10	– 4	16	1.6

$\sum = 5.6$

g) Since the *P*-value = 0.3471 is high, we fail to reject the null hypothesis. There is no evidence that the die is unfair.

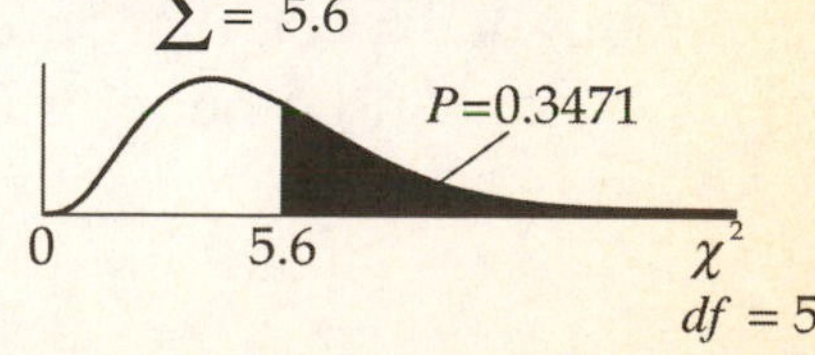

5. **Nuts.**

   a) The weights of the nuts are quantitative. Chi-square goodness-of-fit requires counts.

   b) In order to use a chi-square test, you could count the number of each type of nut. However, it's not clear whether the company's claim was a percentage by number or a percentage by weight.

7. **NYPD and race.**

H_0: The distribution of ethnicities in the police department represents the distribution of ethnicities of the youth of New York City.

H_A: The distribution of ethnicities in the police department does not represent the distribution of ethnicities of the youth of New York City.

Counted data condition: The percentages reported must be converted to counts.
Randomization condition: Assume that the current NYPD is representative of recent departments with respect to ethnicity.
Expected cell frequency condition: The expected counts are all much greater than 5.

(Note: The observed counts should be whole numbers. They are actual policemen. The expected counts may be decimals, since they behave like averages.)

Ethnicity	Observed	Expected
White	16965	7644.852
Black	3796	7383.042
Latino	5001	8247.015
Asian	367	2382.471
Other	52	523.620

Under these conditions, the sampling distribution of the test statistic is χ^2 on 5 – 1 = 4 degrees of freedom. We will use a chi-square goodness-of-fit test.

$$\chi^2 = \sum_{all\,cells} \frac{(Obs - Exp)^2}{Exp} = \frac{(16965 - 7644.852)^2}{7644.852} + \frac{(3796 - 7383.042)^2}{7383.042} + \frac{(5001 - 8247.015)^2}{8247.015}$$

$$+ \frac{(367 - 2382.471)^2}{2382.471} + \frac{(52 - 523.620)^2}{523.620} \approx 16,500$$

With χ^2 of over 16,500, on 4 degrees of freedom, the *P*-value is essentially 0, so we reject the null hypothesis. There is strong evidence that the distribution of ethnicities of NYPD officers does not represent the distribution of ethnicities of the youth of New York City. Specifically, the proportion of white officers is much higher than the proportion of white youth in the community. As one might expect, there are also lower proportions of officers who are black, Latino, Asian, and other ethnicities than we see in the youth in the community.

9. Fruit flies.

a) H_0: The ratio of traits in this type of fruit fly is 9:3:3:1, as genetic theory predicts.

H_A: The ratio of traits in this type of fruit fly is not 9:3:3:1.

trait	Observed	Expected
YN	59	56.25
YS	20	18.75
EN	11	18.75
ES	10	6.25

Counted data condition: The data are counts.
Randomization condition: Assume that these flies are representative of all fruit flies of this type.
Expected cell frequency condition: The expected counts are all greater than 5.

Under these conditions, the sampling distribution of the test statistic is χ^2 on 4 – 1 = 3 degrees of freedom. We will use a chi-square goodness-of-fit test.

$$\chi^2 = \sum_{all\,cells} \frac{(Obs - Exp)^2}{Exp} = \frac{(59-56.25)^2}{56.25} + \frac{(20-18.75)^2}{18.75} + \frac{(11-18.75)^2}{18.75} + \frac{(10-6.25)^2}{6.25} \approx 5.671$$

With $\chi^2 \approx 5.671$, on 3 degrees of freedom, the P-value = 0.1288 is high, so we fail to reject the null hypothesis. There is no evidence that the ratio of traits is different than the theoretical ratio predicted by the genetic model. The observed results are consistent with the genetic model.

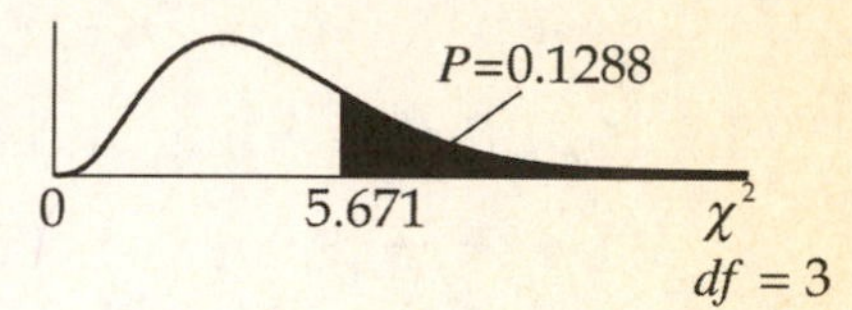

b) With $\chi^2 \approx 11.342$, on 3 degrees of freedom, the P-value = 0.0100 is low, so we reject the null hypothesis. There is strong evidence that the ratio of traits is different than the theoretical ratio predicted by the genetic model. Specifically, there is evidence that the normal wing length occurs less frequently than expected and the short wing length occurs more frequently than expected.

trait	Observed	Expected
YN	118	112.5
YS	40	37.5
EN	22	37.5
ES	20	12.5

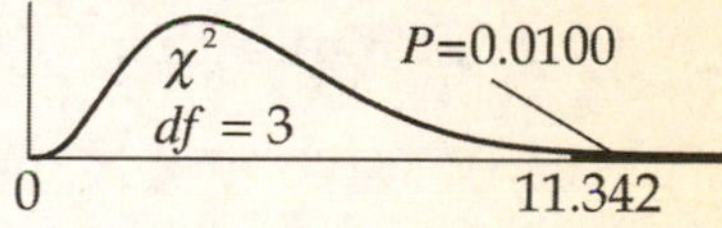

c) At first, this seems like a contradiction. We have two samples with exactly the same ratio of traits. The smaller of the two provides no evidence of a difference, yet the larger one provides strong evidence of a difference. This is explained by the sample size. In general, large samples decrease the proportion of variation from the true ratio. Because of the relatively small sample in the first test, we are unwilling to say that there is a difference. There just isn't enough evidence. But the larger sample allows us to be more certain about the difference.

11. Hurricane frequencies.

a) We would expect 96/16 = 6 hurricanes per time period.

b) We are comparing the distribution of the number of hurricanes, a single variable, to a theoretical distribution. A Chi Square test for Goodness of Fit is appropriate.

c) H_0: The number of large hurricanes remains constant over decades.

H_A: The number of large hurricanes has changed.

d) There are 16 time periods, so there are 16 – 1 = 15 degrees of freedom.

e) $P(\chi^2_{df=15} > 12.67) \approx 0.63$

f) The very high *P*-value means that these data offer no evidence that the number of hurricanes large hurricanes has changed.

g) The final period is only 6 years rather than 10, and already 7 hurricanes have been observed. Perhaps this decade will have an unusually large number of such hurricanes.

13. Childbirth, part 1.

a) There are two variables, breastfeeding and having an epidural, from a single group of births. We will perform a Chi Square test for Independence.

b) H_0: Breastfeeding success is independent of having an epidural.

H_A: There is an association between breastfeeding success and having an epidural.

15. Childbirth, part 2.

a) The table has 2 rows and 2 columns, so there are $(2-1)\times(2-1)=1$ degree of freedom.

b) We expect $\frac{474}{1178} \approx 40.2\%$ of all babies to not be breastfeeding after 6 months, so we expect that 40.2% of the 396 epidural babies, or 159.34, to not be breastfeeding after 6 months.

c) Breastfeeding behavior should be independent for these babies. They are fewer than 10% of all babies, and we assume they are representative of all babies. We have counts, and all the expected cells are at least 5.

17. Childbirth, part 3.

a) $\frac{(Obs-Exp)^2}{Exp} = \frac{(190-159.34)^2}{159.34} = 5.90$ **b)** $P(\chi^2_{df=1} > 14.87) < 0.005$

c) The *P*-value is very low, so reject the null hypothesis. There's strong evidence of an association between having an epidural and subsequent success in breastfeeding.

19. Childbirth, part 4.

a) $c = \frac{Obs - Exp}{\sqrt{Exp}} = \frac{190 - 159.34}{\sqrt{159.34}} = 2.43$

b) It appears that babies whose mothers had epidurals during childbirth are much more likely to be breastfeeding 6 months later.

21. Childbirth, part 5.

These factors would not have been mutually exclusive. There would be yes or no responses for every baby for each.

23. Titanic.

a) $P(\text{crew}) = \frac{885}{2201} \approx 0.402$

b) $P(\text{third and alive}) = \frac{178}{2201} \approx 0.081$

c) $P(\text{alive} \mid \text{first}) = \frac{P(\text{alive and first})}{\text{P(first)}} = \frac{202/2201}{325/2201} = \frac{202}{325} \approx 0.622$

d) The overall chance of survival is $\frac{710}{2201} \approx 0.323$, so we would expect about 32.3% of the crew, or about 285.48 members of the crew, to survive.

e) H_0: Survival was independent of status on the ship.

H_A: Survival depended on status on the ship.

f) The table has 2 rows and 4 columns, so there are $(2-1)\times(4-1) = 3$ degrees of freedom.

g) With $\chi^2 \approx 187.8$, on 3 degrees of freedom, the *P*-value is essentially 0, so we reject the null hypothesis. There is strong evidence survival depended on status. First-class passengers were more likely to survive than any other class or crew.

25. Titanic again.

First class passengers were most likely to survive, while third class passengers and crew were under-represented among the survivors.

27. Cranberry juice.

a) This is an experiment. Volunteers were assigned to drink a different beverage.

b) We are concerned with the proportion of urinary tract infections among three different groups. We will use a chi-square test for homogeneity.

c) H_0: The proportion of urinary tract infection is the same for each group.

H_A: The proportion of urinary tract infection is different among the groups.

d) **Counted data condition:** The data are counts.
Randomization condition: Although not specifically stated, we will assume that the women were randomly assigned to treatments.
Expected cell frequency condition: The expected counts are all greater than 5.

	Cranberry *(Obs / Exp)*	Lactobacillus *(Obs / Exp)*	Control *(Obs / Exp)*
Infection	8 / 15.333	20 / 15.333	18 / 15.333
No infection	42 / 34.667	30 / 34.667	32 / 34.667

e) The table has 2 rows and 3 columns, so there are $(2-1)\times(3-1)=2$ degrees of freedom.

f) $\chi^2 = \sum_{all\,cells} \frac{(Obs - Exp)^2}{Exp} \approx 7.776$

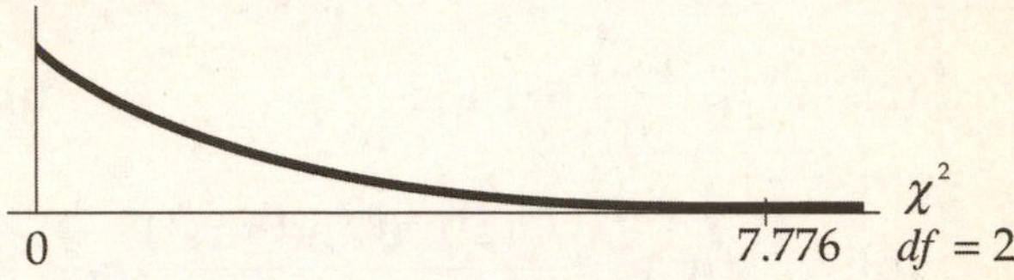

P-value $\approx$ 0.020.

g) Since the P-value is low, we reject the null hypothesis. There is strong evidence of difference in the proportion of urinary tract infections for cranberry juice drinkers, lactobacillus drinkers, and women that drink neither of the two beverages.

h) A table of the standardized residuals is below, calculated by using $c = \frac{Obs - Exp}{\sqrt{Exp}}$.

	Cranberry	Lactobacillus	Control
Infection	–1.87276	1.191759	0.681005
No infection	1.245505	–0.79259	–0.45291

There is evidence that women who drink cranberry juice are less likely to develop urinary tract infections, and women who drank lactobacillus are more likely to develop urinary tract infections.

29. Montana.

a) We have one group, categorized according to two variables, political party and being male or female, so we will perform a chi-square test for independence.

b) H_0: Political party is independent of being male or female in Montana.

H_A: There is an association between political party and being male or female in Montana.

c) **Counted data condition:** The data are counts.
Randomization condition: Although not specifically stated, we will assume that the poll was conducted randomly.
Expected cell frequency condition: The expected counts are all greater than 5.

	Democrat (Obs / Exp)	Republican (Obs / Exp)	Independent (Obs / Exp)
Male	36 / 43.663	45 / 40.545	24 / 20.792
Female	48 / 40.337	33 / 37.455	16 / 19.208

d) Under these conditions, the sampling distribution of the test statistic is χ^2 on 2 degrees of freedom. We will use a chi-square test for independence.

$$\chi^2 = \sum_{all\,cells} \frac{(Obs - Exp)^2}{Exp} \approx 4.851$$

The P-value ≈ 0.0884

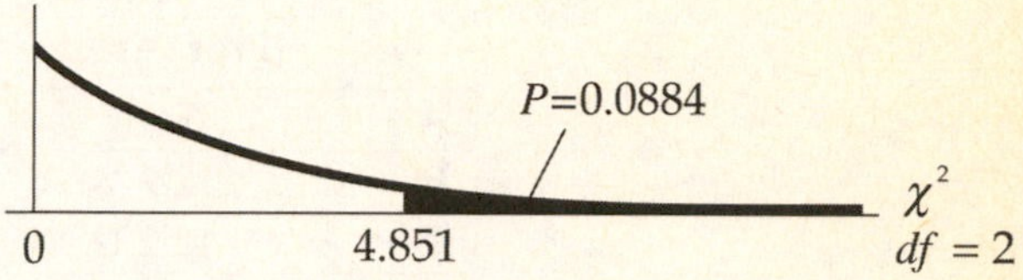

e) Since the P-value ≈ 0.0884 is fairly high, we fail to reject the null hypothesis. There is little evidence of an association between being male or female and political party in Montana.

31. Montana revisited.

H_0: Political party is independent of region in Montana.

H_A: There is an association between political party and region in Montana.

Counted data condition: The data are counts.
Randomization condition: Although not specifically stated, we will assume that the poll was conducted randomly.
Expected cell frequency condition: The expected counts are all greater than 5.

	Democrat (Obs / Exp)	Republican (Obs / Exp)	Independent (Obs / Exp)
West	39 / 28.277	17 / 26.257	12 / 13.465
Northeast	15 / 23.703	30 / 22.01	12 / 11.287
Southeast	30 /32.02	31 / 29.733	16 / 15.248

Under these conditions, the sampling distribution of the test statistic is χ^2 on 4 degrees of freedom. We will use a chi-square test for independence.

$$\chi^2 = \sum_{all\,cells} \frac{(Obs - Exp)^2}{Exp} \approx 13.849,$$

and the P-value ≈ 0.0078

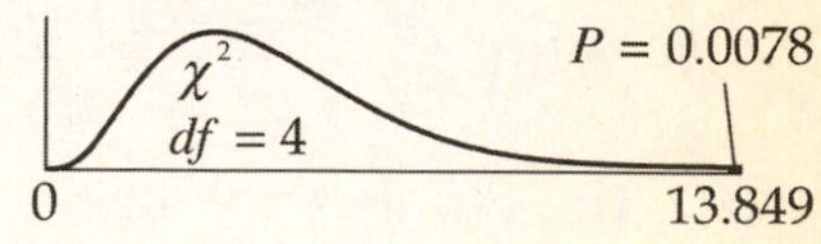

Since the P-value ≈ 0.0078 is low, reject the null hypothesis. There is strong evidence of an association between region and political party in Montana. Residents in the West are more likely to be Democrats than Republicans, and residents in the Northeast are more likely to be Republicans than Democrats.

33. Grades.

a) We have two groups, students of Professor Alpha and students of Professor Beta, and we are concerned with the distribution of one variable, grade. We will perform a chi-square test for homogeneity.

b) H_0: The distribution of grades is the same for the two professors.

H_A: The distribution of grades is different for the two professors.

c) The expected counts are organized in the table below:

	Prof. Alpha	Prof. Beta
A	6.667	5.333
B	12.778	10.222
C	12.222	9.778
D	6.111	4.889
F	2.222	1.778

Since three cells have expected counts less than 5, the chi-square procedures are not appropriate. Cells would have to be combined in order to proceed. (We will do this in another exercise.)

35. Grades again.

a) **Counted data condition:** The data are counts.
Randomization condition: Assume that these students are representative of all students that have ever taken courses from the professors.
Expected cell frequency condition: The expected counts are all greater than 5.

	Prof. Alpha (*Obs / Exp*)	Prof. Beta (*Obs / Exp*)
A	3 / 6.667	9 / 5.333
B	11 / 12.778	12 / 10.222
C	14 / 12.222	8 / 9.778
Below C	12 / 8.333	3 / 6.667

b) Under these conditions, the sampling distribution of the test statistic is χ^2 on 3 degrees of freedom, instead of 4 degrees of freedom before the change in the table. We will use a chi-square test for homogeneity.

c) With $\chi^2 = \sum_{all\,cells} \frac{(Obs - Exp)^2}{Exp} \approx 9.306$, the P-value ≈ 0.0255.

χ^2
$df = 3$
$P = 0.0255$
0
9.306

Since P-value = 0.0255 is low, we reject the null hypothesis. There is evidence that the grade distributions for the two professors are different. Professor Alpha gives fewer As and more grades below C than Professor Beta.

37. Racial steering.

H_0: There is no association between race and section of the complex in which people live.

H_A: There is an association between race and section of the complex in which people live.

Counted data: The data are counts.
Randomization condition: Assume the recently rented apartments are representative of all apartments.
Expected cell frequency condition: The expected counts are all greater than 5.

	White *(Obs / Exp)*	Black *(Obs / Exp)*
Section A	87 / 76.179	8 / 18.821
Section B	83 / 93.821	34 / 23.179

Under these conditions, the sampling distribution of the test statistic is χ^2 on 1 degree of freedom. We will use a chi-square test for independence.

$$\chi^2 = \sum_{all\,cells} \frac{(Obs - Exp)^2}{Exp} \approx \frac{(87-76.179)^2}{76.179} + \frac{(8-18.821)^2}{18.821} + \frac{(83-93.821)^2}{93.821} + \frac{(34-23.179)^2}{23.179}$$

$$\approx 1.5371 + 6.2215 + 1.2481 + 5.0517$$

$$\approx 14.058$$

With $\chi^2 \approx 14.058$, on 1 degree of freedom, the P-value ≈ 0.0002.

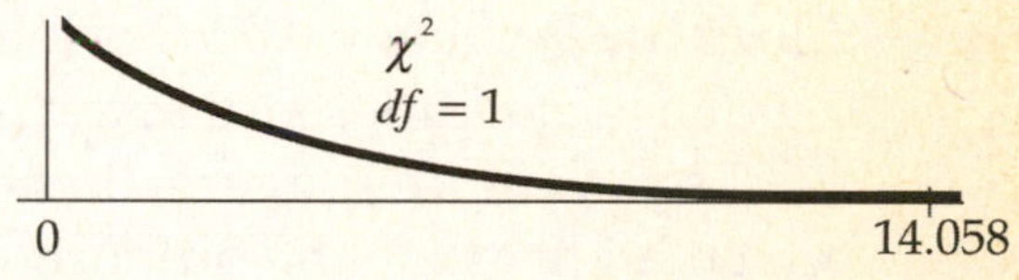

Since the P-value ≈ 0.0002 is low, we reject the null hypothesis. There is strong evidence of an association between race and the section of the apartment complex in which people live. An examination of the components shows us that whites are much more likely to rent in Section A (component = 6.2215), and blacks are much more likely to rent in Section B (component = 5.0517).

39. Steering revisited.

a) H_0: The proportion of whites who live in Section A is the same as the proportion of blacks who live in Section A. $(p_W = p_B \text{ or } p_W - p_B = 0)$

H_A: The proportion of whites who live in Section A is different than the proportion of blacks who live in Section A. $(p_W \neq p_B \text{ or } p_W - p_B \neq 0)$

Independence assumption: Assume that people rent independent of the section.
Independent samples condition: The groups are not associated.
Success/Failure condition: $n\hat{p}$ (white) = 87, $n\hat{q}$ (white) = 83, $n\hat{p}$ (black) = 8, and $n\hat{q}$ (black) = 34. These are not all greater than 10, since the number of black renters in Section A is only 8, but it is close to 10, and the others are large. It should be safe to proceed.

Since the conditions have been satisfied, we will model the sampling distribution of the difference in proportion with a Normal model with mean 0 and standard deviation estimated by

$$SE_{\text{pooled}}\left(\hat{p}_W - \hat{p}_B\right) = \sqrt{\frac{\hat{p}_{\text{pooled}}\hat{q}_{\text{pooled}}}{n_W} + \frac{\hat{p}_{\text{pooled}}\hat{q}_{\text{pooled}}}{n_B}} = \sqrt{\frac{\left(\frac{95}{212}\right)\left(\frac{117}{212}\right)}{170} + \frac{\left(\frac{95}{212}\right)\left(\frac{117}{212}\right)}{42}} \approx 0.0856915.$$

The observed difference between the proportions is:
0.5117647 – 0.1904762 = 0.3212885.

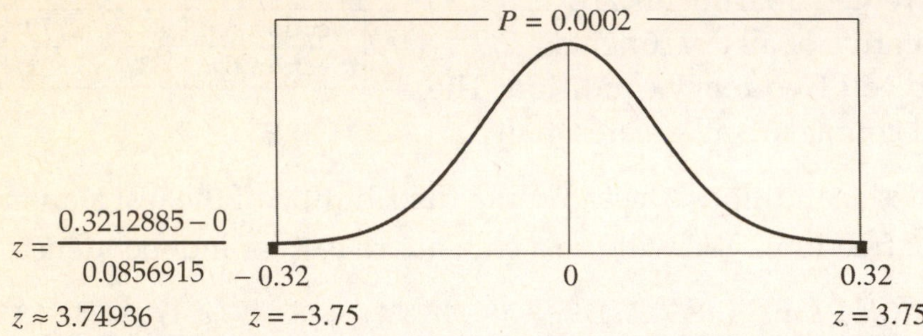

(You have to use a *ridiculous* number of decimal places to get this to come out "right". This is to done to illustrate the point of the question. DO NOT DO THIS! Use technology.)

Since the P-value = 0.0002 is low, we reject the null hypothesis. There is strong evidence of a difference in the proportion of whites and blacks living in Section A. The evidence suggests that the proportion of all whites living in Section A is much higher than the proportion of all black residents living in Section A.

The value of z for this test was approximately 3.74936. $z^2 \approx (3.74936)^2 \approx 14.058$, the same as the value for χ^2 in Exercise 37.

b) The resulting P-values were both approximately 0.0002. The two tests are equivalent.

41. Pregnancies.

H_0: Pregnancy outcome is independent of age.

H_A: There is an association between pregnancy outcome and age.

Counted data condition: The data are counts.
Randomization condition: Assume that these women are representative of all pregnant women.
Expected cell frequency condition: The expected counts are all greater than 5.

	Live Births *(Obs / Exp)*	Fetal losses *(Obs / Exp)*
Under 20	49 / 49.677	13 / 12.323
20 – 29	201 / 193.9	41 / 48.099
30 – 34	88 / 87.335	21 / 21.665
35 and over	49 / 56.087	21 / 13.913

Under these conditions, the sampling distribution of the test statistic is χ^2 on 3 degrees of freedom. We will use a chi-square test for independence.

With $\chi^2 = \sum_{all\,cells} \frac{(Obs - Exp)^2}{Exp} \approx 5.89$, on 3 degrees of freedom, the P-value ≈ 0.1173.

X²=5.8851 P=.1173

Since the P-value is high, we fail to reject the null hypothesis. There is no evidence of an association between pregnancy outcome and age.

Review of Part VI – Learning About the World

1. Crawling.

a) H0: The mean age at which babies begin to crawl is the same whether the babies were born in January or July. $\left(\mu_{Jan} = \mu_{July} \text{ or } \mu_{Jan} - \mu_{July} = 0\right)$

HA: There is a difference in the mean age at which babies begin to crawl, depending on whether the babies were born in January or July. $\left(\mu_{Jan} \neq \mu_{July} \text{ or } \mu_{Jan} - \mu_{July} \neq 0\right)$

Independent groups assumption: The groups of January and July babies are independent.
Randomization condition: Although not specifically stated, we will assume that the babies are representative of all babies.
10% condition: 32 and 21 are less than 10% of all babies.
Nearly Normal condition: We don't have the actual data, so we can't check the distribution of the sample. However, the samples are fairly large. The Central Limit Theorem allows us to proceed.

Since the conditions are satisfied, it is appropriate to model the sampling distribution of the difference in means with a Student's t-model, with 43.68 degrees of freedom (from the approximation formula).

We will perform a two-sample t-test. The sampling distribution model has mean 0, with standard error: $SE(\bar{y}_{Jan} - \bar{y}_{July}) = \sqrt{\frac{7.08^2}{32} + \frac{6.91^2}{21}} \approx 1.9596$.

The observed difference between the mean ages is 29.84 – 33.64 = – 3.8 weeks.

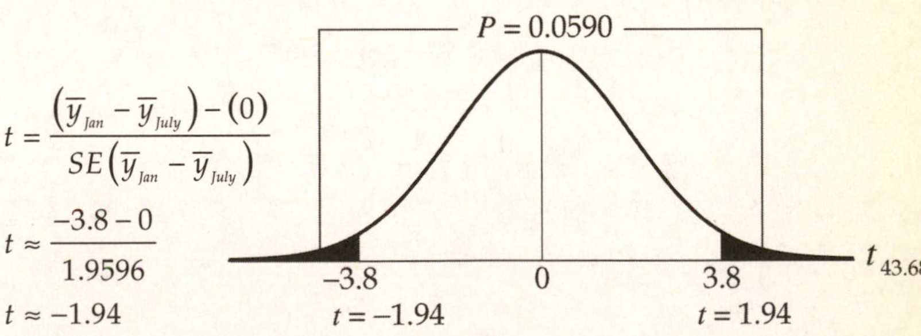

Since the P-value = 0.0590 is fairly low, we reject the null hypothesis. There is some evidence that mean age at which babies crawl is different for January and June babies. June babies appear to crawl a bit earlier than July babies, on average. Since the evidence is not strong, we might want to do some more research into this claim.

b) H_0: The mean age at which babies begin to crawl is the same whether the babies were born in April or October. $(\mu_{Apr} = \mu_{Oct} \text{ or } \mu_{Apr} - \mu_{Oct} = 0)$

H_A: There is a difference in the mean age at which babies begin to crawl. $(\mu_{Apr} \neq \mu_{Oct} \text{ or } \mu_{Apr} - \mu_{Oct} \neq 0)$

The conditions (with minor variations) were checked in part a.

Since the conditions are satisfied, it is appropriate to model the sampling distribution of the difference in means with a Student's *t*-model, with 59.40 degrees of freedom (from the approximation formula).

We will perform a two-sample *t*-test. The sampling distribution model has mean 0, with standard error: $SE(\bar{y}_{Apr} - \bar{y}_{Oct}) = \sqrt{\frac{6.21^2}{26} + \frac{7.29^2}{44}} \approx 1.6404$.

The observed difference between the mean ages is 31.84 – 33.35 = – 1.51 weeks.

Since the *P*-value = 0.3610 is high, we fail to reject the null hypothesis. There is no evidence that mean age at which babies crawl is different for April and October babies.

$$t = \frac{(\bar{y}_{Apr} - \bar{y}_{Oct}) - (0)}{SE(\bar{y}_{Apr} - \bar{y}_{Oct})}$$

$$t \approx \frac{-1.51 - 0}{1.6404}$$

$$t \approx -0.92$$

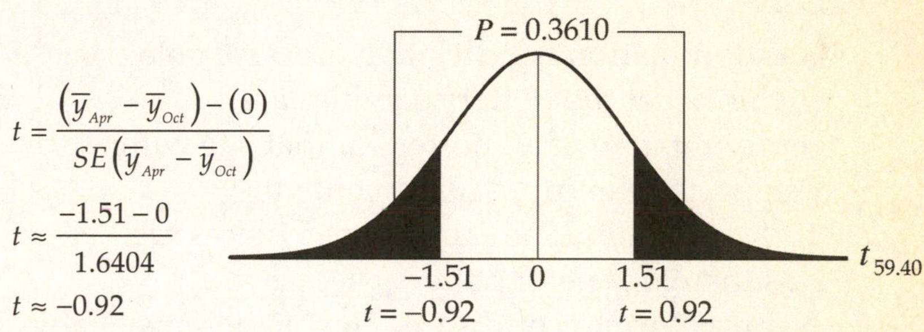

c) These results are not consistent with the researcher's claim. We have slight evidence in one test and no evidence in the other. The researcher will have to do better than this to convince us!

3. **Women.**

H_0 : The percentage of businesses in the area owned by women is 26%. ($p = 0.26$)
H_A : The percentage of businesses owned by women is not 26%. ($p \neq 0.26$)

Random condition: This is a random sample of 410 businesses.
10% condition: The sample of 410 businesses is less than 10% of all businesses.
Success/Failure condition: $np = (410)(0.26) = 106.6$ and $nq = (410)(0.74) = 303.4$ are both greater than 10, so the sample is large enough.

The conditions have been satisfied, so a Normal model can be used to model the sampling distribution of the proportion, with $\mu_{\hat{p}} = p = 0.26$ and

$$\sigma(\hat{p}) = \sqrt{\frac{pq}{n}} = \sqrt{\frac{(0.26)(0.74)}{410}} \approx 0.02166.$$

We can perform a one-proportion z-test. The observed proportion of businesses owned by women is

$$\hat{p} = \frac{115}{410} \approx 0.2805.$$

$$z = \frac{\hat{p} - p_0}{\sqrt{\frac{pq}{n}}}$$

$$z = \frac{0.2805 - 0.26}{\sqrt{\frac{(0.26)(0.74)}{410}}}$$

$$z \approx 0.946$$

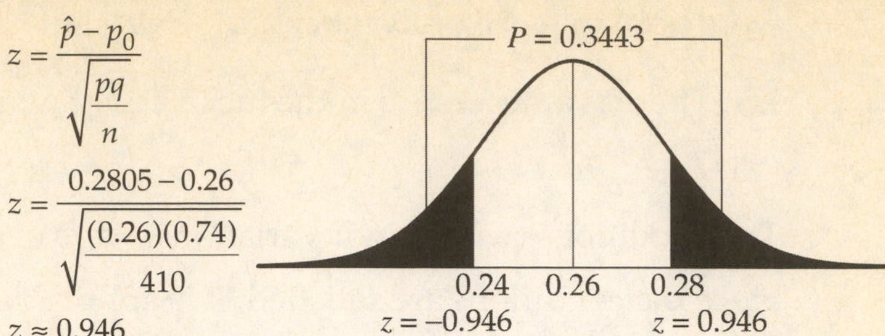

Since the P-value = 0.3443 is high, we fail to reject the null hypothesis. There is no evidence that the proportion of businesses in the Denver area owned by women is any different than the national figure of 26%.

5. Pottery.

Independent groups assumption: The pottery samples are from two different sites.
Randomization condition: It is reasonable to think that the pottery samples are representative of all pottery at that site with respect to aluminum oxide content.
10% condition: The samples of 5 pieces are less than 10% of all pottery pieces.
Nearly Normal condition: The histograms of aluminum oxide content are roughly unimodal and symmetric.

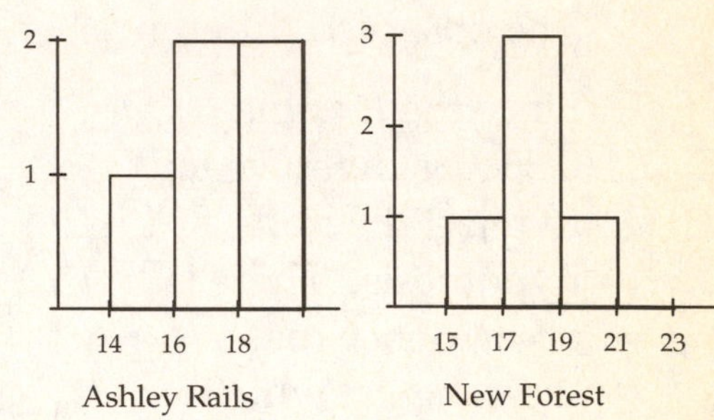

Since the conditions are satisfied, it is appropriate to model the sampling distribution of the difference in means with a Student's t-model, with 7 degrees of freedom (from the approximation formula). We will construct a two-sample t-interval, with 95% confidence.

$$(\bar{y}_{AR} - \bar{y}_{NF}) \pm t^*_{df}\sqrt{\frac{s^2_{AR}}{n_{AR}} + \frac{s^2_{NF}}{n_{NF}}}$$

$$= (17.32 - 18.18) \pm t^*_7\sqrt{\frac{1.65892^2}{5} + \frac{1.77539^2}{5}} \approx (-3.37, 1.65)$$

We are 95% confident that the difference in the mean percentage of aluminum oxide content of the pottery at the two sites is between –3.37% and 1.65%. Since 0 is in the interval, there is no evidence that the aluminum oxide content at the two sites is different. It would be reasonable for the archaeologists to think that the same ancient people inhabited the sites.

7. **Gehrig.**

H₀: The proportion of ALS patients who were athletes is the same as the proportion of patients with other disorders who were athletes.
$(p_{ALS} = p_{Other} \text{ or } p_{ALS} - p_{Other} = 0)$

H_A: The proportion of ALS patients who were athletes is greater than the proportion of patients with other disorders who were athletes.
$(p_{ALS} > p_{Other} \text{ or } p_{ALS} - p_{Other} > 0)$

Random condition: This is NOT a random sample. We must assume that these patients are representative of all patients with neurological disorders.
10% condition: 280 and 151 are both less than 10% of all patients with disorders.
Independent samples condition: The groups are independent.
Success/Failure cond.: $n\hat{p}$(ALS) = (280)(0.38) = 106, $n\hat{q}$(ALS) = (280)(0.72) = 174, $n\hat{p}$(Other) = (151)(0.26) = 39, and $n\hat{q}$(Other) = (151)(0.74) = 112 are all greater than 10, so the samples are both large enough.

Since the conditions have been satisfied, we will model the sampling distribution of the difference in proportion with a Normal model with mean 0 and standard deviation estimated by

$$SE_p(\hat{p}_{ALS} - \hat{p}_{Other}) = \sqrt{\frac{\hat{p}_p\hat{q}_p}{n_{ALS}} + \frac{\hat{p}_p\hat{q}_p}{n_{Other}}} = \sqrt{\frac{(0.336)(0.664)}{280} + \frac{(0.336)(0.664)}{151}} \approx 0.0477$$

The observed difference between the proportions is 0.38 – 0.26 = 0.12.

Since the P-value = 0.0060 is very low, we reject the null hypothesis. There is strong evidence that the proportion of ALS patients who are athletes is greater than the proportion of patients with other disorders who are athletes.

$$z = \frac{0.12 - 0}{0.0477}$$
$$z \approx 2.52$$

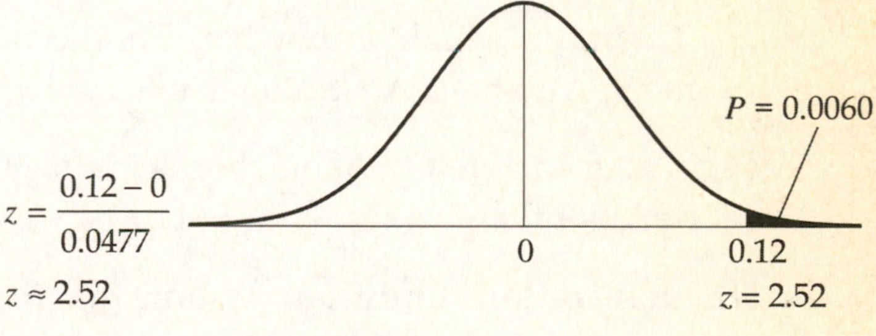

b) This was a retrospective observational study. In order to make the inference, we must assume that the patients studied are representative of all patients with neurological disorders.

9. **Genetics.**

H_0: The proportions of traits are as specified by the ratio 1:3:3:9.
H_A: The proportions of traits are not as specified.

Counted data condition: The data are counts.
Randomization condition: Assume that these students are representative of all people.

Expected cell frequency condition: The expected counts (shown in the table) are all greater than 5.

Under these conditions, the sampling distribution of the test statistic is χ^2 on 4 – 1 = 3 degrees of freedom. We will use a chi-square goodness-of-fit test.

Trait	Observed	Expected	Residual = $(Obs - Exp)$	$(Obs - Exp)^2$	Component = $\frac{(Obs - Exp)^2}{Exp}$
Attached, noncurling	10	7.625	2.375	5.6406	0.73975
Attached, curling	22	22.875	– 0.875	0.7656	0.03347
Free, noncurling	31	22.875	8.125	66.0156	2.8859
Free, curling	59	68.625	– 9.625	92.6406	1.35
					≈ 5.01

$\chi^2 = 5.01$. Since the *P*-value = 0.1711 is high, we fail to reject the null hypothesis.

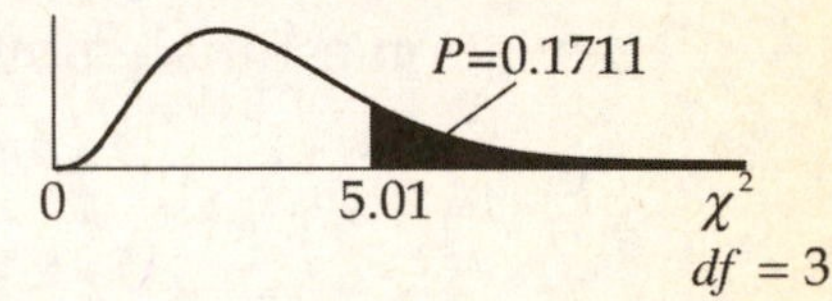

There is no evidence that the proportions of traits are anything other than 1:3:3:9.

11. Babies.

H_0: The mean weight of newborns in the U.S. is 7.41 pounds, the same as the mean weight of Australian babies. $(\mu = 7.41)$

H_A: The mean weight of newborns in the U.S. is not the same as the mean weight of Australian babies. $(\mu \neq 7.41)$

Randomization condition: Assume that the babies at this Missouri hospital are representative of all U.S. newborns. (Given)
10% condition: 112 newborns are less than 10% of all newborns.
Nearly Normal condition: We don't have the actual data, so we cannot look at a graphical display, but since the sample is large, it is safe to proceed.

The babies in the sample had a mean weight of 7.68 pounds and a standard deviation in weight of 1.31 pounds. Since the conditions for inference are satisfied, we can model the sampling distribution of the mean weight of U.S. newborns with a Student's *t* model, with 112 – 1 = 111 degrees of freedom, $t_{111}\left(7.41, \frac{1.31}{\sqrt{112}}\right)$. We will perform a one-sample *t*-test.

Since the *P*-value = 0.0313 is low, we reject the null hypothesis. If we believe that the babies at this Missouri hospital are representative of all U.S. babies, there is evidence to suggest that the mean weight of U.S. babies is different than the mean weight of Australian babies. U.S. babies appear to weigh more on average.

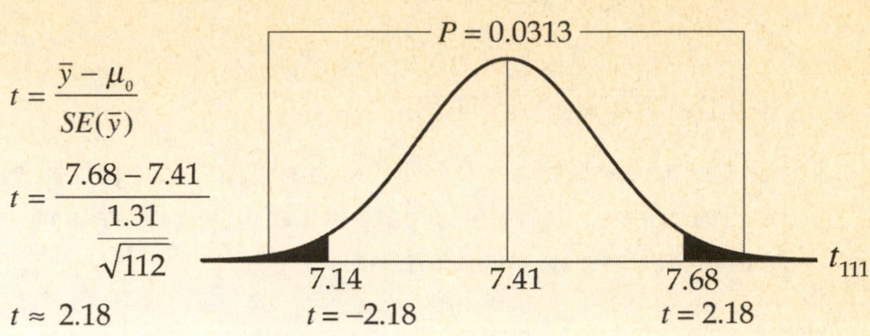

13. Feeding fish.

a) If there is no difference in the average fish sizes, the chance of observing a difference this large, or larger, just by natural sampling variation is 0.1%.

b) There is evidence that largemouth bass that are fed a natural diet are larger. The researchers would advise people who raise largemouth bass to feed them a natural diet.

c) If the advice is incorrect, the researchers have committed a Type I error.

15. Twins.

H_0: There is no association between duration of pregnancy and level of prenatal care.

H_A: There is an association between duration of pregnancy and level of prenatal care.

Counted data condition: The data are counts.
Randomization condition: Assume that these pregnancies are representative of all twin births.
Expected cell frequency condition: The expected counts are all greater than 5.

	Preterm (induced or Cesarean) *(Obs / Exp)*	**Preterm (without procedures)** *(Obs / Exp)*	**Term or postterm** *(Obs / Exp)*
Intensive	18 /16.676	15 / 15.579	28 / 28.745
Adequate	46 / 42.101	43 / 39.331	65 / 72.568
Inadequate	12 / 17.223	13 / 16.090	38 / 29.687

Under these conditions, the sampling distribution of the test statistic is χ^2 on 4 degrees of freedom. We will use a chi-square test for independence.

$$\chi^2 = \sum_{all\,cells} \frac{(Obs - Exp)^2}{Exp} \approx 6.14,$$

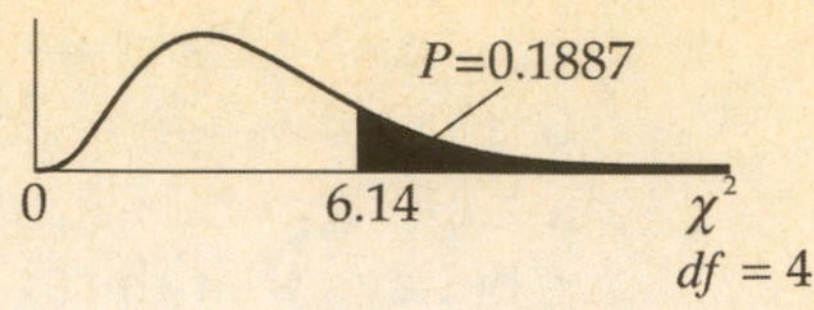

and the P-value ≈ 0.1887.

Since the P-value ≈ 0.1887 is high, we fail to reject the null hypothesis. There is no evidence of an association between duration of pregnancy and level of prenatal care in twin births.

17. Age.

a) **Independent groups assumption:** The group of patients with and without cardiac disease are not related in any way.
Randomization condition: Assume that these patients are representative of all people.
10% condition: 2397 patients without cardiac disease and 450 patients with cardiac disease are both less than 10% of all people.
Normal population assumption: We don't have the actual data, so we will assume that the population of ages of patients is Normal.

Since the conditions are satisfied, it is appropriate to model the sampling distribution of the difference in means with a Student's t-model, with 670 degrees of freedom (from the approximation formula). We will construct a two-sample t-interval, with 95% confidence.

$$(\bar{y}_{Card} - \bar{y}_{None}) \pm t^*_{df}\sqrt{\frac{s^2_{Card}}{n_{Card}} + \frac{s^2_{None}}{n_{None}}} = (74.0 - 69.8) \pm t^*_{670}\sqrt{\frac{7.9^2}{450} + \frac{8.7^2}{2397}} \approx (3.39,\ 5.01)$$

We are 95% confident that the mean age of patients with cardiac disease is between 3.39 and 5.01 years higher than the mean age of patients without cardiac disease.

b) Older patients are at greater risk for a variety of health problems. If an older patient does not survive a heart attack, the researchers will not know to what extent depression was involved, because there will be a variety of other possible variables influencing the death rate. Additionally, older patients may be more (or less) likely to be depressed than younger ones.

19. Back to Montana.

H_0: Political party is independent of income level in Montana.

H_A: There is an association between political party and income level in Montana.

Counted data condition: The data are counts.
Randomization condition: Although not specifically stated, we will assume that the poll was conducted randomly.

Expected cell frequency condition: The expected counts are all greater than 5.

	Democrat (Obs / Exp)	Republican (Obs / Exp)	Independent (Obs / Exp)
Low	30 / 24.119	16 / 22.396	12 / 11.485
Middle	28 / 30.772	24 / 28.574	22 / 14.653
High	26 / 29.109	38 / 27.03	6 / 13.861

Under these conditions, the sampling distribution of the test statistic is χ^2 on 4 degrees of freedom. We will use a chi-square test for independence.

$$\chi^2 = \sum_{\text{all cells}} \frac{(Obs - Exp)^2}{Exp} \approx 17.19$$

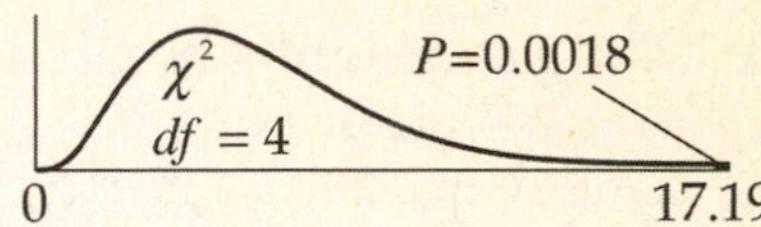

The P-value ≈ 0.0018

Since the P-value ≈ 0.0018 is low, we reject the null hypothesis. There is strong evidence of an association between income level and political party in Montana. An examination of the components shows that Democrats are more likely to have low incomes, Independents are more likely to have middle incomes, and Republicans are more likely to have high incomes.

21. Cesareans.

H_0: The proportion of births involving cesarean deliveries is the same in Vermont and New Hampshire. $(p_{VT} = p_{NH} \text{ or } p_{VT} - p_{NH} = 0)$

H_A: The proportion of births involving cesarean deliveries is different in Vermont and New Hampshire. $(p_{VT} \neq p_{NH} \text{ or } p_{VT} - p_{NH} \neq 0)$

Random condition: Hospitals were randomly selected.
10% condition: 223 and 186 are both less than 10% of all births in these states.
Independent samples condition: Vermont and New Hampshire are different states!
Success/Failure cond.: $n\hat{p}$(VT) = (223)(0.166) = 37, $n\hat{q}$(VT) = (223)(0.834) = 186, $n\hat{p}$(NH) = (186)(0.188) = 35, and $n\hat{q}$(NH) = (186)(0.812) = 151 are all greater than 10, so the samples are both large enough.

Since the conditions have been satisfied, we will model the sampling distribution of the difference in proportion with a Normal model with mean 0 and standard deviation estimated by

$$SE_{\text{pooled}}\left(\hat{p}_{VT} - \hat{p}_{NH}\right) = \sqrt{\frac{\hat{p}_{\text{pooled}}\hat{q}_{\text{pooled}}}{n_{VT}} + \frac{\hat{p}_{\text{pooled}}\hat{q}_{\text{pooled}}}{n_{NH}}}$$

$$= \sqrt{\frac{(0.176)(0.824)}{223} + \frac{(0.176)(0.824)}{186}} \approx 0.03782$$

The observed difference between the proportions is 0.166 – 0.188 = – 0.022.

Since the P-value = 0.5563 is high, we fail to reject the null hypothesis. There is no evidence that the proportion of cesarean births in Vermont is different from the proportion of cesarean births in New Hampshire.

$$z = \frac{-0.22 - 0}{0.03782}$$

$$z \approx -0.59$$

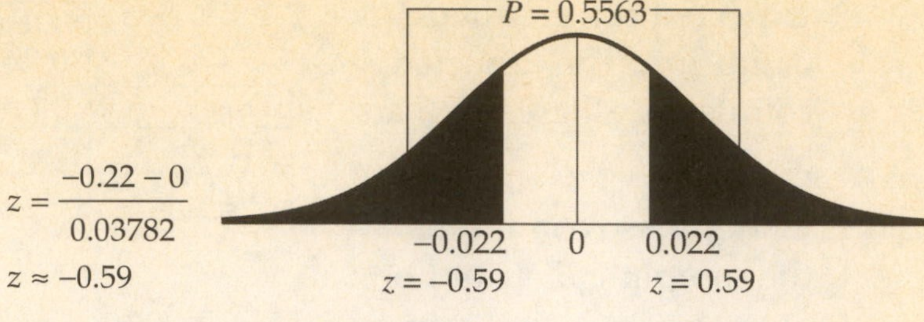

23. Meals.

H0: The college student's mean daily food expense is \$10. $(\mu = 10)$

HA: The college student's mean daily food expense is greater than \$10. $(\mu > 10)$

Randomization condition: Assume that these days are representative of all days.
10% condition: 14 days are less than 10% of all days.
Nearly Normal condition: The histogram of daily expenses is fairly unimodal and symmetric. It is reasonable to think that this sample came from a Normal population.

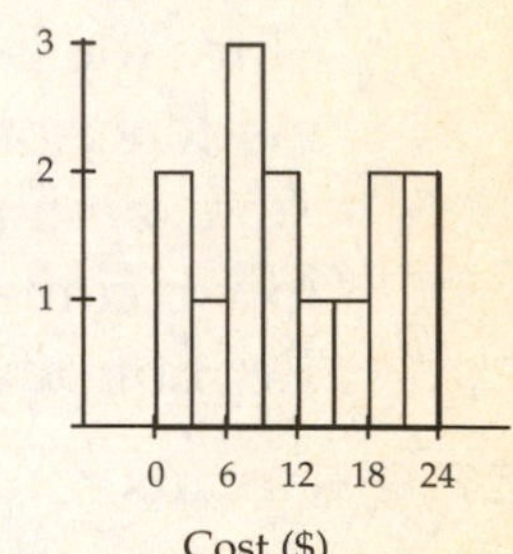

The expenses in the sample had a mean of 11.4243 dollars and a standard deviation of 8.05794 dollars. Since the conditions for inference are satisfied, we can model the sampling distribution of the mean daily expense with a Student's t model, with 14 – 1 = 13 degrees of freedom, $t_{13}\left(10, \frac{8.05794}{\sqrt{14}}\right)$. We will perform a one-sample t-test.

Since the P-value = 0.2600 is high, we fail to reject the null hypothesis. There is no evidence that the student's average spending is more than \$10 per day.

$$t = \frac{\overline{y} - \mu_0}{SE(\overline{y})}$$

$$t = \frac{11.4243 - 10}{\frac{8.05794}{\sqrt{14}}}$$

$$t \approx 0.66$$

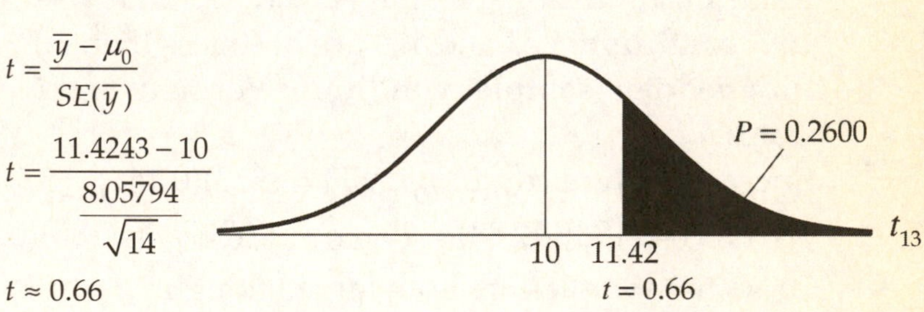

25. Teach for America.

H0: The mean score of students with certified teachers is the same as the mean score of students with uncertified teachers. $(\mu_C = \mu_U \text{ or } \mu_C - \mu_U = 0)$

HA: The mean score of students with certified teachers is greater than as the mean score of students with uncertified teachers. $(\mu_C > \mu_U \text{ or } \mu_C - \mu_U > 0)$

Independent groups assumption: The certified and uncertified teachers are independent groups.
Randomization condition: Assume the students studied were representative of all students.
10% condition: Two samples of size 44 are both less than 10% of the population.
Nearly Normal condition: We don't have the actual data, so we can't look at the graphical displays, but the sample sizes are large, so we can proceed.

Since the conditions are satisfied, it is appropriate to model the sampling distribution of the difference in means with a Student's *t*-model, with 86 degrees of freedom (from the approximation formula).

We will perform a two-sample *t*-test. The sampling distribution model has mean 0, with standard error: $SE(\bar{y}_C - \bar{y}_U) = \sqrt{\frac{9.31^2}{44} + \frac{9.43^2}{44}} \approx 1.9977$.

The observed difference between the mean scores is 35.62 – 32.48 = 3.14.

Since the *P*-value = 0.0598 is fairly high, we fail to reject the null hypothesis. There is little evidence that students with certified teachers had mean scores higher than students with uncertified teachers. However, since the *P*-value is not extremely high, further investigation is recommended.

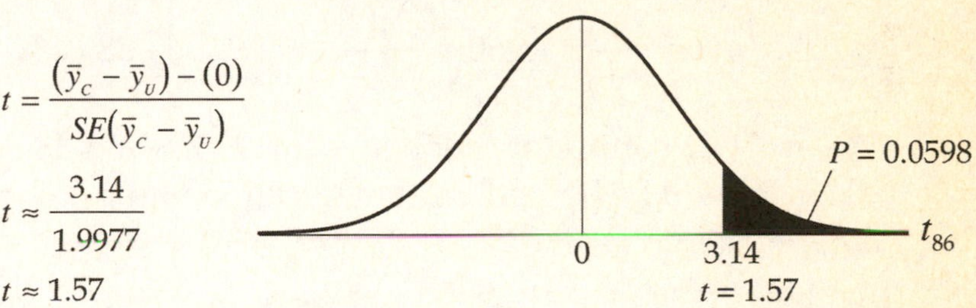

27. Legionnaires' disease.

Paired data assumption: The data are paired by room.
Randomization condition: We will assume that these rooms are representative of all rooms at the hotel.
10% condition: Assume that 8 rooms are less than 10% of the rooms. The hotel must have more than 80 rooms.
Nearly Normal condition: The histogram of differences between before and after measurements is roughly unimodal and symmetric.

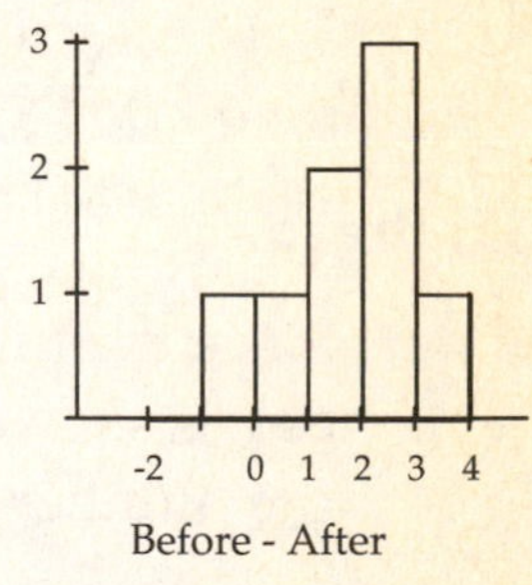

Since the conditions are satisfied, the sampling distribution of the difference can be modeled with a Student's *t*-model with 8 – 1 = 7 degrees of freedom. We will find a paired *t*-interval, with 95% confidence.

$$\bar{d} \pm t^*_{n-1}\left(\frac{s_d}{\sqrt{n}}\right) = 1.6125 \pm t^*_7\left(\frac{1.23801}{\sqrt{8}}\right) \approx (0.58, 2.65)$$

We are 95% confident that the mean difference in the bacteria counts is between 0.58 and 2.65 colonies per cubic foot of air. Since the entire interval is above 0, there is evidence that the new air-conditioning system was effective in reducing average bacteria counts.

29. Bipolar kids.

a) **Random condition:** Assume that the 89 children are representative of all children with bipolar disorder.
10% condition: 89 is less than 10% of all children with bipolar disorder.
Success/Failure condition: $n\hat{p} = 26$ and $n\hat{q} = 63$ both greater than 10.

Since the conditions are met, we can use a one-proportion z-interval to estimate the percentage of children with bipolar disorder who might be helped by this treatment.

$$\hat{p} \pm z^*\sqrt{\frac{\hat{p}\hat{q}}{n}} = \left(\frac{26}{89}\right) \pm 1.960\sqrt{\frac{\left(\frac{26}{89}\right)\left(\frac{63}{89}\right)}{89}} \approx (19.77\%, 38.66\%)$$

We are 95% confident that between 19.77% and 38.66% of children with bipolar disorder will be helped with medication and psychotherapy.

b)

$$ME = z^*\sqrt{\frac{\hat{p}\hat{q}}{n}}$$

$$0.06 = 1.960\sqrt{\frac{\left(\frac{26}{89}\right)\left(\frac{63}{89}\right)}{n}}$$

$$n = \frac{(1.960)^2\left(\frac{26}{89}\right)\left(\frac{63}{89}\right)}{(0.06)^2}$$

$$n \approx 221 \text{ children}$$

In order to estimate the proportion of children helped with medication and psychotherapy to within 6% with 95% confidence, we would need a sample of at least 221 children. All decimals in the final answer must be rounded up, to the next person.
(For a more cautious answer, let $\hat{p} = \hat{q} = 0.5$. This method results in a required sample of 267 children.)

31. Bread.

a) Since the histogram shows that the distribution of the number of loaves sold per day is skewed strongly to the right, we can't use the Normal model to estimate the number of loaves sold on the busiest 10% of days.

b) **Randomization condition:** Assume that these days are representative of all days.
10% condition: 100 days are less than 10% of all days.
Nearly Normal condition: The histogram is skewed strongly to the right. However, since the sample size is large, the Central Limit Theorem guarantees that the distribution of averages will be approximately Normal.

The days in the sample had a mean of 103 loaves sold and a standard deviation of 9 loaves sold. Since the conditions are satisfied, the sampling distribution of the mean can be modeled by a Student's *t*- model, with 103 – 1 = 103 degrees of freedom. We will use a one-sample *t*-interval with 95% confidence for the mean number of loaves sold. (By hand, use $t_{50}^{*} \approx 2.403$ from the table.)

c) $\bar{y} \pm t_{n-1}^{*}\left(\frac{s}{\sqrt{n}}\right) = 103 \pm t_{102}^{*}\left(\frac{9}{\sqrt{103}}\right) \approx (101.2,\ 104.8)$

We are 95% confident that the mean number of loaves sold per day at the Clarksburg Bakery is between 101.2 and 104.8.

d) We know that in order to cut the margin of error in half, we need to a sample four times as large. If we allow a margin of error that is twice as wide, that would require a sample only one-fourth the size. In this case, our original sample is 100 loaves; so 25 loaves would be a sufficient number to estimate the mean with a margin of error twice as wide.

e) Since the interval is completely above 100 loaves, there is strong evidence that the estimate was incorrect. The evidence suggests that the mean number of loaves sold per day is greater than 100. This difference is statistically significant, but may not be practically significant. It seems like the owners made a pretty good estimate!

33. Insulin and diet.

a) H_0: People with high dairy consumption have IRS at the same rate as those with low dairy consumption. $\left(p_{High} = p_{Low} \text{ or } p_{High} - p_{Low} = 0\right)$

H_A: People with high dairy consumption have IRS at a different rate than those with low dairy consumption. $\left(p_{High} \neq p_{Low} \text{ or } p_{High} - p_{Low} \neq 0\right)$

Random condition: Assume the people studied are representative of all people.
10% condition: 102 and 190 are both less than 10% of all people.
Independent samples condition: The two groups are not related.
Success/Failure condition: $n\hat{p}$ (high) = 24, $n\hat{q}$ (high) = 78, $n\hat{p}$ (low) = 85, and $n\hat{q}$ (low) = 105 are all greater than 10, so the samples are both large enough.

Since the conditions have been satisfied, we will model the sampling distribution of the difference in proportion with a Normal model with mean 0 and standard deviation estimated by

$$SE_{\text{pooled}}\left(\hat{p}_{High} - \hat{p}_{Low}\right) = \sqrt{\frac{\hat{p}_{\text{pooled}}\hat{q}_{\text{pooled}}}{n_{High}} + \frac{\hat{p}_{\text{pooled}}\hat{q}_{\text{pooled}}}{n_{Low}}}$$

$$= \sqrt{\frac{(0.373)(0.627)}{102} + \frac{(0.373)(0.627)}{190}} \approx 0.05936$$

The observed difference between the proportions is 0.2352 – 0.4474 = – 0.2122.

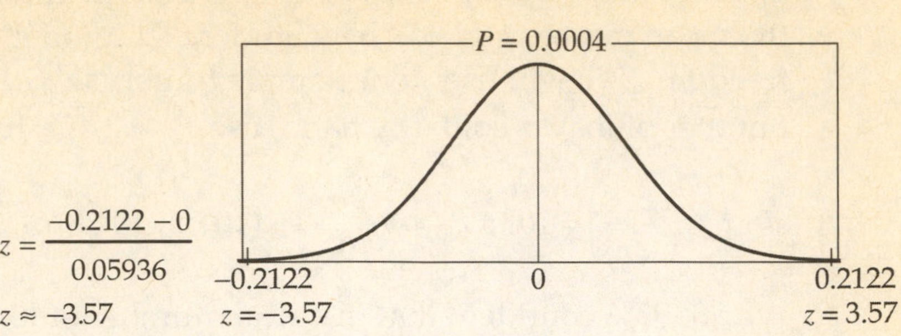

Since the P-value = 0.0004 is very low, we reject the null hypothesis. There is strong evidence that the proportion of people with IRS is different for those who with high dairy consumption compared to those with low dairy consumption. People who consume dairy products more than 35 times per week appear less likely to have IRS than those who consume dairy products fewer than 10 times per week.

b) There is evidence of an association between the low consumption of dairy products and IRS, but that does not prove that dairy consumption influences the development of IRS. This is an observational study, and a controlled experiment is required to prove cause and effect.

35. Rainmakers?

Independent groups assumption: The two groups of clouds are independent.
Randomization condition: Researchers randomly assigned clouds to be seeded with silver iodide or not seeded.
10% condition: We are testing cloud seeding, not clouds themselves, so this condition doesn't need to be checked.
Nearly Normal condition: We don't have the actual data, so we can't look at the distributions, but the means of group are significantly higher than the medians. This is an indication that the distributions are skewed to the right, with possible outliers. The samples sizes of 26 each are fairly large, so it should be safe to proceed, but we should be careful making conclusions, since there may be outliers.

Since the conditions are satisfied, it is appropriate to model the sampling distribution of the difference in means with a Student's t-model, with 33.86 degrees of freedom (from the approximation formula). We will construct a two-sample t-interval, with 95% confidence.

$$(\bar{y}_S - \bar{y}_U) \pm t^*_{df}\sqrt{\frac{s_S^2}{n_S} + \frac{s_U^2}{n_U}}$$

$$= (441.985 - 164.588) \pm t^*_{33.86}\sqrt{\frac{650.787^2}{26} + \frac{278.426^2}{26}} \approx (-4.76, 559.56)$$

We are 95% confident the mean amount of rainfall produced by seeded clouds is between 4.76 acre-feet less than and 559.56 acre-feet more than the mean amount of rainfall produced by unseeded clouds.

Since the interval contains 0, there is little evidence that the mean rainfall produced by seeded clouds is any different from the mean rainfall produced by unseeded clouds. However, we shouldn't place too much faith in this conclusion. It is based on a procedure that is sensitive to outliers, and there may have been outliers present.

37. Color or text?

a) By randomizing the order of the cards (shuffling), the researchers are attempting to avoid bias that may result from volunteers remembering the order of the cards from the first task. Although mentioned, hopefully the researchers are randomizing the order in which the tasks are performed. For example, if each volunteer performs the color task first, they may all do better on the text task, simply because they have had some practice in memorizing cards. By randomizing the order, that bias is controlled.

b) H0: The mean difference between color and word scores is zero. $(\mu_d = 0)$

HA: The mean difference between color and word scores is not zero. $(\mu_d \neq 0)$

c) **Paired data assumption:** The data are paired by volunteer.
Randomization condition: Hopefully, the volunteers were randomized with respect to the order in which they performed the tasks, and the cards were shuffled between tasks.
10% condition: We are testing the format of the task (color/word), not the people, so we don't need to check this condition.
Nearly Normal condition: The histogram of the differences is roughly unimodal and symmetric.

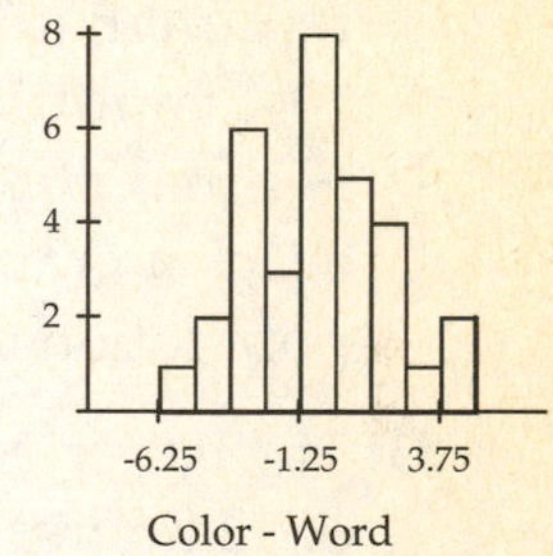

d) Since the conditions are satisfied, the sampling distribution of the difference can be modeled with a Student's *t*-model with 32 – 1 = 31 degrees of freedom, $t_{31}\left(0, \frac{2.50161}{\sqrt{32}}\right)$.

We will use a paired *t*-test, (color – word) with $\bar{d} = -0.75$.

e) Since the *P*-value = 0.0999 is high, we fail to reject the null hypothesis. There is no evidence that the mean difference in score is different from zero. It is reasonable to think that neither color nor word dominates perception. However, if the order in which volunteers took the test was not randomized, making conclusions from this test may be risky.

$$t = \frac{\bar{d} - 0}{\frac{s_d}{\sqrt{n}}}$$

$$t \approx \frac{-0.75 - 0}{\frac{2.50161}{\sqrt{32}}}$$

$$t \approx -1.70$$

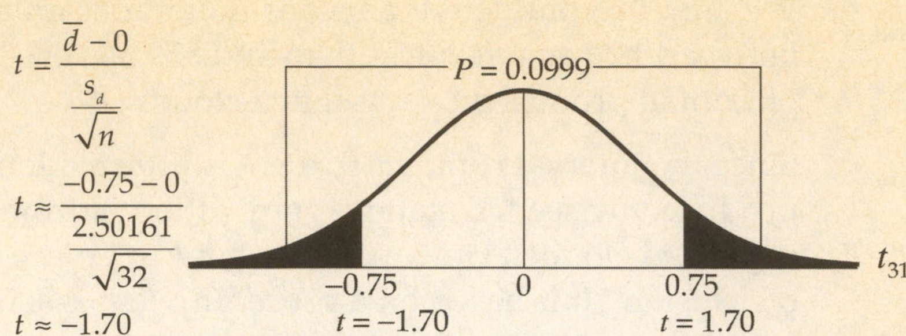

39. Batteries.

a) Different samples have different means. Since this is a fairly small sample, the difference may be due to natural sampling variation. Also, we have no idea how to quantify "a lot less" with out considering the variation as measured by the standard deviation.

b) H_0: The mean life of a battery is 100 hours. $(\mu = 100)$

H_A: The mean life of a battery is less than 100 hours. $(\mu < 100)$

c) **Randomization condition:** It is reasonable to think that these 16 batteries are representative of all batteries of this type
10% condition: 16 batteries are less than 10% of all batteries.
Normal population assumption: Since we don't have the actual data, we can't check a graphical display, and the sample is not large. Assume that the population of battery lifetimes is Normal.

d) The batteries in the sample had a mean life of 97 hours and a standard deviation of 12 hours. Since the conditions for inference are satisfied, we can model the sampling distribution of the mean battery life with a Student's *t* model, with 16 − 1 = 15 degrees of freedom, $t_{15}\left(100, \frac{12}{\sqrt{16}}\right)$, or $t_{15}(100, 3)$

We will perform a one-sample *t*-test.

Since the *P*-value = 0.1666 is greater than $\alpha = 0.05$, we fail to reject the null hypothesis. There is no evidence to suggest that the mean battery life is less than 100 hours.

$$t = \frac{\bar{y} - \mu_0}{SE(\bar{y})}$$

$$t = \frac{97 - 100}{\frac{12}{\sqrt{16}}}$$

$$t = -1$$

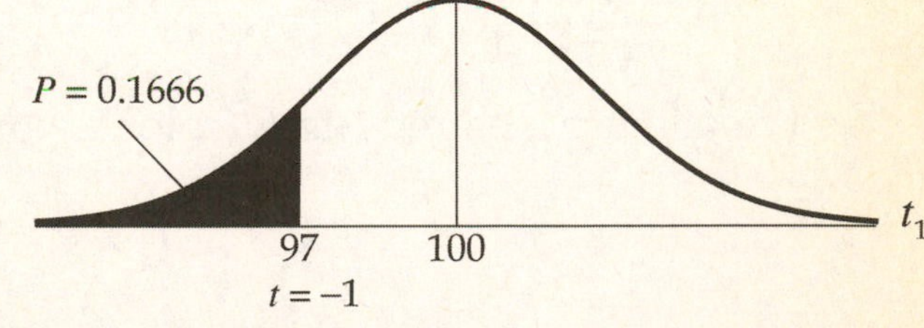

e) If the mean life of the company's batteries is only 98 hours, then the mean life is less than 100, and the null hypothesis is false. We failed to reject a false null hypothesis, making a Type II error.

41. Family planning.

H_0: Unplanned pregnancies and education level are independent.

H_A: There is an association between unplanned pregnancies and education level.

Counted data condition: The percentages must be converted to counts.
Randomization condition: Assume that these women are representative of all women.
Expected cell frequency condition: The expected counts are all greater than 5.

Under these conditions, the sampling distribution of the test statistic is χ^2 on 1 degree of freedom. We will use a chi-square test for independence.

$\chi^2 = \sum_{all\,cells} \frac{(Obs - Exp)^2}{Exp} \approx 40.71$, and the *P*-value is essentially 0.

Since the *P*-value is essentially 0, we reject the null hypothesis. There is strong evidence of an association between unplanned pregnancies and education level. More educated women tend to have fewer unplanned pregnancies.

Chapter 27 – Inferences for Regression

1. Hurricane predictions.

a) The equation of the line of best fit for these data points is $\widehat{Error} = 453.22 - 8.37(Year)$, where *Year* is measured in years since 1970.

According to the linear model, the error made in predicting a hurricane's path was about 453 nautical miles, on average, in 1970. It has been declining at rate of about 8.37 nautical miles per year.

b) H_0: There has been no change in prediction accuracy. $(\beta_1 = 0)$

H_A: There has been a change in prediction accuracy. $(\beta_1 \neq 0)$

c) Assuming the conditions have been met, the sampling distribution of the regression slope can be modeled by a Student's *t*-model with (34 – 2) = 32 degrees of freedom. We will use a regression slope *t*-test.

The value of $t = -6.92$. The *P*-value ≤ 0.0001 means that the association we see in the data is unlikely to occur by chance. We reject the null hypothesis, and conclude that there is strong evidence that the prediction accuracies have in fact been changing during the time period.

d) 58.8% of the variation in the prediction accuracy is accounted for by the linear model based on year.

3. Movie budgets.

a) $\widehat{Budget} = -63.9981 + 1.02648(RunTime)$. The model suggests that each additional minute of run time for a movie costs about $1,026,000.

b) A negative intercept makes no sense, but the *P*-value of 0.07 indicates that we can't discern a difference between our estimated value and zero. The statement that a movie of zero length should cost $0 makes sense.

c) Amounts by which movie costs differ from predictions made by this model vary, with a standard deviation of about $33 million.

d) The standard error of the slope is 0.1541 million dollars per minute.

e) If we constructed other models based on different samples of movies, we'd expect the slopes of the regression lines to vary, with a standard deviation of about $154,000 per minute.

5. Movie budgets, the sequel.

a) **Straight enough condition:** The scatterplot is straight enough, and the residuals plot looks unpatterned.
Independence assumption: The residuals plot shows no evidence of dependence.

Does the plot thicken? condition: The residuals plot shows no obvious trends in the spread.
Nearly Normal condition, Outlier condition: The histogram of residuals is unimodal and symmetric, and shows no outliers.

b) Since conditions have been satisfied, the sampling distribution of the regression slope can be modeled by a Student's t-model with (120 – 2) = 118 degrees of freedom.

$b_1 \pm t^*_{n-2} \times SE(b_1) = 1.02648 \pm (t^*_{118}) \times 0.1541 \approx (0.72, 1.33)$

We are 95% confident that the cost of making longer movies increases at a rate of between 0.72 and 1.33 million dollars per minute.

7. Hot dogs.

a) H_0: There's no association between calories and sodium content of all-beef hot dogs. $(\beta_1 = 0)$

H_A: There is an association between calories and sodium content. $(\beta_1 \neq 0)$

b) Assuming the conditions have been met, the sampling distribution of the regression slope can be modeled by a Student's t-model with (13 – 2) = 11 degrees of freedom. We will use a regression slope t-test. The equation of the line of best fit for these data points is: $\widehat{Sodium} = 90.9783 + 2.29959(Calories)$

The value of $t = 4.10$. The P-value of 0.0018 means that the association we see in the data is very unlikely to occur by chance alone. We reject the null hypothesis, and conclude that there is evidence of a linear association between the number of calories in all-beef hotdogs and their sodium content. Because of the positive slope, there is evidence that hot dogs with more calories generally have higher sodium contents.

9. Second frank.

a) Among all-beef hot dogs with the same number of calories, the sodium content varies, with a standard deviation of about 60 mg.

b) The standard error of the slope of the regression line is 0.5607 milligrams of sodium per calorie.

c) If we tested many other samples of all-beef hot dogs, the slopes of the resulting regression lines would be expected to vary, with a standard deviation of about 0.56 mg of sodium per calorie.

11. Last dog.

$$b_1 \pm t^*_{n-2} \times SE(b_1) = 2.29959 \pm (2.201) \times 0.5607 \approx (1.03, 3.57)$$

We are 95% confident that for every additional calorie, all-beef hot dogs have, on average, between 1.03 and 3.57 mg more sodium.

13. Marriage age 2007.

a) H_0: The difference in age between men and women at first marriage has not been decreasing since 1975. $(\beta_1 = 0)$

H_A: The difference in age between men and women at first marriage has been decreasing since 1975. $(\beta_1 < 0)$

b) **Straight enough condition:** The scatterplot is straight enough.
Independence assumption: We are examining a relationship over time, so there is reason to be cautious, but the residuals plot shows no evidence of dependence.
Does the plot thicken? condition: The residuals plot shows no obvious trends in the spread.
Nearly Normal condition, Outlier condition: The histogram is not particularly unimodal and symmetric, but shows no obvious skewness or outliers.

c) Since conditions have been satisfied, the sampling distribution of the regression slope can be modeled by a Student's t-model with (31 – 2) = 29 degrees of freedom. We will use a regression slope t-test. The equation of the line of best fit for these data points is: $(\widehat{Men - Women}) = 65.3103 - 0.031725(Year)$

The value of $t = -9.11$. The P-value of less than 0.0001 (even though this is the value for a two-tailed test, it is still very small) means that the association we see in the data is unlikely to occur by chance. We reject the null hypothesis, and conclude that there is strong evidence of a negative linear relationship between difference in age at first marriage and year. The difference in marriage age between men and women appears to be decreasing over time.

15. Marriage age 2007, again.

$$b_1 \pm t^*_{n-2} \times SE(b_1) = -0.031725 \pm (2.045) \times 0.0035 \approx (-0.039, -0.025)$$

We are 95% confident that the mean difference in age between men and women at first marriage decreases by between 0.025 and 0.039 years in age for each year that passes.

17. Fuel economy.

a) H_0: There is no linear relationship between the weight of a car and its mileage.
$(\beta_1 = 0)$

H_A: There is a linear relationship between the weight of a car and its mileage.
$(\beta_1 \neq 0)$

b) **Straight enough condition:** The scatterplot is straight enough to try a linear model.
Independence assumption: The residuals plot is scattered.
Does the plot thicken? condition: The residuals plot indicates some possible "thickening" as the predicted values increases, but it's probably not enough to worry about.
Nearly Normal condition, Outlier condition: The histogram of residuals is unimodal and symmetric, with one possible outlier. With the large sample size, it is okay to proceed.

Since conditions have been satisfied, the sampling distribution of the regression slope can be modeled by a Student's *t*-model with (50 – 2) = 48 degrees of freedom. We will use a regression slope *t*-test. The equation of the line of best fit for these data points is: $\widehat{MPG} = 48.7393 - 8.21362(Weight)$, where *Weight* is measured in thousands of pounds.

The value of $t = -12.2$. The *P*-value of less than 0.0001 means that the association we see in the data is unlikely to occur by chance. We reject the null hypothesis, and conclude that there is strong evidence of a linear relationship between weight of a car and its mileage. Cars that weigh more tend to have lower gas mileage.

19. Fuel economy, part II.

a) Since conditions have been satisfied in Exercise 7, the sampling distribution of the regression slope can be modeled by a Student's *t*-model with (50 – 2) = 48 degrees of freedom. (Use $t^*_{45} = 2.014$ from the table.) We will use a regression slope *t*-interval, with 95% confidence.

$$b_1 \pm t^*_{n-2} \times SE(b_1) = -8.21362 \pm (2.014) \times 0.6738 \approx (-9.57, -6.86)$$

b) We are 95% confident that the mean mileage of cars decreases by between 6.86 and 9.57 miles per gallon for each additional 1000 pounds of weight.

21. Fuel economy, part III.

a) The regression equation predicts that cars that weigh 2500 pounds will have a mean fuel efficiency of $48.7393 - 8.21362(2.5) = 28.20525$ miles per gallon.

$$\hat{y}_\nu \pm t^*_{n-2}\sqrt{SE^2(b_1)\cdot(x_\nu - \bar{x})^2 + \frac{s_e^2}{n}}$$

$$= 28.20525 \pm (2.014)\sqrt{0.6738^2 \cdot (2.5 - 2.8878)^2 + \frac{2.413^2}{50}} \approx (27.34,\ 29.07)$$

We are 95% confident that cars weighing 2500 pounds will have mean fuel efficiency between 27.34 and 29.07 miles per gallon.

b) The regression equation predicts that cars that weigh 3450 pounds will have a mean fuel efficiency of $48.7393 - 8.21362(3.45) = 20.402311$ miles per gallon.

$$\hat{y}_\nu \pm t^*_{n-2}\sqrt{SE^2(b_1)\cdot(x_\nu - \bar{x})^2 + \frac{s_e^2}{n} + s_e^2}$$

$$= 20.402311 \pm (2.014)\sqrt{0.6738^2 \cdot (3.45 - 2.8878)^2 + \frac{2.413^2}{50} + 2.413^2} \approx (15.44,\ 25.37)$$

We are 95%confident that a car weighing 3450 pounds will have fuel efficiency between 15.44 and 25.37 miles per gallon.

23. Cereal.

a) H_0: There is no linear relationship between the number of calories and the sodium content of cereals. $(\beta_1 = 0)$

H_A: There is a linear relationship. $(\beta_1 \neq 0)$

Since these data were judged acceptable for inference, the sampling distribution of the regression slope can be modeled by a Student's *t*-model with (77 – 2) = 75 degrees of freedom. We will use a regression slope *t*-test. The equation of the line of best fit for these data points is: $\widehat{Sodium} = 21.4143 + 1.29357(Calories)$.

The value of $t = 2.73$. The *P*-value of 0.0079 means that the association we see in the data is unlikely to occur by chance. We reject the null hypothesis, and conclude that there is strong evidence of a linear relationship between the number of calories and sodium content of cereals. Cereals with higher numbers of calories tend to have higher sodium contents.

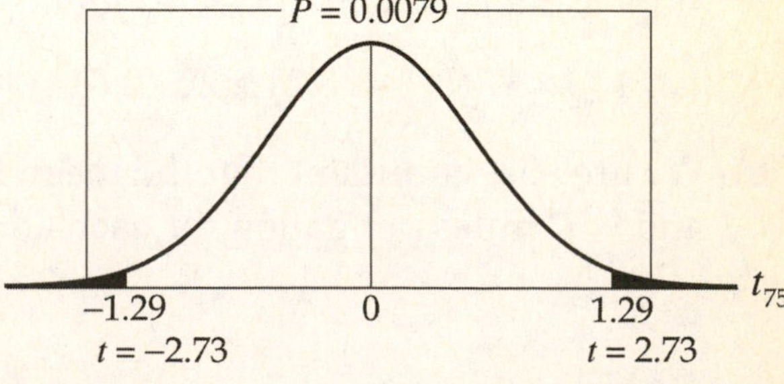

b) Only 9% of the variability in sodium content can be explained by the number of calories. The residual standard deviation is 80.49 mg, which is pretty large when you consider that the range of sodium content is only 320 mg. Although there is strong evidence of a linear association, it is too weak to be of much use. Predictions would tend to be very imprecise.

25. Another bowl.

Straight enough condition: The scatterplot is not straight.
Independence assumption: The residuals plot shows a curved pattern.
Does the plot thicken? condition: The spread of the residuals is not consistent. The residuals plot "thickens" as the predicted values increase.
Nearly Normal condition, Outlier condition: The histogram of residuals is skewed to the right, with an outlier.

These data are not appropriate for inference.

27. Acid rain.

a) H_0: There is no linear relationship between BCI and pH. $(\beta_1 = 0)$

H_A: There is a linear relationship between BCI and pH. $(\beta_1 \neq 0)$

b) Assuming the conditions for inference are satisfied, the sampling distribution of the regression slope can be modeled by a Student's *t*-model with (163 – 2) = 161 degrees of freedom. We will use a regression slope *t*-test. The equation of the line of best fit for these data points is: $\widehat{BCI} = 2733.37 - 197.694(pH)$.

c) The value of $t \approx -7.73$. The *P*-value (two-sided!) of essentially 0 means that the association we see in the data is unlikely to occur by chance. We reject the null hypothesis, and conclude that there is strong evidence of a linear relationship between BCI and pH. Streams with higher pH tend to have lower BCI.

$$t = \frac{b_1 - \beta_1}{SE(b_1)}$$

$$t = \frac{-197.694 - 0}{25.57}$$

$$t \approx -7.73$$

29. Ozone.

a) H_0: There is no linear relationship between population and ozone level. $(\beta_1 = 0)$

H_A: There is a positive linear relationship between population and ozone level. $(\beta_1 > 0)$

Assuming the conditions for inference are satisfied, the sampling distribution of the regression slope can be modeled by a Student's *t*-model with (16 – 2) = 14 degrees of freedom. We will use a regression slope *t*-test. The equation of the line of best fit for these data points is: $\widehat{Ozone} = 18.892 + 6.650(Population)$, where ozone level is measured in parts per million and population is measured in millions.

The value of $t \approx 3.48$. The P-value of 0.0018 means that the association we see in the data is unlikely to occur by chance. We reject the null hypothesis, and conclude that there is strong evidence of a positive linear relationship between ozone level and population. Cities with larger populations tend to have higher ozone levels.

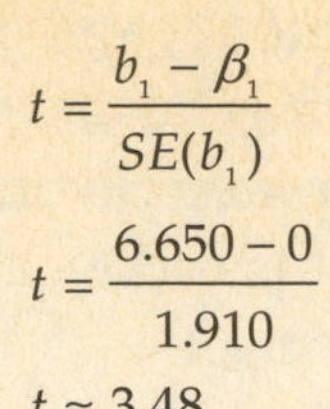

$$t = \frac{b_1 - \beta_1}{SE(b_1)}$$

$$t = \frac{6.650 - 0}{1.910}$$

$$t \approx 3.48$$

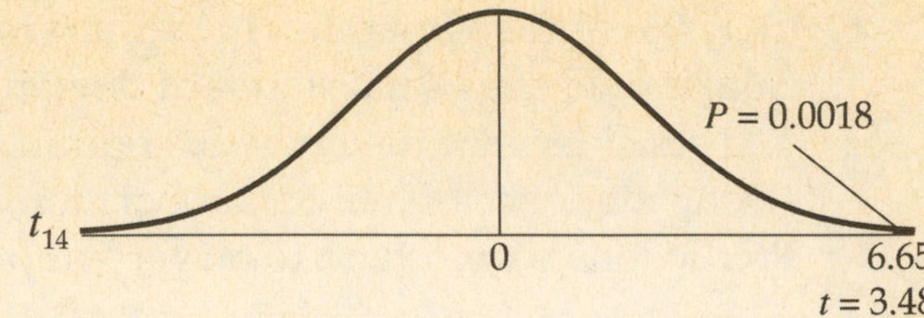

b) City population is a good predictor of ozone level. Population explains 84% of the variability in ozone level and s is just over 5 parts per million.

31. Ozone, again

a) $b_1 \pm t^*_{n-2} \times SE(b_1) = 6.65 \pm (1.761) \times 1.910 \approx (3.29, 10.01)$

We are 90% confident that each additional million people will increase mean ozone levels by between 3.29 and 10.01 parts per million.

b) The regression equation predicts that cities with a population of 600,000 people will have ozone levels of 18.892 + 6.650(0.6) = 22.882 parts per million.

$$\hat{y}_\nu \pm t^*_{n-2}\sqrt{SE^2(b_1)\cdot(x_\nu - \bar{x})^2 + \frac{s_e^2}{n}}$$

$$= 22.882 \pm (1.761)\sqrt{1.91^2 \cdot (0.6 - 1.7)^2 + \frac{5.454^2}{16}} \approx (18.47, 27.29)$$

We are 90% confident that the mean ozone level for cities with populations of 600,000 will be between 18.47 and 27.29 parts per million.

33. Start the car!

a) Since there are 33 – 2 = 31 degrees of freedom, there were 33 batteries tested.

b) **Straight enough condition:** The scatterplot is roughly straight, but scattered.
Independence assumption: The residuals plot shows no pattern.
Does the plot thicken? condition: The spread of the residuals is consistent.
Nearly Normal condition: The Normal probability plot of residuals is reasonably straight.

c) H$_0$: There is no linear relationship between cost and power. $(\beta_1 = 0)$

H$_A$: There is a positive linear relationship between cost and power. $(\beta_1 > 0)$

Since the conditions for inference are satisfied, the sampling distribution of the regression slope can be modeled by a Student's t-model with (33 – 2) = 31 degrees of freedom. We will use a regression slope t-test. The equation of the line of best fit for these data points is: $\widehat{Power} = 384.594 + 4.14649(Cost)$, with power measured in cold cranking amps, and cost measured in dollars.

The value of $t \approx 3.23$. The P-value of 0.0015 means that the association we see in the data is unlikely to occur by chance. We reject the null hypothesis, and conclude that there is strong evidence of a positive linear relationship between cost and power. Batteries that cost more tend to have more power.

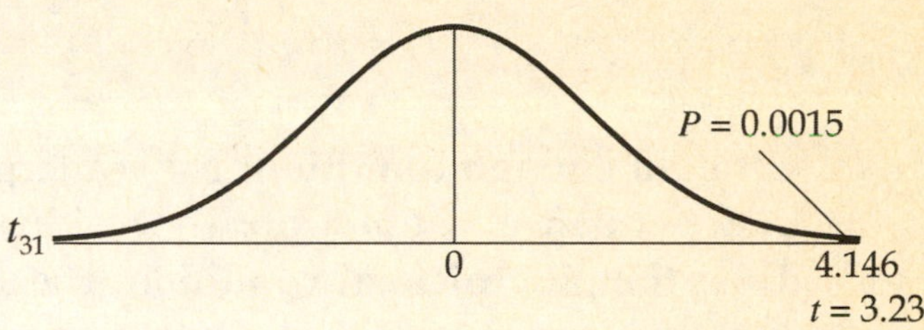

d) Since $R^2 = 25.2\%$, only 25.2% of the variability in power can be accounted for by cost. The residual standard deviation is 116 amps. That's pretty large, considering the range battery power is only about 400 amps. Although there is strong evidence of a linear association, it is too weak to be of much use. Predictions would tend to be very imprecise.

e) The equation of the line of best fit for these data points is: $\widehat{Power} = 384.594 + 4.14649(Cost)$, with power measured in cold cranking amps, and cost measured in dollars.

f) There are 31 degrees of freedom, so use $t^*_{30} = 1.697$ as a conservative estimate from the table.

$$b_1 \pm t^*_{n-2} \times SE(b_1) = 4.14649 \pm (1.697) \times 1.282 \approx (1.97, 6.32)$$

g) We are 95% confident that the mean power increases by between 1.97 and 6.32 cold cranking amps for each additional dollar in cost.

35. Body fat.

a) H$_0$: There is no linear relationship between waist size and percent body fat. $(\beta_1 = 0)$

H$_A$: There is a linear relationship between waist size and percent body fat. $(\beta_1 \neq 0)$

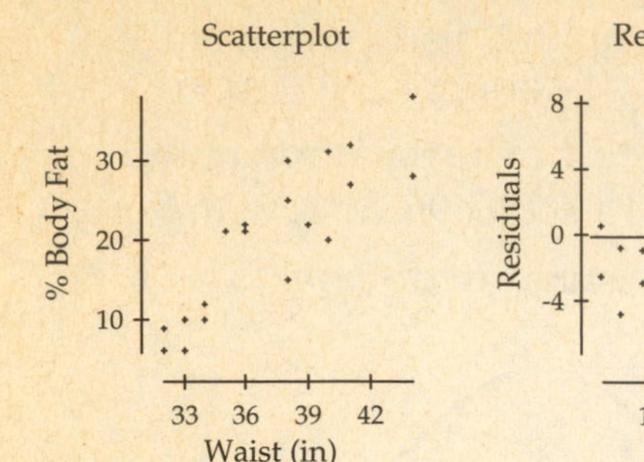

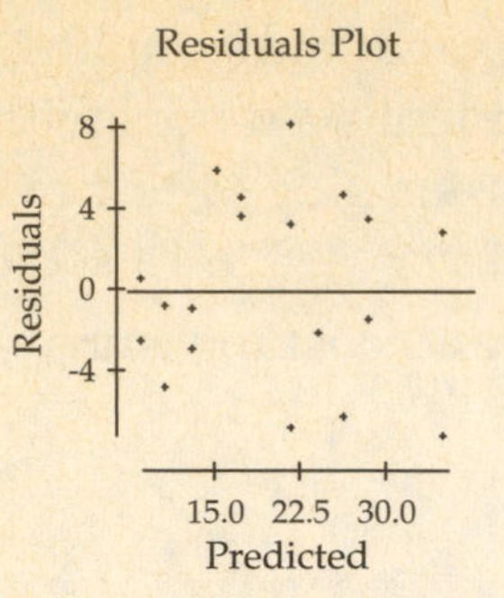

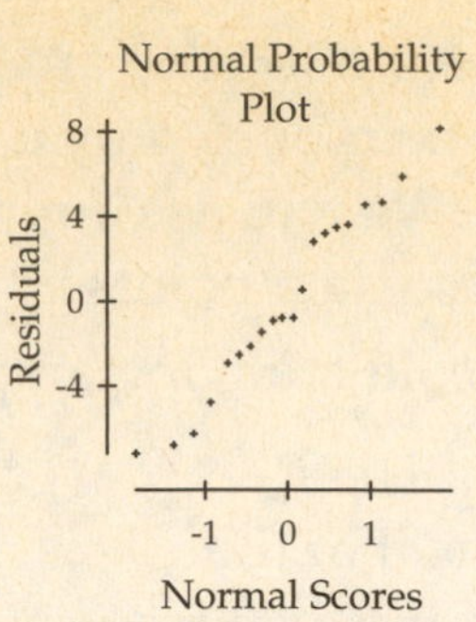

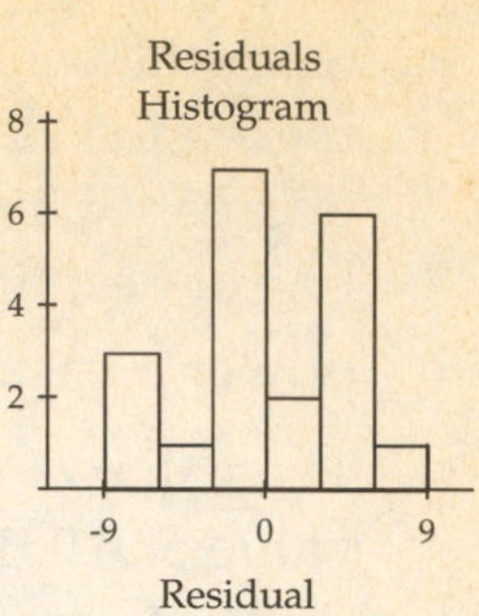

Straight enough condition: The scatterplot is straight enough.
Independence assumption: The residuals plot shows no pattern.
Does the plot thicken? condition: The spread of the residuals is consistent.
Nearly Normal condition, Outlier condition: The Normal probability plot of residuals is straight, and the histogram of the residuals is unimodal and symmetric with no outliers.

Since the conditions for inference are inference are satisfied, the sampling distribution of the regression slope can be modeled by a Student's *t*-model with (20 – 2) = 18 degrees of freedom. We will use a regression slope *t*-test.

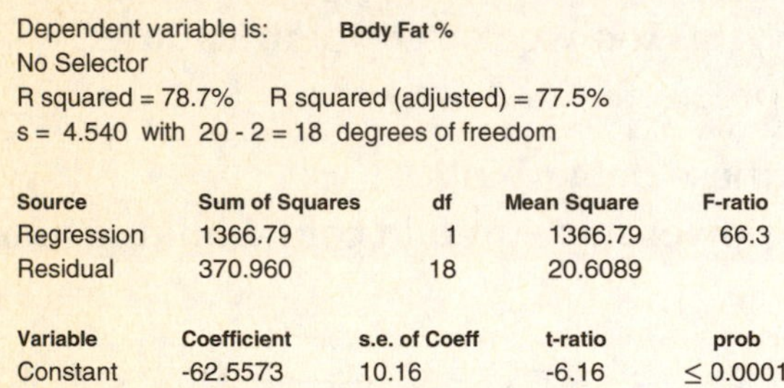

Dependent variable is: **Body Fat %**
No Selector
R squared = 78.7% R squared (adjusted) = 77.5%
s = 4.540 with 20 - 2 = 18 degrees of freedom

Source	Sum of Squares	df	Mean Square	F-ratio
Regression	1366.79	1	1366.79	66.3
Residual	370.960	18	20.6089	

Variable	Coefficient	s.e. of Coeff	t-ratio	prob
Constant	-62.5573	10.16	-6.16	≤ 0.0001
Waist (in)	2.22152	0.2728	8.14	≤ 0.0001

The equation of the line of best fit for these data points is:

$$\widehat{\%BodyFat} = -62.5573 + 2.22152(Waist).$$

The value of $t \approx 8.14$. The *P*-value of essentially 0 means that the association we see in the data is unlikely to occur by chance. We reject the null hypothesis, and conclude that there is strong evidence of a linear relationship between waist size and percent body fat. People with larger waists tend to have a higher percentage of body fat.

b) The regression equation predicts that people with 40-inch waists will have $-62.5573 + 2.22152(40) = 26.3035\%$ body fat. The average waist size of the people sampled was approximately 37.05 inches.

$$\hat{y}_v \pm t^*_{n-2}\sqrt{SE^2(b_1)\cdot(x_v - \bar{x})^2 + \frac{s_e^2}{n}}$$

$$= 26.3035 \pm (2.101)\sqrt{0.2728^2 \cdot (40 - 37.05)^2 + \frac{4.54^2}{20}}$$

$$\approx (23.58,\ 29.03)$$

We are 95% confident that the mean percent body fat for people with 40-inch waists is between 23.58% and 29.03%.

37. Grades.

a) The regression output is to the right. The model is:

$\widehat{Midterm2} = 12.005 + 0.721(Midterm1)$

Dependent variable is: **Midterm 2**
No Selector
R squared = 19.9% R squared (adjusted) = 18.6%
s = 16.78 with 64 - 2 = 62 degrees of freedom

Source	Sum of Squares	df	Mean Square	F-ratio
Regression	4337.14	1	4337.14	15.4
Residual	17459.5	62	281.604	

Variable	Coefficient	s.e. of Coeff	t-ratio	prob
Constant	12.0054	15.96	0.752	0.4546
Midterm 1	0.720990	0.1837	3.92	0.0002

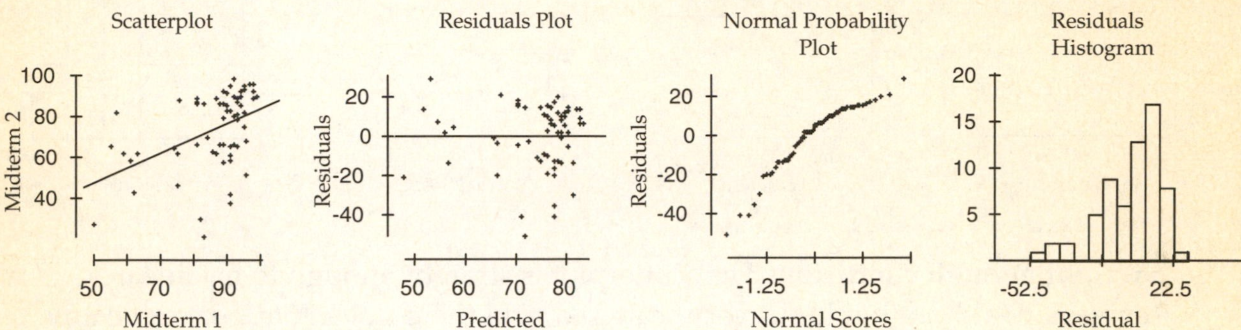

b) **Straight enough condition:** The scatterplot shows a weak, positive relationship between Midterm 2 score and Midterm 1 score. There are several outliers, but removing them only makes the relationship slightly stronger. The relationship is straight enough to try linear regression.
Independence assumption: The residuals plot shows no pattern..
Does the plot thicken? condition: The spread of the residuals is consistent.
Nearly Normal condition, Outlier condition: The histogram of the residuals is unimodal, slightly skewed with several possible outliers. The Normal probability plot shows some slight curvature.

Since we had some difficulty with the conditions for inference, we should be cautious in making conclusions from these data. The small *P*-value of 0.0002 for the slope would indicate that the slope is statistically distinguishable from zero, but the R^2 value of 0.199 suggests that the relationship is weak. Midterm 1 isn't a useful predictor of Midterm 2.

c) The student's reasoning is not valid. The R^2 value is only 0.199 and the value of s is 16.8 points. Although correlation between Midterm 1 and Midterm 2 may be statistically significant, it isn't of much practical use in predicting Midterm 2 scores. It's too weak.

39. Strike two.

H0: The effectiveness of the video is independent of the player's initial ability. $(\beta_1 = 0)$

HA: The effectiveness of the video depends on the player's initial ability. $(\beta_1 \neq 0)$

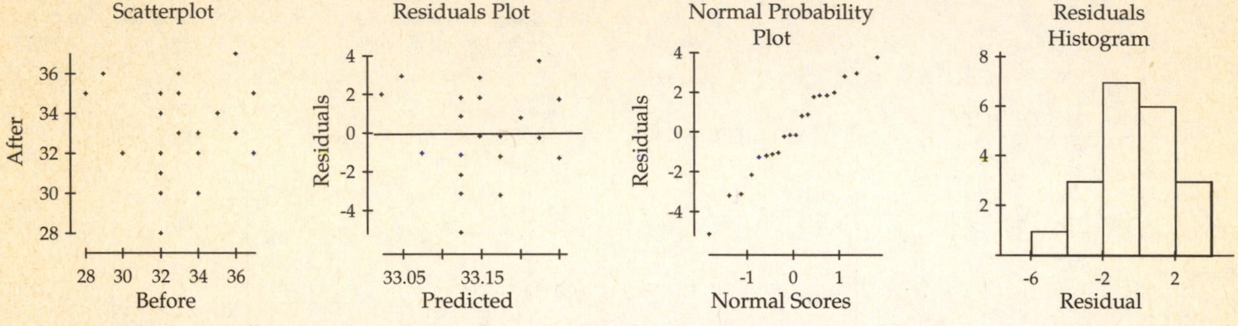

Straight enough condition: The scatterplot is straight enough to try linear regression, although it looks very scattered, and there doesn't appear to be any association.
Independence assumption: The residuals plot shows no pattern.
Does the plot thicken? condition: The spread of the residuals is consistent.
Nearly Normal condition, Outlier condition: The Normal probability plot of residuals is very straight, and the histogram of the residuals is unimodal and symmetric with no outliers.

Since the conditions for inference are inference are satisfied, the sampling distribution of the regression slope can be modeled by a Student's t-model with $(20 - 2) = 18$ degrees of freedom. We will use a regression slope t-test.

Dependent variable is: **After**
No Selector
R squared = 0.1% R squared (adjusted) = -5.5%
s = 2.386 with 20 - 2 = 18 degrees of freedom

Source	Sum of Squares	df	Mean Square	F-ratio
Regression	0.071912	1	0.071912	0.013
Residual	102.478	18	5.69323	

Variable	Coefficient	s.e. of Coeff	t-ratio	prob
Constant	32.3161	7.439	4.34	0.0004
Before	0.025232	0.2245	0.112	0.9118

The equation of the line of best fit for these data points is:
$\widehat{After} = 32.3161 + 0.025232(Before)$, where we are counting the number of strikes thrown before and after the training program.

The value of $t \approx 0.112$. The P-value of 0.9118 means that the association we see in the data is quite likely to occur by chance. We fail to reject the null hypothesis, and conclude that there is no evidence of a linear relationship between the number of strikes thrown before the training program and the number of strikes thrown after the program. The effectiveness of the program does not appear to depend on the initial ability of the player.

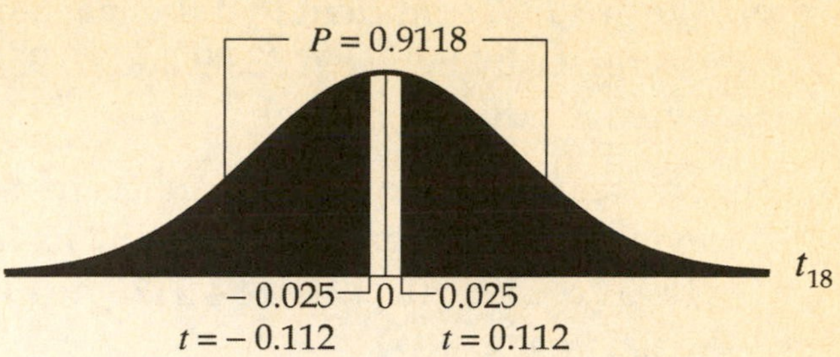

41. Education and mortality.

a) **Straight enough condition:** The scatterplot is straight enough to try linear regression.
Independence assumption: The residuals plot shows no pattern. If these cities are representative of other cities, we can generalize our results.
Does the plot thicken? condition: The spread of the residuals is consistent.
Nearly Normal condition, Outlier condition: The histogram of the residuals is unimodal and symmetric with no outliers.

b) H0: There is no linear relationship between education and mortality. $(\beta_1 = 0)$

HA: There is a linear relationship between education and mortality. $(\beta_1 \neq 0)$

Since the conditions for inference are inference are satisfied, the sampling distribution of the regression slope can be modeled by a Student's t-model with $(58 - 2) = 56$ degrees of freedom. We will use a regression slope t-test. The equation of the line of best fit for these data points is:
$\widehat{Mortality} = 1493.26 - 49.9202(Education)$.

The value of $t \approx -6.24$. The P-value of essentially 0 means that the association we see in the data is unlikely to occur by chance. We reject the null hypothesis, and conclude that there is strong evidence of a linear relationship between the level of education in a city and its mortality rate. Cities with lower education levels tend to have higher mortality rates.

$$t = \frac{b_1 - \beta_1}{SE(b_1)}$$
$$t = \frac{-49.9202 - 0}{8.000}$$
$$t \approx -6.24$$

c) We cannot conclude that getting more education is likely to prolong your life. Association does not imply causation. There may be lurking variables involved.

d) For 95% confidence, $t^*_{56} \approx 2.00327$.

$b_1 \pm t^*_{n-2} \times SE(b_1) = -49.9202 \pm (2.003) \times 8.000 \approx (-65.95, -33.89)$

e) We are 95% confident that the mean number of deaths per 100,000 people decreases by between 33.89 and 65.95 deaths for an increase of one year in average education level.

f) The regression equation predicts that cities with an adult population with an average of 12 years of school will have a mortality rate of $1493.26 - 49.9202(12) = 894.2176$ deaths per 100,000. The average education level was 11.0328 years.

$$\hat{y}_v \pm t^*_{n-2}\sqrt{SE^2(b_1)\cdot(x_v - \bar{x})^2 + \frac{s_e^2}{n}}$$

$$= 894.2176 \pm (2.003)\sqrt{8.00^2 \cdot (12 - 11.0328)^2 + \frac{47.92^2}{58}} \approx (874.239, 914.196)$$

We are 95% confident that the mean mortality rate for cities with an average of 12 years of schooling is between 874.239 and 914.196 deaths per 100,000 residents.

Chapter 28 – Analysis of Variance

1. Popcorn

a) H_0: The mean number of unpopped kernels is the same for all four brands of popcorn. $(\mu_1 = \mu_2 = \mu_3 = \mu_4)$

H_A: The mean number of unpopped kernels is not the same for all four brands of popcorn.

b) MS_T has $k-1=4-1=3$ degrees of freedom.
MS_E has $N\text{-}k=16-4=12$ degrees of freedom.

c) The F-statistic is 13.56 with 3 and 12 degrees of freedom, resulting in a P-value of 0.00037. We reject the null hypothesis and conclude that there is strong evidence that the mean number of unpopped kernels is not the same for all four brands of popcorn.

d) To check the Similar Variance condition, look at side-by-side boxplots of the treatment groups to see whether they have similar spreads. To check the Nearly Normal condition, look to see if a normal probability plot of the residuals is straight, look to see that a histogram of the residuals is nearly normal, and look to see if a residuals plot shows no pattern, and no systematic change in spread.

3. Gas mileage.

a) H_0: The mean gas mileage is the same for each muffler. $(\mu_1 = \mu_2 = \mu_3)$

H_A: The mean gas mileages for each muffler are not all the same.

b) MS_T has $k-1=3-1=2$ degrees of freedom.
MS_E has $N\text{-}k=24-3=21$ degrees of freedom.

c) The F-statistic is 2.35 with 2 and 21 degrees of freedom, resulting in a P-value of 0.1199. We fail to reject the null hypothesis and conclude that there is no evidence to suggest that gas mileage associated with any single muffler is different than the others.

d) To check the Similar Variance condition, look at side-by-side boxplots of the treatment groups to see whether they have similar spreads. To check the Nearly Normal condition, look to see if a normal probability plot of the residuals is straight, look to see that a histogram of the residuals is nearly normal, and look to see if a residuals plot shows no pattern, and no systematic change in spread.

e) By failing to notice that one of the mufflers resulted in significantly different gas mileage, you have committed a Type II error.

5. Activating baking yeast.

a) H_0: The mean activation time is the same for all four recipes. $(\mu_1 = \mu_2 = \mu_3 = \mu_4)$

H_A: The mean activation times for each recipe are not all the same.

b) The *F*-statistic is 44.7392 with 3 and 12 degrees of freedom, resulting in a *P*-value less than 0.0001. We reject the null hypothesis and conclude that there is strong evidence that the mean activation times for each recipe are not all the same.

c) Yes, it would be appropriate to follow up with multiple comparisons, because we have rejected the null hypothesis.

7. Eye and hair color.

An analysis of variance is not appropriate, because eye color is a categorical variable. The students could consider a chi-square test of independence.

9. Fuel economy revisited.

a) H_0: The mean mileage is the same for each engine type (number of cylinders).
$(\mu_4 = \mu_5 = \mu_6 = \mu_8)$

H_A: The mean mileages for each engine type are not all the same.

b) The Similar Variance condition is not met, because the boxplots show distributions with radically different spreads. A re-expression of the response variable may equalize the spreads, allowing us to proceed with an analysis of variance.

11. Tellers.

a) H_0: The mean time to serve a customer is the same for each teller.
$(\mu_1 = \mu_2 = \mu_3 = \mu_4 = \mu_5 = \mu_6)$

H_A: The mean times to serve a customer for each teller are not all the same.

b) The *F*-statistic is 1.508 with 5 and 134 degrees of freedom, resulting in a *P*-value equal to 0.1914. We fail to reject the null hypothesis and conclude that there is not enough evidence that the mean times to serve a customer for each teller are not all the same.

c) No, it would not be appropriate to follow up with multiple comparisons, because we have failed to reject the null hypothesis.

13. Yogurt.

a) MS_T has $k-1=3-1=2$ degrees of freedom.

$$MS_T = \frac{SS_T}{df_T} = \frac{17.300}{2} = 8.65$$

MS_E has $N\text{-}k = 9-3 = 6$ degrees of freedom.

$$MS_E = \frac{SS_E}{df_E} = \frac{0.4600}{6} \approx 0.0767$$

b) $F\text{-statistic} = \frac{MS_T}{MS_E} = \frac{8.65}{0.0767} \approx 112.78$

c) With a *P*-value equal to 0.000017, there is very strong evidence that the mean taste test scores for each method of preparation are not all the same.

d) We have assumed that the experimental runs were performed in random order, that the variances of the treatment groups are equal, and that the errors are Normal.

e) To check the Similar Variance condition, look at side-by-side boxplots of the treatment groups to see whether they have similar spreads. To check the Nearly Normal condition, look to see if a normal probability plot of the residuals is straight, look to see that a histogram of the residuals is nearly normal, and look to see if a residuals plot shows no pattern, and no systematic change in spread.

f) $s_p = \sqrt{MS_E} = \sqrt{0.0767} \approx 0.277$ points

15. Eggs.

a) H_0: The mean taste test scores are the same for both real and substitute eggs. $(\mu_R = \mu_S)$

H_A: The mean taste test scores are different. $(\mu_R \neq \mu_S)$

b) The *F*-statistic is 31.0712 with 1 and 6 degrees of freedom, resulting in a *P*-value equal to 0.0014. We reject the null hypothesis and conclude that there is strong evidence that the mean taste test scores for real and substitute eggs are different. The real eggs have a higher mean taste test score.

c) The Similar Variance assumption is a bit of a problem. The spread of the distribution of taste test scores for real eggs looks greater than the spread of the distribution of taste test scores for substitute eggs. Additionally, we cannot check the Nearly Normal condition because we don't have a residuals plot. Caution should be used in making any conclusions.

d) H$_0$: The mean taste test score of brownies made with real eggs is the same as the mean taste test score of brownies made with substitute eggs.
$(\mu_R = \mu_S \text{ or } \mu_R - \mu_S = 0)$

H$_A$: The mean taste test score of brownies made with real eggs is different than the mean taste test score of brownies made with substitute eggs.
$(\mu_R \neq \mu_S \text{ or } \mu_R - \mu_S \neq 0)$

Independent groups assumption: Taste test scores for each type of brownie should be independent.
Randomization condition: Brownies were tasted in random order.
10% condition: 4 and 4 are less than 10% of all possible batches.
Nearly Normal condition: We don't have the actual data, so we can't check the distribution of the sample. We will need to assume that the distribution of taste test scores is unimodal. The boxplots show distributions that are at least symmetric.

Since the conditions are satisfied, it is appropriate to model the sampling distribution of the difference in means with a Student's *t*-model, with 6 degrees of freedom ($n_R + n_S - 2 = 6$ for a pooled *t*-test). We will perform a two-sample *t*-test.

The pooled sample variance is $s_p^2 = \dfrac{(4-1)(0.651)^2 + (4-1)(0.395)^2}{(4-1)+(4-1)} \approx 0.2899$

The sampling distribution model has mean 0, with standard error:

$$SE_{pooled}(\bar{y}_R - \bar{y}_S) = \sqrt{\frac{0.2899}{4} + \frac{0.2899}{4}} \approx 0.3807.$$

The observed difference between the mean scores is 6.78 – 4.66 = 2.12.

$$t = \frac{(\bar{y}_R - \bar{y}_T) - (0)}{SE_{pooled}(\bar{y}_R - \bar{y}_T)} \approx \frac{2.12 - 0}{0.3807} \approx 5.5687$$

With a $t = 5.5687$ and 6 degrees of freedom, the 2-sided p-value is 0.0014. Since the *P*-value is low, we reject the null hypothesis. There is strong evidence that the mean taste test score for brownies made with real eggs is different from the mean taste test score for brownies made with substitute eggs. In fact, there is evidence that the brownies made with real eggs taste better.

The *P*-value for the 2-sample pooled *t*-test, 0.0014, which is the same as the p-value for the analysis of variance test. The *F*-statistic, 31.0712, was approximately the same as $t^2 = (5.5727)^2 \approx 31.05499$. Also, the pooled estimate of the variance, 0.2899, is approximately equal to MS_E, the mean squared error. This is because an analysis of variance for two samples is equivalent to a 2-sample pooled *t*-test.

17. School system.

a) H_0: The mean math test score is the same at each of the 15 schools.
$(\mu_A = \mu_B = \ldots = \mu_O)$

H_A: The mean math test scores are not all the same at each of the 15 schools.

b) The *F*-statistic is 1.0735 with 14 and 105 degrees of freedom, resulting in a *P*-value equal to 0.3899. We fail to reject the null hypothesis and conclude that there is not enough evidence to suggest that the mean math scores are not all the same for the 15 schools.

c) This does not match our findings in part b. Because the intern performed so many *t*-tests, he may have committed several Type I errors. (Type I error is the probability of rejecting the null hypothesis when there is actually no difference between the schools.) Some tests would be expected to result in Type I error due to chance alone. The overall Type I error rate is higher for multiple *t*-tests than for an analysis of variance, or a multiple comparisons method. Because we failed to reject the null hypothesis in the analysis of variance, we should not do multiple comparison tests.

19. Cereals.

a) H_0: The mean sugar content is the same for each of the 3 shelves. $(\mu_1 = \mu_2 = \mu_3)$

H_A: The mean sugar contents are not all the same for each of the 3 shelves.

b) The *F*-statistic is 7.3345 with 2 and 74 degrees of freedom, resulting in a *P*-value equal to 0.0012. We reject the null hypothesis and conclude that there is strong evidence to suggest that the mean sugar content of cereal on at least one shelf differs from that on other shelves.

c) We cannot conclude that cereals on shelf two have a higher mean sugar content than cereals on shelf three, or that cereals on shelf two have a higher mean sugar content than cereals on shelf one. We can conclude only that the mean sugar contents are not all equal.

d) We can conclude that the mean sugar content on shelf two is significantly different from the mean sugar contents on shelves one and three. In fact, there is evidence that the mean sugar content on shelf two is greater than that of shelf one and shelf three.

21. Downloading.

a) H_0: The mean download time is the same for each of the three times of day.
$(\mu_{Early} = \mu_{Evening} = \mu_{Late})$

H_A: The mean download times are not all the same for each of the three times of day.

b)

Analysis of Variance

Source	DF	Sum of Squares	Mean Square	F-ratio	P-Value
Time of Day	2	204641	102320	46.035	<0.0001
Error	45	100020	2222.67		
Total	47	304661			

The *F*-statistic is 46.035 with 2 and 45 degrees of freedom, resulting in a *P*-value less than 0.0001. We reject the null hypothesis and conclude that there is strong evidence to suggest that the mean download time is different in at least one of the three times of day.

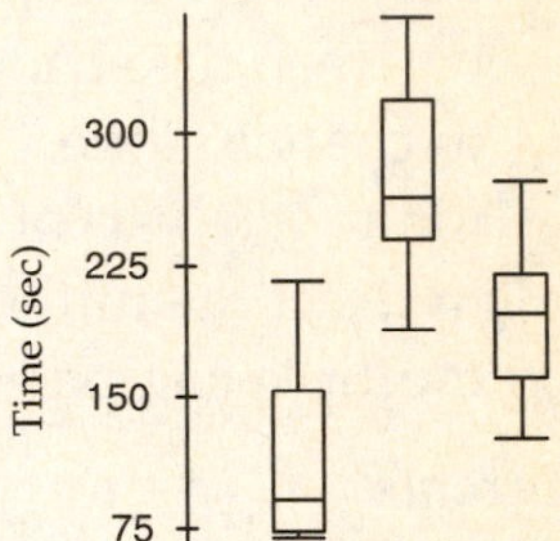

c) **Randomization condition:** The runs were not randomized, but it is likely that they are representative of all download times at these times of day.
Similar Variance condition: The boxplots show similar spreads for the distributions of download times for the different times of day.
Nearly Normal condition: The normal probability plot of residuals is reasonably straight.

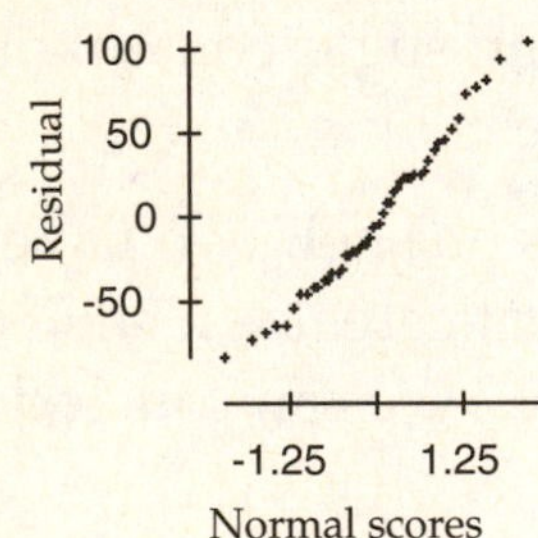

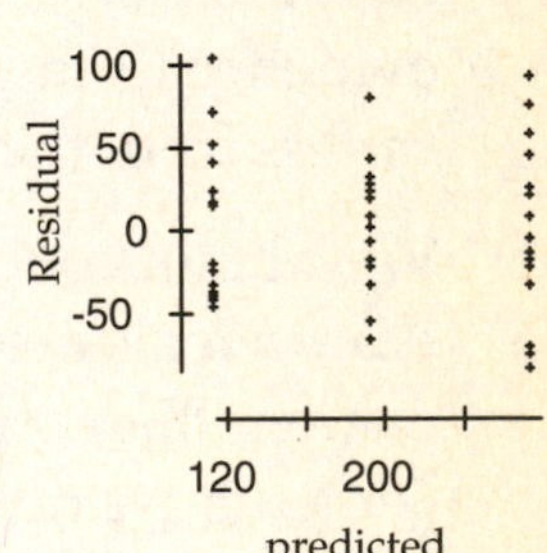

d) A Bonferroni test shows that all three pairs are different from each other at $\alpha = 0.05$.

Chapter 29 – Multifactor Analysis of Variance

1. Popcorn revisited.

a) H_0: The effect due to power level is the same for each level. $(\gamma_{Low} = \gamma_{Med} = \gamma_{High})$

H_A: Not all of the power levels have the same effect on popcorn popping.

H_0: The effect due to popping time is the same for each time. $(\tau_3 = \tau_4 = \tau_5)$

H_A: Not all of the popping times have the same effect on popcorn popping.

b) The *power* sum of squares has 3 – 1 = 2 degrees of freedom.
The *popping time* sum of squares has 3 – 1 = 2 degrees of freedom.
The error sum of squares has (9 – 1) – 2 – 2 = 4 degrees of freedom.

c) Because the experiment did not include replication, there are no degrees of freedom left for the interaction term. The interaction term would require 2(2) = 4 degrees of freedom, leaving none for the error term, making any tests impossible.

3. Popcorn again.

a) The *F*-statistic for *power* is 13.56 with 2 and 4 degrees of freedom, resulting in a p-value equal to 0.0165. The *F*-statistic for *time* is 9.36 with 2 and 4 degrees of freedom, resulting in a *P*-value equal to 0.0310.

b) With a *P*-value equal to 0.0165 we reject the null hypothesis that *power* has no effect and conclude that the mean number of uncooked kernels is not equal across all 3 *power* levels. With a *P*-value equal to 0.0310 we reject the null hypothesis that *time* has no effect and conclude that the mean number of uncooked kernels is not equal across all 3 *time* levels.

c) **Randomization condition:** The bags should be randomly assigned to treatments.
Similar Variance condition: The side-by-side boxplots should have similar spreads. The residuals plot should show no pattern, and no change in spread.
Nearly Normal condition: The Normal probability plot of the residuals should be straight, the histogram of the residuals should be unimodal and symmetric.

5. Crash analysis.

a) H_0: The effect on head injury severity is the same for both seats. $(\gamma_D = \gamma_P)$

H_A: The effects on head injury severity are different for the two seats.

H_0: The size of the vehicle has no effect on head injury severity.
$(\tau_1 = \tau_2 = \tau_3 = \tau_4 = \tau_5 = \tau_6)$

H_A: The size of the vehicle does have an effect on head injury severity.

b) **Randomization condition:** Assume that the cars are representative of all cars.
Additive Enough condition: The interaction plot is reasonably parallel.
Similar Variance condition: The side-by-side boxplots have similar spreads. The residuals plot shows no pattern, and no systematic change in spread.
Nearly Normal condition: The Normal probability plot of the residuals should be straight, the histogram of the residuals should be unimodal and symmetric.

c) With a *P*-value equal to 0.838 the interaction term between *seat* and *vehicle size* is not significant. With a *P*-value less than 0.0001 for both *seat* and *car size* we reject both null hypotheses and conclude that both *seat* and *car size* affect the severity of head injury. By looking at one of the partial boxplots we see that the mean head injury severity is higher for the driver's side. The effect of driver's *seat* seems to be roughly the same for all six cars.

7. Baldness and Heart Disease.

a) A two-factor ANOVA must have a quantitative response variable. Here the response is whether they exhibited baldness or not, which is a categorical variable. A two-factor ANOVA is not appropriate.

b) We could use a chi-square analysis to test whether baldness and heart disease are independent.

9. Baldness and heart disease again.

a) A chi-square test of independence gives a chi-square statistic of 14.510 with a *P*-value equal to 0.0023. We reject the hypothesis that baldness and heart disease are independent.

b) No, the fact that these are not independent does not mean that one causes the other. There could be a lurking variable (such as age) that influences both.

11. Basketball shots.

a) H_0: The time of day has no effect on the number of shots made. $(\gamma_M = \gamma_N)$

H_A: The time of day does have an effect on the number of shots made.

H_0: The type of shoe has no effect on the number of shots made. $(\tau_F = \tau_O)$

H_A: The type of shoe does have an effect on the number of shots made.

b)

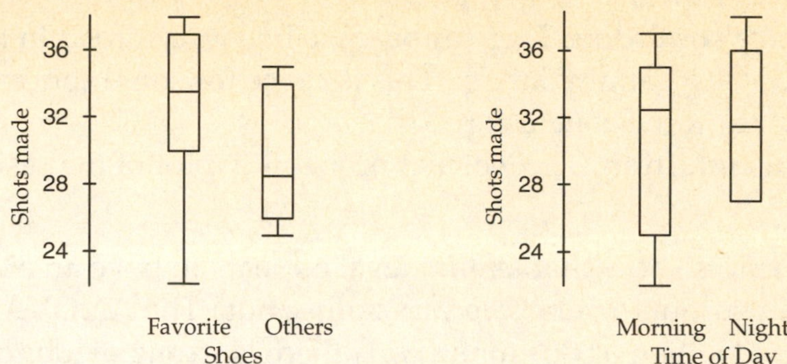

The partial boxplots show little effect of either *time of day* or *shoes* on the number of shots made.

Randomization condition: We assume that the number of shots made were independent from one treatment condition to the next.
Similar Variance condition: The side-by-side boxplots have similar spreads. The residuals plot shows no pattern, and no systematic change in spread.
Nearly Normal condition: The Normal probability plot of the residuals is straight.

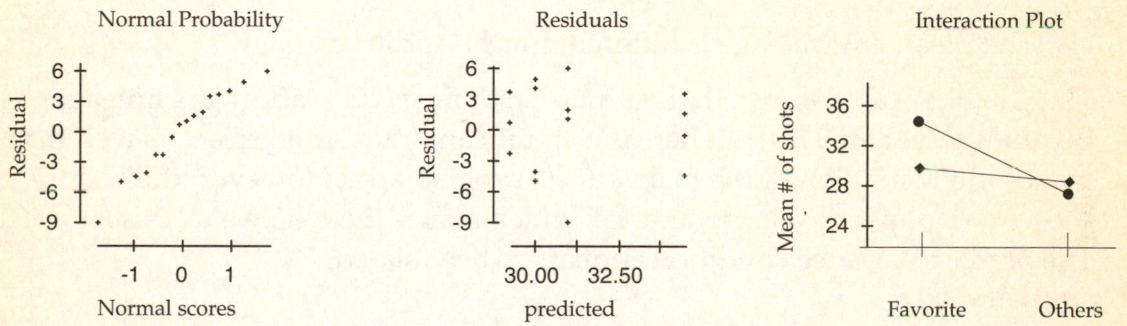

The interaction plot shows a possible interaction effect. It looks as though the favorite shoes may make more of a difference at night. However, we fail to reject the null hypothesis that there is no interaction effect. In fact, none of the effects appears to be significant. It looks as though she cannot conclude that either *shoes* or *time of day* affect her mean free throw percentage.

13. Sprouts again.

a) H0: The temperature level has no effect on the number of sprouts. $(\gamma_{32} = \gamma_{34} = \gamma_{36})$

HA: The temperature level does have an effect on the number of sprouts.

H0: The salinity level has no effect on the number of sprouts. $(\tau_0 = \tau_4 = \tau_8 = \tau_{12})$

HA: The salinity level does have an effect on the number of sprouts.

b) **Randomization condition:** We assume that the sprouts were representative of all sprouts.
Similar Variance condition: There appears to be more spread in the number of sprouts for the lower salinity levels. This is cause for some concern, but most likely does not affect the conclusions.
Nearly Normal condition: The Normal probability plot of the residuals should be straight.

The partial boxplots show that *salinity* level appears to have an effect on the number of bean sprouts while *temperature* does not. The ANOVA supports this. With a p-value less than 0.0001 for *salinity*, there is strong evidence that the salinity level affects the number of bean sprouts. However, it appears that neither the interaction term nor *temperature* have a significant effect on the number of bean sprouts. The interaction term has a *P*-value equal to 0.7549 and *temperature* has a *P*-value equal to 0.3779.

15. Gas additives.

H_0: The car type has no effect on gas mileage. $(\gamma_H = \gamma_M = \gamma_S)$

H_A: The car type does have an effect on gas mileage.

H_0: The mean gas mileage is the same for both additives. $(\tau_G = \tau_R)$

H_A: The mean gas mileage is different for the additives.

The ANOVA table shows that both *car type* and *additive* affect gas mileage, with *P*-values less than 0.0001. There is a significant interaction effect as well that makes interpretation of the main effects problematic. However, the residual plot shows a strong increase in variance, which makes the whole analysis suspect. The Similar Variance condition appears to be violated.

17. Gas additives again.

After the re-expression of the response, gas mileage, the Normal probability plot of residuals looks straight and the residuals plot shows constant spread over the predicted values.

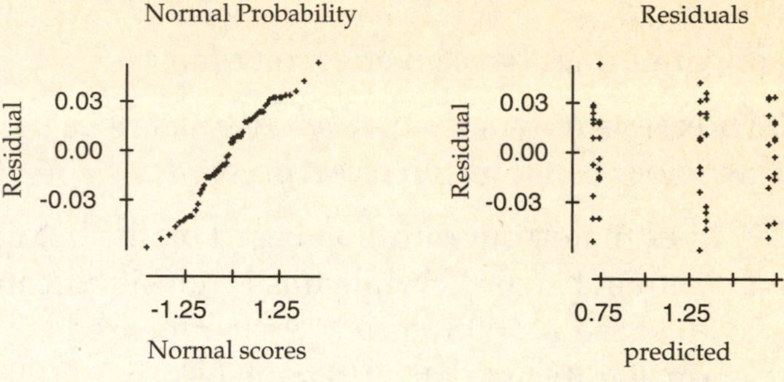

Analysis of Variance

Source	**DF**	**Sum of Squares**	**Mean Square**	**F-ratio**	**P-Value**
Type	2	10.1254	5.06268	5923.1	<0.0001
Additive	1	0.026092	0.0260915	30.526	<0.0001
Interaction	2	7.57E-05	3.78E-05	0.044265	0.9567
Error	54	0.046156	8.55E-04		
Total	59	10.1977			

After the re-expression, the ANOVA table only shows the main effects to be significant, while the interaction term is not. We can conclude that both the *car type* and *additive* have an effect on mileage and that the effects are constant (in log(*mpg*)) over the values of the various levels of the other factor.

19. Batteries again.

a) H_0: The mean times are the same under both environments. $(\gamma_C = \gamma_{RT})$

H_A: The mean times are different under the two environments.

H_0: The brand of battery has no effect on time. $(\tau_A = \tau_B = \tau_C = \tau_D)$

H_A: The brand of batter does have an effect on time.

b) From the partial boxplots it appears that *environment* does have an effect on time, but it is unclear whether *brand* has an effect on time.

c) Yes, the ANOVA does match our intuition based on the boxplots. The *brand* effect has a *P*-value equal to 0.099. While this is not significant at the $\alpha = 0.05$ level, it is significant at the $\alpha = 0.10$ level. As it appeared on the boxplot, the *environment* is clearly significant with a *P*-value less than 0.0001.

d) There is also an interaction, however, which makes the statement about *brands* problematic. Not all *brands* are affected by the environment in the same way. Brand C, which works best in the cold, performs worst at room temperature.

e) I would be uncomfortable recommending brand C because it performs the worst of the four at room temperature.

21. Batteries once more.

In this one-way ANOVA, we can see that the means vary across treatments. (However, boxplots with only 2 observations are not appropriate.) By looking closely, it seems obvious that the four flashlights at room temperature lasted much longer than the ones in the cold. It is much harder to see whether the means of the four brands are different, or whether they differ by the same amounts across both environmental conditions. The two-way ANOVA with interaction makes these distinctions clear.

Chapter 30 – Multiple Regression

1. Interpretations.

a) There are two problems with this interpretation. First, the other predictors are not mentioned. Secondly, the prediction should be stated in terms of a mean, not a precise value.

b) This is a correct interpretation.

c) This interpretation attempts to predict in the wrong direction. This model cannot predict *lotsize* from *price.*

d) R^2 concerns the fraction of variability accounted for by the regression model, not the fraction of data values.

3. Predicting final exams.

a) $\widehat{final} = -6.72 + 0.2560(Test1) + 0.3912(Test2) + 0.9015(Test3)$

b) $R^2 = 77.7\%$, which means that 77.7% of the variation in *final* grade is accounted for by the multiple regression model.

c) According to the multiple regression model, each additional point on *Test3* is associated with an average increase of 0.9015 points on the final, for students with given *Test1* and *Test2* scores.

d) Test scores are probably collinear. If we are only concerned about predicting the final exam score, *Test1* may not add much to the regression. However, we would expect it to be associated with the final exam score.

5. Home prices.

a) $\widehat{price} = -152037 + 9530(baths) + 139.87(sqft)$

b) $R^2 = 71.1\%$, which means that 71.1% of the variation in asking *price* is accounted for by the multiple regression model.

c) According to the multiple regression model, the asking *price* increases, on average, by about $139.87 for each additional square foot, for homes with the same number of bathrooms.

d) The number of bathrooms is probably correlated with the size of the house, even after considering the square footage of the bathroom itself. This correlation may account for the coefficient of *baths* not being discernibly different from 0. Moreover, the regression model does not predict what will happen when a house is modified, for example, by converting existing space into a bathroom.

7. Predicting finals II.

Straight enough condition: The plot of residuals versus predicted values looks curved, rising in the middle, and falling on both ends. This is a potential difficulty.
Randomization condition: It is reasonable to think of this class as a representative sample of all classes.
Nearly Normal condition: The Normal probability plot and the histogram of residuals suggest that the highest five residuals are extraordinarily high.
Does the plot thicken? condition: The spread is not consistent over the range of predicted values.

These data may benefit from a re-expression.

9. Secretary performance.

a) $\widehat{salary} = 9.788 + 0.11(service) + 0.053(education) + 0.071(score) + 0.004(speed) + 0.065(dictation)$

b) $\widehat{salary} = 9.788 + 0.11(120) + 0.053(9) + 0.071(50) + 0.004(60) + 0.065(30) \approx \$29{,}200$

c) H_0: When including the other potential predictors, typing speed does not increase our ability to predict salary. $(\beta_4 = 0)$

H_A: When including the other potential predictors, typing speed increases our ability to predict salary. $(\beta_4 \neq 0)$

With $t = 0.013$, and $30 - 6 = 24$ degrees of freedom, the *P*-value equals 0.9897. Since the *P*-value is so large, we fail to reject the null hypothesis. There is no evidence to suggest that the coefficient for typing *speed* is anything other than 0.

d) Omitting typing *speed* would simplify the model, and probably result in a model that was almost as good. Other predictors might also be omitted, but we cannot make that decision from the information given.

e) *Age* may be collinear with other predictors in the model. In particular, it is likely to be highly associated with months of *service*.

11. Body fat revisited.

a) H_0: There is no linear relationship between weight and percent body fat. $(\beta_W = 0)$

H_A: There is a linear relationship between weight and percent body fat. $(\beta_W \neq 0)$

With $t = 12.4$, and $250 - 2 = 248$ degrees of freedom, the *P*-value is less than 0.0001. Since the *P*-value is so small, we reject the null hypothesis. There is strong evidence of a linear relationship between *weight* and *%body fat*.

b) According to the linear model, each pound of *weight* is associated with a 0.189% increase in *%body fat*.

c) After removing the linear effects of *waist* and *height*, each pound of *weight* is associated, on average, with a decrease of 0.10% in *%body fat*. The change in coefficient and sign is a result of including the other predictors. We expect *weight* to be correlated with both *waist* and *height*. It may be collinear with them.

d) The *P*-value of 0.1567 says that if the coefficient of *height* in this model is truly 0, we could expect to observe a sample regression coefficient at least as far from 0 as the one we have here about 15.7% of the time.

13. Body fat again.

a) H_0: There is no linear relationship between chest size and percent body fat.

$(\beta_C = 0)$

H_A: There is a linear relationship between chest size and percent body fat.

$(\beta_C \neq 0)$

With $t = 15.5$, and $250 - 2 = 248$ degrees of freedom, the p-value is less than 0.0001. Since the p-value is so small, we reject the null hypothesis. There is strong evidence of a linear relationship between *chest* and *%body fat*.

b) According to the linear model, each additional inch in *chest* is associated with an average increase of 0.71272% in *%body fat*.

c) After allowing for the linear effects of *waist* and *height*, the multiple regression model predicts an average decrease of 0.233531% in *%body fat* for each additional inch *chest* size.

d) Each of the variables appears to contribute to the model. There does not appear to be an advantage to removing any of them.

15. Fifty states 2008.

a) The only model that seems to do poorly is the one that omits *murder*. The other three are hard to choose among.

$\widehat{Lifeexp} = 74.781 - 0.39266(Murder) + 0.007670(HSgrad) + 0.00010773(Income)$

$R^2 = 68.7\%$

$\widehat{Lifeexp} = 77.332 - 0.45525(Murder) + 0.01538(HSgrad) + 5.060(Illiteracy)$

$R^2 = 57.6\%$

$\widehat{Lifeexp} = 75.3472 - 0.42699(Murder) + 0.00010727(Income) + 1.335(Illiteracy)$

$R^2 = 68.5\%$

b) Each of the models has at least one coefficient with a large *P*-value. This predictor variable could be omitted to simplify the model without degrading it too much.

c) No. Regression models cannot be interpreted that way. Association is not the same thing as causation.

d) Plots of the residuals highlight some states as possible outliers. You may want to consider setting them aside to see if the model changes.

17. Burger King revisited.

a) With an $R^2 = 100\%$, the model should make excellent predictions.

b) The value of s, 3.140 calories, is very small compared to the initial standard variation of *calories*. This means that the model fits the data quite well, leaving very little variation unaccounted for.

c) No, the residuals are not all 0. Indeed, we know that their standard deviation is $s = 3.140$ calories. They are very small compared with the original values. The true value of R^2 was rounded up to 100%.

Chapter 31 – Multiple Regression Wisdom

1. Global warming.

a) The distribution of Studentized residuals might be bimodal. We should check for evidence of 2 groups in the data.

b) The pattern of Studentized residuals over time suggests that carbon dioxide levels changed in character in the late 1970s, and began a more rapid increase than had been seen earlier in the 20th century.

c) The largest residuals may not be outliers, but rather the result of a change in the underlying pattern. It might be better to fit different regression models to the data up to 1977, and to the data after 1977.

3. Healthy breakfast.

a) The slope of a partial regression plot is the coefficient of the corresponding predictor – in this case, – 1.020.

b) Quaker oatmeal makes the slope more strongly negative. It appears to have substantial influence on this slope.

c) Not surprisingly, omitting Quaker oatmeal changes the coefficient of *fiber*. It is now positive (although not significantly different from 0). This second regression model has a higher R^2, suggesting that it fits the data better. Without the influential point, the second regression is probably the better model.

d) The coefficient of *fiber* is not discernibly different from 0. We have no evidence that it contributes significantly to calories.

5. A nutritious breakfast.

a) After allowing for the effects of *sodium* and *sugars*, the model predicts a decrease of 0.019 calories for each additional gram of *potassium*.

b) Those points pull the slope of the relationship down. Omitting them should increase the value of the coefficient of *potassium*. It would likely become positive, since the remaining points show a positive slope in the partial regression plot.

c) These appear to be influential points. They have both high leverage and large residuals, and the partial regression plot shows their influence.

d) If our goal is to understand the relationships among these variables, then it might be best to omit these cereals because they seem to behave as if they are from a separate subgroup.

7. Traffic delays.

a) This set of indicators uses *medium* size as its base. The coefficients of *small, large,* and *very large* estimate the average change in amount of *delay/person* relative to the amount of *delay/person* for *medium* size cities found for each of the other three sizes. If there were an indicator variable for *medium* as well, the four indicators would be collinear, so the coefficients could not be estimated.

b) On average, the total *delay/person* is 5.01 hours longer for people in *large* cities, after allowing for the effects of *arterial road speed* and *highway road speed.*

9. More traffic.

a) An assumption required for indicator variables to be useful is that the regression models fit for the different groups identified by the indicators are parallel, i.e. they have the same slope. These lines are not parallel.

b) The coefficient of $Am \times Sml$ adjusts the slope of the regression model fit for the small cities. We would say that the slope of *delay/person* on *arterial mph* for *small* cities (after allowing for the linear effects of the other variables in the model) is $-2.60848 + 3.81461 = 1.20613$.

c) The regression model seems to do a good job. The R^2 shows that 80.7% of the variability in *delay/person* is accounted for by the model. Most of the *P*-values for the coefficients are small. The coefficients concerning the very large cities have larger *P*-values, but that may be due to having a relatively small number of such cities. It may still be wise to keep those predictors in the model.

11. Influential traffic?

Colorado Springs has high leverage, but it does not have a particularly high Studentized residual. It appears that the point has influence but does not exert it. Removing this case from the regression may not result in a large change in the model, so the case is probably not influential.

Review of Part VII – Inference When Variables Are Related

1. **Tableware.**

   a) Since there are 57 degrees of freedom, there were 59 different products in the analysis.

   b) 84.5% of the variation in retail price is explained by the polishing time.

   c) Assuming the conditions have been met, the sampling distribution of the regression slope can be modeled by a Student's *t*-model with (59 – 2) = 57 degrees of freedom. We will use a regression slope *t*-interval. For 95% confidence, use $t^*_{57} \approx 2.0025$, or estimate from the table $t^*_{50} \approx 2.009$.

   $b_1 \pm t^*_{n-2} \times SE(b_1) = 2.49244 \pm (2.0025) \times 0.1416 \approx (2.21, 2.78)$

   d) We are 95% confident that the average price increases between \$2.21 and \$2.78 for each additional minute of polishing time.

3. **Mutual funds.**

   a) **Paired data assumption:** These data are paired by mutual fund.
   Randomization condition: Assume that these funds are representative of all large cap mutual funds.
   10% condition: 15 mutual funds are less than 10% of all large cap mutual funds.
   Nearly Normal condition: The histogram of differences is unimodal and symmetric.

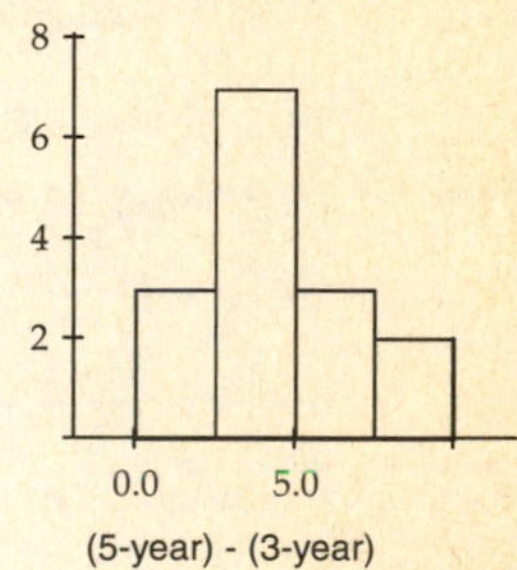

   Since the conditions are satisfied, the sampling distribution of the difference can be modeled with a Student's *t*-model with 15 – 1 = 14 degrees of freedom. We will find a paired *t*-interval, with 95% confidence.

   $$\bar{d} \pm t^*_{n-1}\left(\frac{s_d}{\sqrt{n}}\right) = 4.54 \pm t^*_{14}\left(\frac{2.50508}{\sqrt{15}}\right) \approx (3.15, 5.93)$$

   Provided that these mutual funds are representative of all large cap mutual funds, we are 95% confident that, on average, 5-year yields are between 3.15% and 5.93% higher than 3-year yields.

b) H_0: There is no linear relationship between 3-year and 5-year rates. $(\beta_1 = 0)$

H_A: There is a linear relationship between 3-year and 5-year rates. $(\beta_1 \neq 0)$

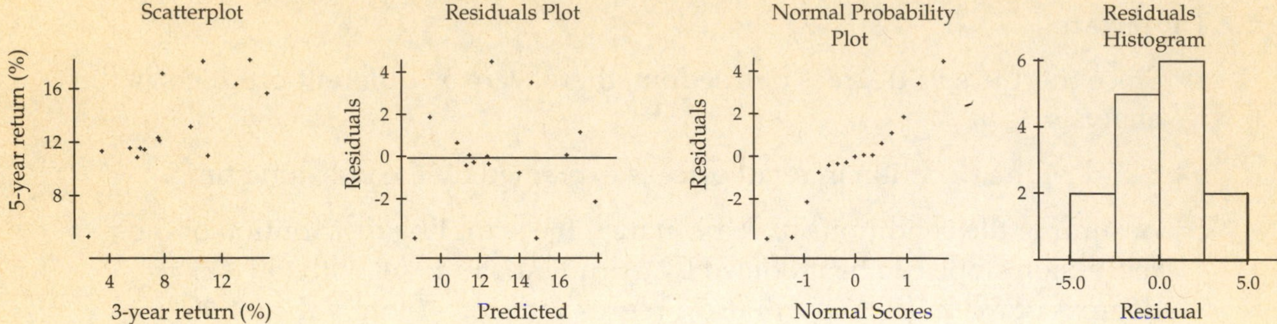

Straight enough condition: The scatterplot is straight enough for regression.
Independence assumption: The residuals plot shows no pattern.
Does the plot thicken? condition: The spread of the residuals is consistent.
Nearly Normal condition: The Normal probability plot of residuals isn't very straight. However, the histogram of residuals is unimodal and symmetric. With a sample size of 15, it is probably okay to proceed.

Since the conditions for inference are satisfied, the sampling distribution of the regression slope can be modeled by a Student's *t*-model with (15 – 2) = 13 degrees of freedom. We will use a regression slope *t*-test.

Dependent variable is: 5-year
No Selector
R squared = 58.4% R squared (adjusted) = 55.2%
s = 2.360 with 15 - 2 = 13 degrees of freedom

Source	Sum of Squares	df	Mean Square	F-ratio
Regression	101.477	1	101.477	18.2
Residual	72.3804	13	5.56773	

Variable	Coefficient	s.e. of Coeff	t-ratio	prob
Constant	6.92904	1.557	4.45	0.0007
3-year	0.719157	0.1685	4.27	0.0009

The equation of the line of best fit for these data points is:
$(5\widehat{year}) = 6.92904 + 0.719157(3year)$.

The value of $t \approx 4.27$. The *P*-value of 0.0009 means that the association we see in the data is unlikely to occur by chance. We reject the null hypothesis, and conclude that there is strong evidence of a linear relationship between the rates of return for 3-year and 5-year periods. Provided that these mutual funds are representative of all large cap mutual funds, mutual funds with higher 3-year returns tend to have higher 5-year returns.

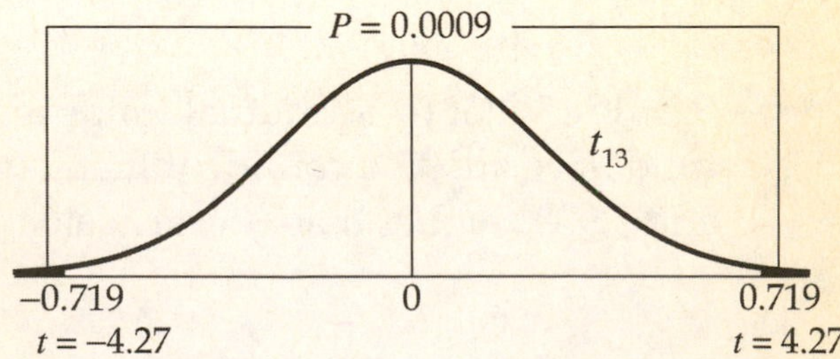

5. Football.

a) He should have performed a one-way ANOVA and an F-test.

b) H_0: The mean distance thrown is the same for each grip. $(\mu_1 = \mu_2 = \mu_3 = \mu_4)$

H_A: The mean distances thrown are not all the same.

c) With a P-value equal to 0.0032, we reject the null hypothesis and conclude that the distance thrown is not the same for all grips, on average.

d) **Randomization condition:** The grips used were randomized.
Similar Variance condition: The boxplots should show similar spread.
Nearly Normal condition: The histogram of the residuals should be unimodal and symmetric.

e) He might want to perform a multiple comparison test to see which grip is best.

7. Wild horses.

a) Since there are 36 degrees of freedom, 38 herds of wild horses were studied.

b) **Straight enough condition:** The scatterplot is straight enough to try linear regression.
Independence assumption: The residuals plot shows no pattern.
Does the plot thicken? condition: The spread of the residuals is consistent.
Nearly Normal condition: The histogram of residuals is unimodal and symmetric.

c) Since the conditions for inference are satisfied, the sampling distribution of the regression slope can be modeled by a Student's t-model with (38 – 2) = 36 degrees of freedom. We will use a regression slope t-interval, with 95% confidence. Use $t_{35}^{*} \approx 2.030$ as an estimate.

$$b_1 \pm t_{n-2}^{*} \times SE(b_1) = 0.153969 \pm (2.030) \times 0.0114 \approx (0.131,\ 0.177)$$

d) We are 95% confident that the mean number of foals in a herd increases by between 0.131 and 0.177 foals for each additional adult horse.

e) The regression equation predicts that herds with 80 adults will have $-1.57835 + 0.153969(80) = 10.73917$ foals. The average size of the herds sampled is 110.237 adult horses. Use $t_{36}^{*} \approx 1.6883$, or use an estimate of $t_{35}^{*} \approx 1.690$, from the table.

$$\hat{y}_v \pm t_{n-2}^{*} \sqrt{SE^2(b_1) \cdot (x_v - \bar{x})^2 + \frac{s_e^2}{n} + s_e^2}$$

$$= 10.73917 \pm (1.6883) \sqrt{0.0114^2 \cdot (80 - 110.237)^2 + \frac{4.941^2}{38} + 4.941^2}$$

$$\approx (2.26,\ 19.21)$$

We are 95% confident that number of foals in a herd of 80 adult horses will be between 2.26 and 19.21. This prediction interval is too wide to be of much use.

9. Horses again, but less wild.

a) *Sterilized* is an indicator variable.

b) According to the model, there were, on average, 6.4 fewer foals in herds in which some of the stallions were sterilized, after allowing for the number of adults in the herd.

c) The p-value for *sterilized* is equal to 0.096. That is not significant at the $\alpha = 0.05$ level, but it does seem to indicate some effect.

11. Video racing.

a) H0: The mean time is the same for all three types of mouse. $(\gamma_{Ergo} = \gamma_{\text{Reg}} = \gamma_{Cord})$

HA: The mean times are not all the same for the three types of mouse.

H0: The mean time is the same whether the lights are on or off. $(\tau_{On} = \tau_{Off})$

HA: The mean times are not the same for lights on and lights off.

b) The *mouse* sum of squares has 3 – 1 = 2 degrees of freedom.
The *light* sum of squares has 2 – 1 = 1 degrees of freedom.
The error sum of squares has (6 – 1) – 2 – 1 = 2 degrees of freedom.

c) No, he should not fit an interaction term. The interaction term would use 2 degrees of freedom. This would not leave any degrees of freedom for the error sum of squares and thus would not allow any interpretation of the factors.

13. Paper airplanes.

a) It is reasonable to think that the flight distances are independent of one another. The histogram of flight distances (given) is unimodal and symmetric. Since the conditions are satisfied, the sampling distribution of the mean can be modeled by a Student's t model, with 11 – 1 = 10 degrees of freedom. We will use a one-sample t -interval with 95% confidence for the mean flight distance.

$$\bar{y} \pm t^*_{n-1}\left(\frac{s}{\sqrt{n}}\right) = 48.3636 \pm t^*_{10}\left(\frac{18.0846}{\sqrt{11}}\right) \approx (36.21, 60.51)$$

We are 95% confident that the mean distance the airplane may fly is between 36.21 and 60.51 feet.

b) Since 40 feet is contained within our 95% confidence interval, it is plausible that the mean distance is 40 feet.

c) A 99% confidence interval would be wider. Intervals with greater confidence are less precise.

d) In order to cut the margin of error in half, she would need a sample size four times as large, or 44 flights.

15. Barbershop music.

a) With an R^2 of 90.9%, your friend is right about being able to predict singing scores.

b) H0: When including the other predictor, performance does not change our ability to predict singing scores. $(\beta_{Prs} = 0)$

HA: When including the other predictor, performance changes our ability to predict singing scores. $(\beta_{Prs} \neq 0)$

With $t = 8.13$, and 31 degrees of freedom, the *P*-value is less than 0.0001, so we reject the null hypothesis and conclude that *performance* is useful in predicting singing scores.

c) Based on the multiple regression, both *performance* and *music* (even with a *P*-value equal to 0.0766) are useful in predicting singing scores. According to the residuals plot the spread is constant across the predicted values. The histogram of residuals is unimodal and symmetric. Based on the information provided we have met the Similar Variance and Nearly Normal conditions.

17. Study habits.

a) H0: The mean hours studied is the same for both sexes. $(\gamma_F = \gamma_M)$

HA: The mean hours studied not the same for both sexes.

H0: The mean hours studied is the same for all classes. $(\tau_{Fr} = \tau_{So} = \tau_{Jr} = \tau_{Sr})$

HA: The mean hours studied are not the same for all classes.

b) Based on the interaction plot, the lines are not parallel, so the Additive Enough condition is not met. The interaction term should be fit.

c) With all three p-values so large, none of the effects appear to be significant.

d) There are a few outliers. Based on the residuals plot, the Similar Variance condition appears to be met, but we do not have a Normal probability plot of residuals. The main concern is that we may not have enough power to detect differences between groups. We would need more data to increase the power.

19. Lost baggage.

These are not counts, so the **Counted data condition** is not met. We cannot use a chi-square goodness-of-fit test.

21. Togetherness.

a) H0: There is no linear relationship number of meals eaten as a family and grades. $(\beta_1 = 0)$

HA: There is a linear relationship. $(\beta_1 \neq 0)$

Since the conditions for inference are satisfied (given), the sampling distribution of the regression slope can be modeled by a Student's *t*-model with (142 – 2) = 140 degrees of freedom. We will use a regression slope *t*-test. The equation of the line of best fit for these data points is: $\widehat{GPA} = 2.7288 + 0.1093(Meals / Week)$.

$$t = \frac{b_1 - \beta_1}{SE(b_1)}$$
$$t = \frac{0.1093 - 0}{0.0263}$$
$$t \approx 4.16$$

The value of $t \approx 4.16$. The *P*-value of less than 0.0001 means that the association we see in the data is unlikely to occur by chance. We reject the null hypothesis, and conclude that there is strong evidence of a linear relationship between grades and the number of meals eaten as a family. Students whose families eat together relatively frequently tend to have higher grades than those whose families don't eat together as frequently.

b) This relationship would not be particularly useful for predicting a student's grade point average. $R^2 = 11.0\%$, which means that only 11% of the variation in GPA can be explained by the number of meals eaten together per week.

c) These conclusions are not contradictory. There is strong evidence that the slope is not zero, and that means strong evidence of a linear relationship. This does not mean that the relationship itself is strong, or useful for predictions.

23. Lefties and music.

H0: The proportion of right-handed people who can match the tone is the same as the proportion of left-handed people who can match the tone. $(p_L = p_R \text{ or } p_L - p_R = 0)$

HA : The proportion of right-handed people who can match the tone is different from the proportion of left-handed people who can match the tone. $(p_L \neq p_R \text{ or } p_L - p_R \neq 0)$

Random condition: Assume that the people tested are representative of all people.
10% condition: 76 and 53 are both less than 10% of all people.
Independent samples condition: The groups are not associated.
Success/Failure condition: $n\hat{p}$ (right) = 38, $n\hat{q}$ (right) = 38, $n\hat{p}$ (left) = 33, and $n\hat{q}$ (left) = 20 are all greater than 10, so the samples are both large enough.

Since the conditions have been satisfied, we will model the sampling distribution of the difference in proportion with a Normal model with mean 0 and standard deviation estimated by

$$SE_{\text{pooled}}\left(\hat{p}_L - \hat{p}_R\right) = \sqrt{\frac{\hat{p}_{\text{pooled}}\hat{q}_{\text{pooled}}}{n_L} + \frac{\hat{p}_{\text{pooled}}\hat{q}_{\text{pooled}}}{n_R}} = \sqrt{\frac{\left(\frac{71}{129}\right)\left(\frac{58}{129}\right)}{53} + \frac{\left(\frac{71}{129}\right)\left(\frac{58}{129}\right)}{76}} \approx 0.089.$$

The observed difference between the proportions is: $0.6226 - 0.5 = 0.1226$.

Since the P-value = 0.1683 is high, we fail to reject the null hypothesis. There is no evidence that the proportion of people able to match the tone differs between right-handed and left-handed people.

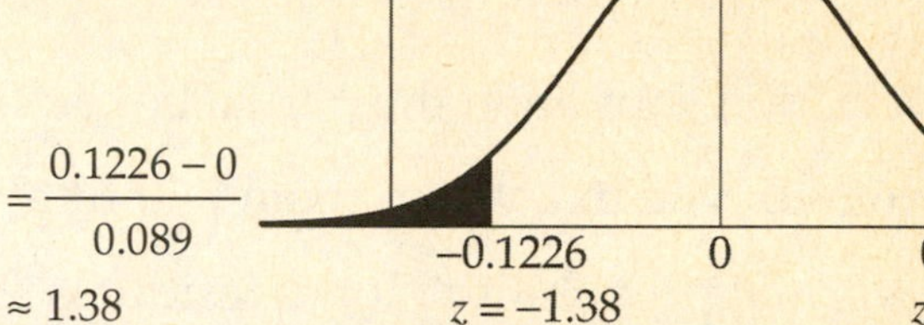

25. Teen traffic deaths 2007.

a) **Straight enough condition:** The scatterplots of the response versus each predicted value should be reasonably straight.
Independence: These data are measured over time. We should plot the data and the residuals against time to look for failures of independence.
Nearly Normal condition: The Normal probability plot should be straight, and the histogram of residuals should be unimodal and symmetric.
Similar Variance condition: The spread of the residuals plot is constant.

b) $R^2 = 57.4\%$, so 57.4% of the variation in female teen traffic deaths is accounted for by the linear model.

c) $\widehat{deaths} = 45074.0 - 21.6006(year)$

d) The model predicts that female teen traffic deaths have been declining at a rate of approximately 21.6 per year.

27. Typing.

a) H_0: Typing speed is the same whether the gloves are on or off. $(\gamma_{On} = \gamma_{Off})$
H_A: Typing speeds are not the same when the gloves are on and off.

H_0: Typing speed is the same at hot or cold temperatures. $(\tau_H = \tau_C)$
H_A: Typing speeds are not the same at hot and cold temperatures.

b) Based on the boxplots, it appears that both the effects of *temperature* and *gloves* affect typing speed.

c) The interaction plot is not parallel so the additive enough condition is not met. Therefore, an interaction term should be fit.

d) Yes, I think it will be significant because a difference in typing speed due to temperature seems to be significant only when he is not wearing the gloves.

e) The p-values for all three effects: *gloves, temperature* and the interaction term, are all very small, so all three effects are significant.

f) Gloves decreased typing speed at both temperatures. A hot temperature increases typing speed with the gloves off, but has little effect with the gloves on.

g) $s = \sqrt{\frac{58.75}{28}} = 1.45$ words per minute. Yes, this seems consistent with the size of the variation shown on the partial boxplots.

h) Tell him to type in a warm room without wearing his gloves.

i) The Normal probability plot of residuals should be straight. The residuals plot shows constant spread across the predicted values. The Similar Variance condition appears to be met. For the levels of the factors that he used, it seems clear that a warm room and not wearing gloves are beneficial for his typing.

29. New York Marathon.

a) The number of finishers per minute appears to increase by about 0.519037 finishers per minute.

b) There is a definite pattern in the residuals. This is a clear violation of the linearity assumption. While there does seem to be a strong association between time and number of finishers, it is not linear.

31. Depression and the Internet.

a) H_0: There is no linear relationship between depression and Internet usage. $(\beta_1 = 0)$

H_A: There is a linear relationship between depression and Internet usage. $(\beta_1 \neq 0)$

Since the conditions for inference are satisfied (given), the sampling distribution of the regression slope can be modeled by a Student's *t*-model with (162 – 2) = 160 degrees of freedom. We will use a regression slope *t*-test. The equation of the line of best fit for these data points is:
$\widehat{DepressionAfter} = 0.565485 + 0.019948(InternetUsage)$.

The value of $t \approx 2.76$. The *P*-value of 0.0064 means that the association we see in the data is unlikely to occur by chance. We reject the null hypothesis, and conclude that there is strong evidence of a linear relationship between depression and Internet usage. Those with high levels of Internet usage tend to have high levels of depression. It should be noted, however, that although the evidence is strong, the association is quite weak, with $R^2 = 4.6\%$. The regression analysis only explains 4.6% of the variation in depression level.

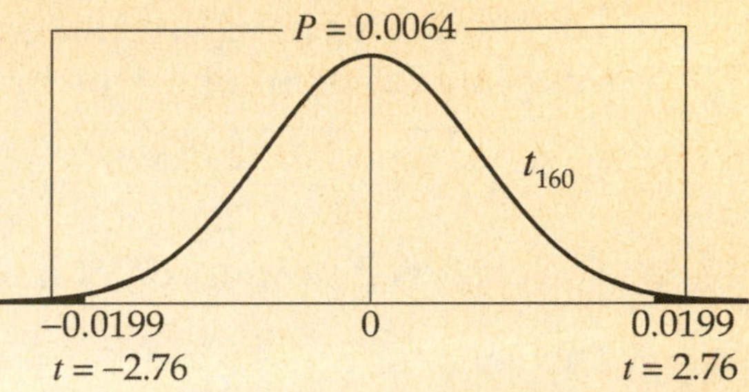

b) The study says nothing about causality, merely association. Furthermore, there are almost certainly other factors involved. In fact, if 4.6% of the variation in depression level is related to Internet usage, the other 95.4% of the variation must be related to something else!

c) H$_0$: The mean difference in depression before and after the experiment is zero. $(\mu_d = 0)$

H$_A$: The mean difference in depression is greater than zero. $(\mu_d > 0)$

Since the conditions are satisfied (given), the sampling distribution of the difference can be modeled with a Student's *t*-model with 162 – 1 = 161 degrees of freedom, $t_{161}\left(0, \frac{0.552417}{\sqrt{162}}\right)$.

We will use a paired *t*-test, with $\bar{d} = -0.118457$.

Since the *P*-value = 0.9965 is very high, we fail to reject the null hypothesis. There is no evidence that the mean depression level increased over the course of the experiment. In fact, these data suggest that depression levels actually decreased.

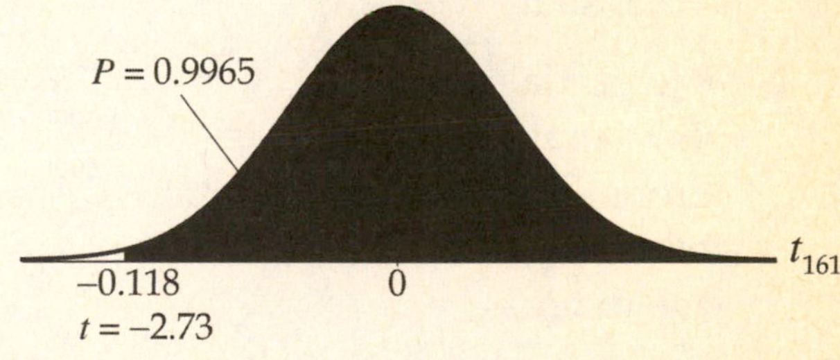

33. Pesticides.

H$_0$: The percentage of males born to workers at the plant is 51.2%.($p = 0.512$)
H$_A$: The percentage of males is less than 51.2%. ($p < 0.512$)

Independence assumption: It is reasonable to think that the births are independent.
Success/Failure Condition: $np = (227)(0.512) = 116$ and $nq = (227)(0.488) = 111$ are both greater than 10, so the sample is large enough.

The conditions have been satisfied, so a Normal model can be used to model the sampling distribution of the proportion, with $\mu_{\hat{p}} = p = 0.512$ and

$\sigma(\hat{p}) = \sqrt{\frac{pq}{n}} = \sqrt{\frac{(0.512)(0.488)}{227}} \approx 0.0332$. We can perform a one-proportion z-test.

The observed proportion of males is $\hat{p} = 0.40$.

The value of $z \approx -3.35$, meaning that the observed proportion of males is over 3 standard deviations below the expected proportion. The P-value associated with this z score is approximately 0.0004.

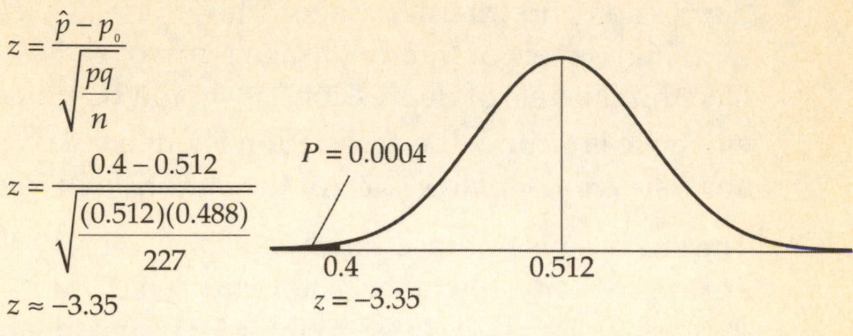

With a P-value this low, we reject the null hypothesis. There is strong evidence that the percentage of males born to workers is less than 51.2%. This provides evidence that human exposure to dioxin may result in the birth of more girls.

35. Video pinball.

a) H0: Pinball score is the same whether the tilt is on or off. ($\gamma_{On} = \gamma_{Off}$)

HA: Pinball scores are different when the tilt is on and when it is off.

H0: Pinball score is the same whether both eyes are open or the right is closed. ($\tau_B = \tau_R$)

HA: Pinball scores are different when both eyes are open and when the right is closed.

b) The partial boxplots show that the *eye* effect is strong, but there appears to be little or no effect due to *tilt*.

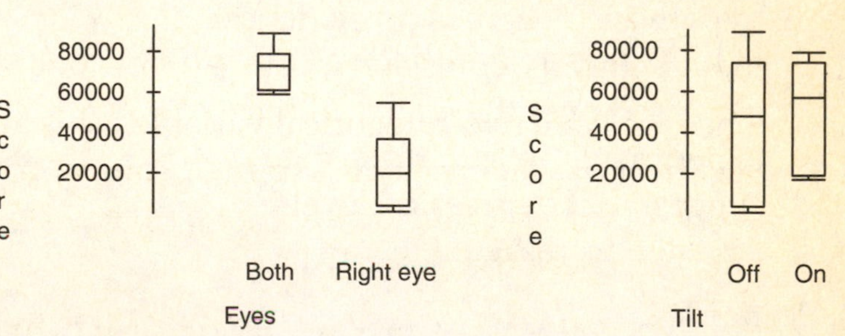

Randomization condition: The experiment was performed in random order.
Similar Variance condition: The side-by-side boxplots have similar spreads. The residuals plot shows no pattern, and no systematic change in spread.
Nearly Normal condition: The Normal probability plot of the residuals is reasonably straight.

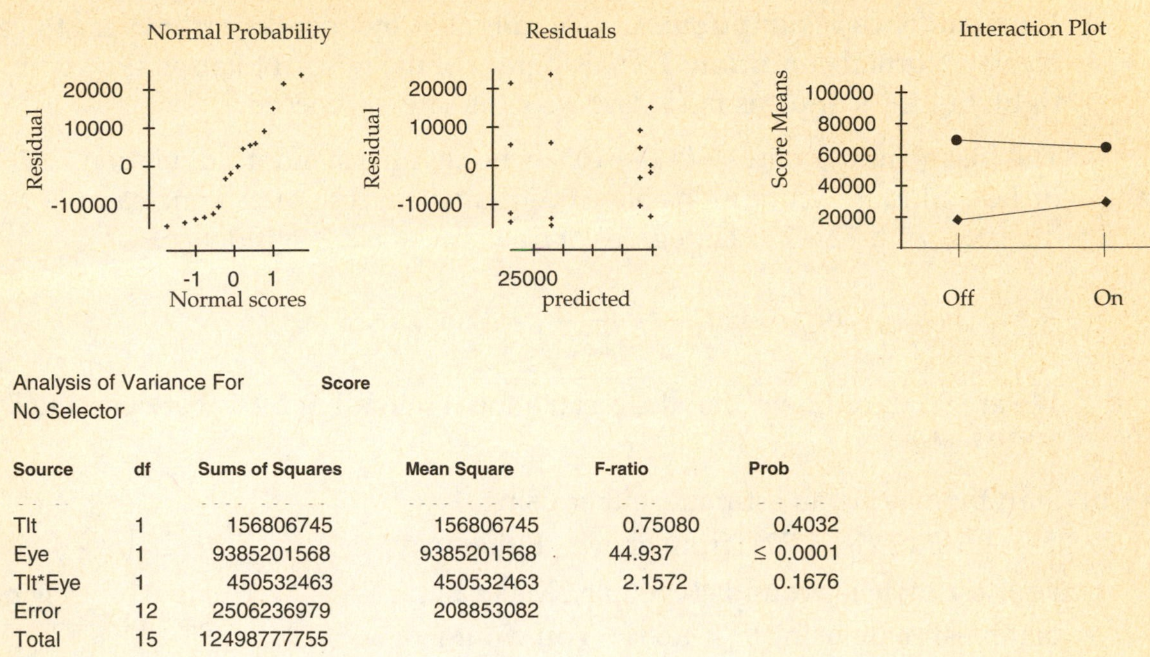

Analysis of Variance For **Score**
No Selector

Source	**df**	**Sums of Squares**	**Mean Square**	**F-ratio**	**Prob**
Tlt	1	156806745	156806745	0.75080	0.4032
Eye	1	9385201568	9385201568	44.937	≤ 0.0001
Tlt*Eye	1	450532463	450532463	2.1572	0.1676
Error	12	2506236979	208853082		
Total	15	12498777755			

According to the ANOVA, only eye effect is significant with a p-value less than 0.0001. Both the *tilt* and interaction effects are not significant. We conclude that keeping both eyes open improves score but that using the tilt does not.

37. Eye and hair color.

a) This is an attempt at linear regression. Regression inference is meaningless here, since eye and hair color are categorical variables.

b) This is an analysis based upon a chi-square test for independence.

H_0: Eye color and hair color are independent.

H_A: There is an association between eye color and hair color.

Since we have two categorical variables, this analysis seems appropriate. However, if you check the expected counts, you will find that 4 of them are less than 5. We would have to combine several cells in order to perform the analysis. (Always check the conditions!)

Since the value of chi-square is so high, it is likely that we would find an association between eye and hair color, even after the cells were combined. There are many cells of interest, but some of the most striking differences that would not be affected by cell combination involve people with fair hair. Blonds are likely to have blue eyes, and not likely to have brown eyes. Those with red hair are not likely to have brown eyes. Additionally, those with black hair are much more likely to have brown eyes than blue.

39. LA rainfall.

a) Independence assumption: Annual rainfall is independent from year to year.
Nearly Normal condition: The histogram of the rainfall totals is skewed to the right, but the sample is fairly large, so it is safe to proceed.

The mean annual rainfall is 14.5165 inches, with a standard deviation 7.82044 inches. Since the conditions have been satisfied, construct a one-sample t-interval, with 22 – 1 = 21 degrees of freedom, at 90% confidence.

$$\bar{y} \pm t^*_{n-1}\left(\frac{s}{\sqrt{n}}\right) = 14.5164 \pm t^*_{21}\left(\frac{7.82044}{\sqrt{22}}\right) \approx (11.65, 17.39)$$

We are 90% confident that the mean annual rainfall in LA is between 11.65 and 17.39 inches.

b) Start by making an estimate, either using $z^* = 1.645$ or $t^*_{21} = 1.721$ from above. Either way, your estimate is around 40 people. Make a better estimate using $t^*_{40} = 1.684$. You would need about 44 years' data to estimate the annual rainfall in LA to within 2 inches.

$$ME = t^*_{40}\left(\frac{s}{\sqrt{n}}\right)$$

$$2 = 1.684\left(\frac{7.82044}{\sqrt{n}}\right)$$

$$n = \frac{(2.064)^2(7.82044)^2}{(2)^2}$$

$$n \approx 44 \text{ years}$$

c) H0: There is no linear relationship between year and annual LA rainfall. $(\beta_1 = 0)$
HA: There is a linear relationship between year and annual LA rainfall. $(\beta_1 \neq 0)$

Straight enough condition: The scatterplot is straight enough to try linear regression, although there is no apparent pattern.
Independence assumption: The residuals plot shows no pattern.
Does the plot thicken? condition: The spread of the residuals is consistent.

Nearly Normal condition: The Normal probability plot of residuals is not straight, and the histogram of the residuals is skewed to the right, but a sample of 22 years is large enough to proceed.

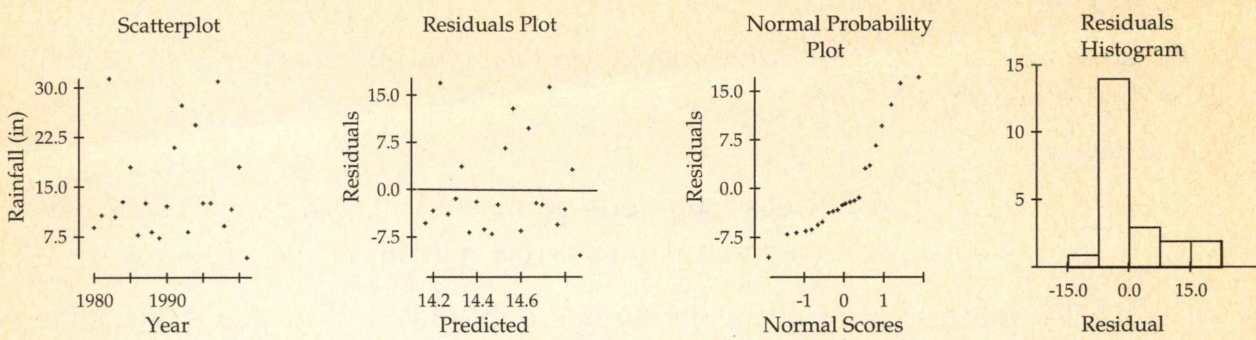

Since the conditions for inference are satisfied, the sampling distribution of the regression slope can be modeled by a Student's t-model with $(22 - 2) = 20$ degrees of freedom. We will use a regression slope t-test.

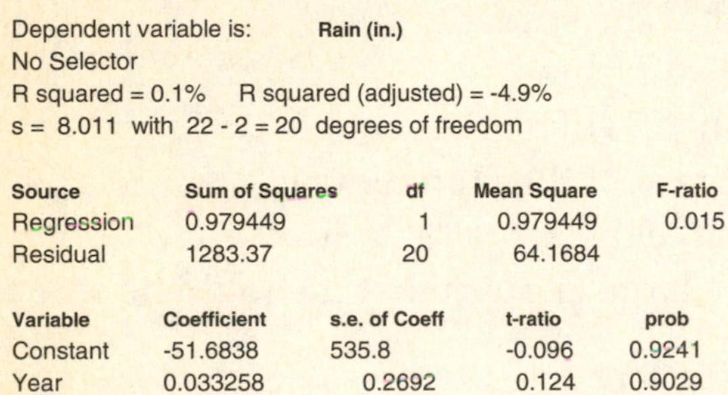

Dependent variable is: **Rain (in.)**
No Selector
R squared = 0.1% R squared (adjusted) = -4.9%
s = 8.011 with 22 - 2 = 20 degrees of freedom

Source	Sum of Squares	df	Mean Square	F-ratio
Regression	0.979449	1	0.979449	0.015
Residual	1283.37	20	64.1684	

Variable	Coefficient	s.e. of Coeff	t-ratio	prob
Constant	-51.6838	535.8	-0.096	0.9241
Year	0.033258	0.2692	0.124	0.9029

The equation of the line of best fit for these data points is:

$\widehat{Rain} = -51.6838 + 0.033258(Year)$.

The value of $t \approx 0.124$. The P-value of 0.92029 means that the association we see in the data is quite likely to occur by chance. We fail to reject the null hypothesis, and conclude that there is no evidence of a linear relationship between the annual rainfall in LA and the year.

41. Weight and athletics.

a) H_0: Mean weight is the same for all 3 groups. $(\mu_1 = \mu_2 = \mu_3)$

H_A: Mean weights are not the same for all 3 groups.

b) According to the boxplots the spread appears constant for all three groups. There is one outlier. We do not have a Normal probability plot, but we suspect that the data may be Normal enough.

c) With a *P*-value equal to 0.0042, we reject the null hypothesis, there is evidence that the mean weights are not the same for all three groups. It may be that more men are involved with athletics, which might explain the weight differences. On the other hand, it may simply be that those who weigh more are more likely to be involved with sports.

d) It seems that differences are evident even when the outlier is removed. It seems that conclusion is valid.

43. Cramming.

a) H_0: The mean score of week-long study group students is the same as the mean score of overnight cramming students. $(\mu_1 = \mu_2 \text{ or } \mu_1 - \mu_2 = 0)$

H_A: The mean score of week-long study group students is greater than the mean score of overnight cramming students. $(\mu_1 > \mu_2 \text{ or } \mu_1 - \mu_2 > 0)$

Independent Groups Assumption: Scores of students from different classes should be independent.
Randomization Condition: Assume that the students are assigned to each class in a representative fashion.
10% Condition: 45 and 25 are less than 10% of all students.
Nearly Normal Condition: The histogram of the crammers is unimodal and symmetric. We don't have the actual data for the study group, but the sample size is large enough that it should be safe to proceed.

$$\bar{y}_1 = 43.2 \qquad \bar{y}_2 = 42.28$$
$$s_1 = 3.4 \qquad s_2 = 4.43020$$
$$n_1 = 45 \qquad n_2 = 25$$

Since the conditions are satisfied, it is appropriate to model the sampling distribution of the difference in means with a Student's *t*-model, with 39.94 degrees of freedom (from the approximation formula). We will perform a two-sample *t*-test. The sampling distribution model has mean 0, with standard error:

$$SE(\bar{y}_1 - \bar{y}_2) = \sqrt{\frac{3.4^2}{45} + \frac{4.43020^2}{25}} \approx 1.02076\,.$$

The observed difference between the mean scores is $43.2 - 42.28 = 0.92$.

Since the P-value = 0.1864 is high, we fail to reject the null hypothesis. There is no evidence that students with a week to study have a higher mean score than students who cram the night before.

$$t = \frac{(\bar{y}_1 - \bar{y}_2) - (0)}{SE(\bar{y}_1 - \bar{y}_2)}$$
$$t \approx \frac{0.92}{1.02076}$$
$$t \approx 0.90$$

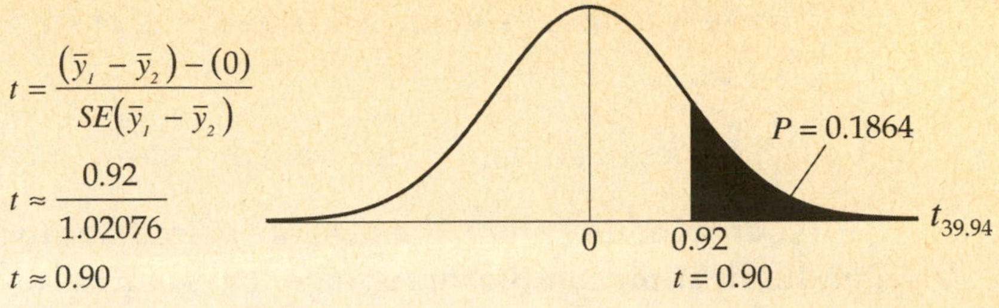

b) H0: The proportion of study group students who will pass is the same as the proportion of crammers who will pass. $(p_1 = p_2 \text{ or } p_1 - p_2 = 0)$

HA : The proportion of study group students who will pass is different from the proportion of crammers who will pass. $(p_1 \neq p_2 \text{ or } p_1 - p_2 \neq 0)$

Random condition: Assume students are assigned to classes in a representative fashion.
10% condition: 45 and 25 are both less than 10% of all students.
Independent samples condition: The groups are not associated.
Success/Failure condition: $n_1\hat{p}_1 = 15$, $n_1\hat{q}_1 = 30$, $n_2\hat{p}_2 = 18$, and $n_2\hat{q}_2 = 7$ are not all greater than 10, since only 7 crammers didn't pass. However, if we check the pooled value, $n_2\hat{p}_{\text{pooled}} = (25)(0.471) = 11.775$. All of the samples are large enough.

Since the conditions have been satisfied, we will model the sampling distribution of the difference in proportion with a Normal model with mean 0 and standard deviation estimated by

$$SE_{\text{pooled}}(\hat{p}_1 - \hat{p}_2) = \sqrt{\frac{\hat{p}_{\text{pooled}}\hat{q}_{\text{pooled}}}{n_1} + \frac{\hat{p}_{\text{pooled}}\hat{q}_{\text{pooled}}}{n_2}} = \sqrt{\frac{\left(\frac{33}{70}\right)\left(\frac{37}{70}\right)}{45} + \frac{\left(\frac{33}{70}\right)\left(\frac{37}{70}\right)}{25}} \approx 0.1245.$$

The observed difference between the proportions is:
$0.3333 - 0.72 = -0.3867$.

Since the P-value = 0.0019 is low, we reject the null hypothesis. There is strong evidence to suggest a difference in the proportion of passing grades for study group participants and overnight crammers. The crammers generally did better.

$$z = \frac{-0.3867 - 0}{0.1245}$$
$$z \approx -3.11$$

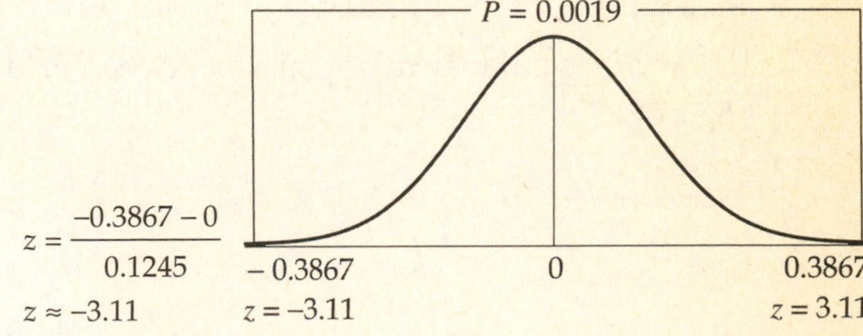

c) H0: There is no mean difference in the scores of students who cram, after 3 days. $(\mu_d = 0)$

HA: The scores of students who cram decreases, on average, after 3 days. $(\mu_d > 0)$

Paired data assumption: The data are paired by student.
Randomization condition: Assume that students are assigned to classes in a representative fashion.
10% condition: 25 students are less than 10% of all students.
Nearly Normal condition: The histogram of differences is roughly unimodal and symmetric.

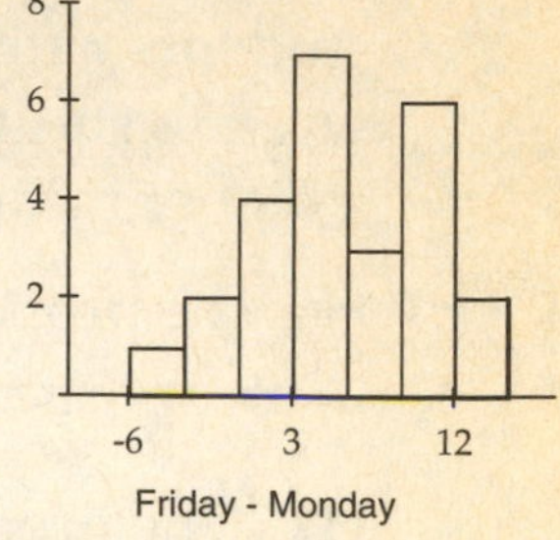

Since the conditions are satisfied, the sampling distribution of the difference can be modeled with a Student's t-model with 25 – 1 = 24 degrees of freedom, $t_{24}\left(0, \frac{4.8775}{\sqrt{25}}\right)$. We will use a paired t-test, with $\bar{d} = 5.04$.

Since the P-value is less than 0.0001, we reject the null hypothesis. There is strong evidence that the mean difference is greater than zero. Students who cram seem to forget a significant amount after 3 days.

$$t = \frac{\bar{d} - 0}{\frac{s_d}{\sqrt{n}}}$$

$$t = \frac{5.04 - 0}{\frac{4.8775}{\sqrt{25}}}$$

$$t \approx 5.17$$

d) $\bar{d} \pm t^*_{n-1}\left(\frac{s_d}{\sqrt{n}}\right) = 5.04 \pm t^*_{24}\left(\frac{4.8775}{\sqrt{25}}\right) \approx (3.03, 7.05)$

We are 95% confident that students who cram will forget an average of 3.03 to 7.05 words in 3 days.

e) H0: There is no linear relationship between Friday score and Monday score. $(\beta_1 = 0)$

HA: There is a linear relationship between Friday score and Monday score. $(\beta_1 \neq 0)$

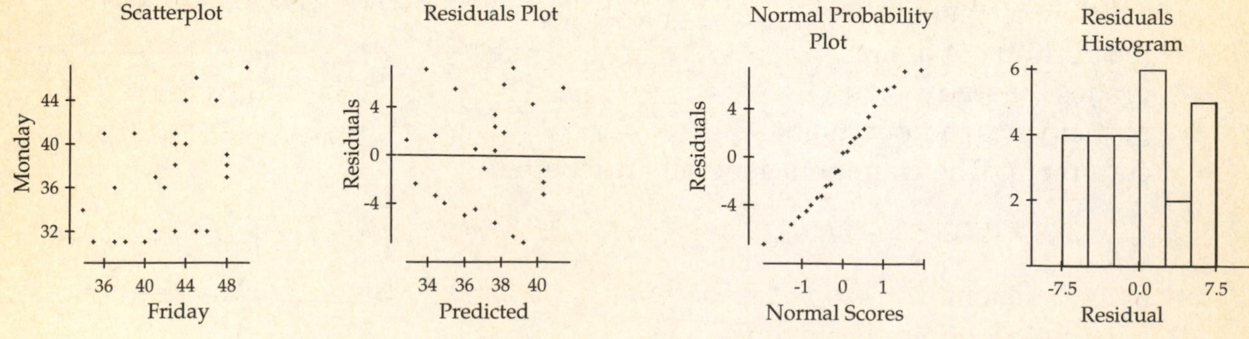

Straight enough condition: The scatterplot is straight enough for regression.
Independence assumption: The residuals plot shows no pattern.
Does the plot thicken? condition: The spread of the residuals is consistent.
Nearly Normal condition: The Normal probability plot of residuals is reasonably straight, and the histogram of the residuals is roughly unimodal and symmetric.

Since the conditions for inference are satisfied, the sampling distribution of the regression slope can be modeled by a Student's t-model with (25 – 2) = 23 degrees of freedom. We will use a regression slope t-test.

Dependent variable is: **Monday**
No Selector
R squared = 22.4%   R squared (adjusted) = 19.0%
s = 4.518 with 25 - 2 = 23 degrees of freedom

Source	Sum of Squares	df	Mean Square	F-ratio
Regression	135.159	1	135.159	6.62
Residual	469.401	23	20.4087	

Variable	Coefficient	s.e. of Coeff	t-ratio	prob
Constant	14.5921	8.847	1.65	0.1127
Friday	0.535666	0.2082	2.57	0.0170

The equation of the line of best fit for these data points is: $\widehat{Monday} = 14.5921 + 0.535666(Friday)$.

The value of $t \approx 2.57$. The P-value of 0.0170 means that the association we see in the data is unlikely to occur by chance. We reject the null hypothesis, and conclude that there is strong evidence of a linear relationship between Friday score and Monday score. Students who do better in the first place tend to do better after 3 days. However, since R^2 is only 22.4%, Friday score is not a very good predictor of Monday score.

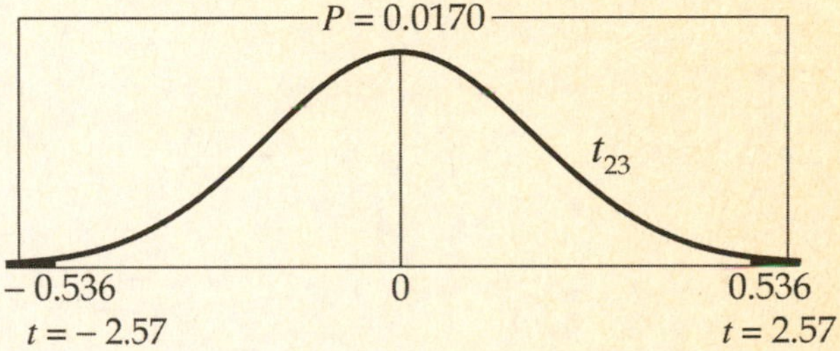

45. Airport screening.

a) No, this is not a good model for incidence of false information. It only has an R^2=8.7%.

b) **Similar Variance condition:** The residuals plot shows no pattern, and no systematic change in spread.
Nearly Normal condition: The histogram of the residuals is unimodal and symmetric.

Isolating the outlier with an indicator variable improved the model dramatically. We can predict incidents of false information well (R^2=86.8%) with 1988 removed from the data:

$$\widehat{false\ info} = 47.2152 + 0.750819(long\ guns) - 0.023583(handguns) - 0.089368(exp).$$

c) The coefficient for *handgun* discoveries has a large P-value and might be removed from the model to simplify it.